Withdrawn

After Steve

After Steve

How Apple Became a Trillion-Dollar
Company and Lost Its Soul

Tripp Mickle

Wm

WILLIAM MORROW
An Imprint of HarperCollinsPublishers

HarperCollins books may be purchased for educational, business, or sales promotional use. For information, please email the Special Markets Department at SPsales@harpercollins.com.

FIRST EDITION

Designed by Nancy Singer

Library of Congress Cataloging-in-Publication Data has been applied for.

ISBN 978-0-06-300981-3

22 23 24 25 26 LSC 10 9 8 7 6 5 4 3 2 1

For my wife, Amanda,

and my parents, Marilynn and Russ

An institution is the lengthened shadow of one man.

—*Ralph Waldo Emerson*

The reasonable man adapts himself to the world; the unreasonable one persists in trying to adapt the world to himself. Therefore, all progress depends on the unreasonable man.

—*George Bernard Shaw*

Contents

Author's Note   xi

Cast of Characters   xiii

Prologue   1

Chapter 1:   One More Thing   5

Chapter 2:   The Artist   21

Chapter 3:   The Operator   42

Chapter 4:   Keep Him   64

Chapter 5:   Intense Determination   92

Chapter 6:   Fragile Ideas   111

Chapter 7:   Possibilities   129

Chapter 8:   Can't Innovate   143

Chapter 9:   The Crown   169

Chapter 10:   Deals   192

Chapter 11:   Blowout   210

Chapter 12:   Pride   222

Chapter 13:   Out of Fashion   232

Chapter 14:   Fuse   251

Chapter 15: Accountants   266

Chapter 16: Security   280

Chapter 17: Hawaii Days   297

Chapter 18: Smoke   307

Chapter 19: The Jony 50   323

Chapter 20: Power Moves   335

Chapter 21: Not Working   358

Chapter 22: A Billion Pockets   376

Chapter 23: Yesterday   392

             Epilogue   404

             Acknowledgments   415

             A Note on Sources   419

             Notes   421

             Bibliography   477

             Index   481

Author's Note

This is a work of nonfiction, based primarily on firsthand reporting over the course of five years, including four years covering Apple for the *Wall Street Journal*. More than two hundred current and former Apple employees spoke with me, providing perspective from every level of the company. I also interviewed their family members, friends, suppliers, competitors, and government officials. Many of them spoke with me for several hours on separate occasions over a span of many months. Most agreed to help on the condition that I not identify them as sources, noting that Apple has a history of making legal threats against people who speak about its business. I also drew on decades of news articles, books, court filings, and other published material. Those sources are detailed at the end of the book.

The dialogue that appears in the book is taken from video or audio recordings, or is reconstructed based on the recollections of people familiar with the events described. When faced with conflicting accounts of situations, I used the most plausible descriptions and provided endnotes detailing other recollections if warranted.

Cast of Characters

Tim Cook: Chief Executive Officer (2011–present), Senior Vice President for Worldwide Operations, COO (1998–2011)

Jony Ive: Chief Design Officer (2015–2019), Senior Vice President of Design, member of design team (1992–2015)

EXECUTIVES

Angela Ahrendts: Senior Vice President, Retail (2014–2019)

Katie Cotton: Vice President of Worldwide Communications (1996–2014)

Eddy Cue: Senior Vice President, Services (2011–present, joined Apple in 1989)

Steve Dowling: Vice President, Communications (2015–2019, joined Apple in 2003)

Tony Fadell: Senior Vice President, iPod Division (2005–2008, joined Apple in 2001)

Scott Forstall: Senior Vice President, iOS (2007–2012, joined Apple in 1997)

Greg Joswiak: Senior Vice President, Worldwide Marketing (2020–present, joined Apple in 1986)

Luca Maestri: Senior Vice President and Chief Financial Officer (2014–present, joined Apple in 2013)

Bob Mansfield: Senior Vice President, Hardware Engineering (2005–2012, joined Apple in 1999, remained as an adviser on future projects after 2012)

Deirdre O'Brien: Senior Vice President, Retail + People (2019–present, joined Apple in 1988)

Peter Oppenheimer: Senior Vice President and Chief Financial Officer (2004–2014, joined Apple in 1996)

Dan Riccio: Senior Vice President, Hardware Engineering (2012–2021, joined Apple in 1998)

Jon Rubinstein: Senior Vice President, Hardware Engineering and iPod Division (1997–2006)

Phil Schiller: Senior Vice President, Worldwide Marketing (1997–2020, joined Apple in 1987 and 1997)

Bruce Sewell: Senior Vice President and General Counsel (2009–2017)

Jeff Williams: Chief Operating Officer (2015–present, joined Apple in 1998)

INDUSTRIAL DESIGN

Bart Andre: Designer (1992–present)

Robert Brunner: Director of Industrial Design (1990–1996)

Danny Coster: Designer (1994–2016)

Daniele De Iuliis: Designer (1992–2018)

Julian Hönig: Designer (2010–2019)

Richard Howarth: Designer (1996–present)

Duncan Kerr: Designer (1999–present)

Marc Newson: Designer (2014–2019), LoveFrom (2019–present)

Tim Parsey: Manager, Industrial Design Studio (1991–1996)

Doug Satzger: Designer (1996–2008)

Christopher Stringer: Designer (1995–2017)

Eugene Whang: Designer (1999–2021)

Rico Zorkendorfer: Designer (2003–2019)

SOFTWARE TEAM

Imran Chaudhri: Designer (1995–2016)

Greg Christie: Vice President, Human Interface Design (1996–2015)

Alan Dye: Vice President, Human Interface Design (2012–present); Creative Director (2006–2012)

Henri Lamiraux: Vice President, Software Engineering (2009–2013, joined Apple in 1990)

Richard Williamson: Designer (2001–2012)

MARKETERS

Hiroki Asai: Vice President, Global Marketing Communications (2010–2016, joined Apple in 2000)

Paul Deneve: Sales & Marketing Manager, Apple Europe (1990–1997); Vice President, Special Projects (2013–2017)

Duncan Milner: Chief Creative Officer, TBWA\Media Arts Lab (2000–2016)

James Vincent: CEO, TBWA\Media Arts Lab; Managing Director, TBWA\Chiat\Day, Apple (2000–2006)

MUSIC MEN

Dr. Dre: Cofounder, Beats

Jimmy Iovine: Cofounder, Beats

Trent Reznor: Chief Creative Officer, Beats

Jeff Robbin: Vice President, Consumer Applications

ENGINEERS

Jeff Dauber: Engineer (1999–2014, Senior Director, Silicon, architecture and technologies)

Eugene Kim: Engineer (2001–present, named Vice President, 2018)

OPERATIONS

Tony Blevins: Vice President, Procurement (~2000–present)

Nick Forlenza: Vice President, Manufacturing Design (2002–2020)

SERVICES

Jamie Erlicht: Head of Worldwide Video (2017–present)

Peter Stern: Vice President, Cloud Services (2016–present)

Zack Van Amburg: Head of Worldwide Video (2017–present)

BOARD OF DIRECTORS

James Bell (2015–present)

Mickey Drexler (1999–2015)

Al Gore (2003–present)

Bob Iger (2011–2019)

Susan Wagner (2014–present)

FASHION

Andrew Bolton: Head Curator, Metropolitan Museum of Art's
        Costume Institute

Karl Lagerfeld: Designer

Anna Wintour: Editor in Chief, Vogue

COLLEAGUES

Martin Darbyshire: Cofounder & CEO, Tangerine

Jim Dawton: Designer, Tangerine, classmate at Newcastle Polytechnic

Clive Grinyer: Cofounder, Tangerine

Peter Phillips: Partner, Tangerine

PARENTS OF TIM COOK AND JONY IVE

Donald Cook

Geraldine Cook

Mike Ive

Pam Ive

Prologue

The artist loitered in the dimly lit corridor of a San Jose theater, waiting for his cue. He knew his lines, understood what was expected. Knowing that others were studying him, he wore a face that betrayed nothing.

It was early June 2019, and Jony Ive's presence was required at a product demonstration event after one of Apple's annual gatherings, the ritual performances where the secretive company unveiled its newest wonders, all of which Ive had been instrumental in designing. Dressed rich casual in loose-fitting linen pants, a T-shirt, and a woven cardigan, he was fifty-two years old now and had nothing left to prove. It was no exaggeration to say that his way of seeing, his love of pure, simple lines, had already redrawn the world. Yet he was never satisfied with his own creations, noticing imperfections invisible to others such as a watch he considered a millimeter too thick or the infinitesimal gap where iPhone parts intersected. He saw poetry inside the machine. He found inspiration in the curve of flowers and the color of tropical waters. He considered imitation to be lazy theft, not flattery. When he stood among the members of his team, they felt as though any problem were solvable, any breakthrough possible.

Yet here he was, waiting like a bit player for his moment under the lights, passing time before an oak table that held a newly made

Mac Pro. He knew its every detail. He had been in the studio as his design team had discussed the holes of deep-sea coral that brought ocean reefs to life. He had watched as that conversation helped create an aluminum computer frame with a series of overlapping holes that could breathe air and heat into and out of the machine. The result was a computer that looked unlike any other in the world.

Standing before his latest marvel, Ive looked bored.

Then a buzz rippled from the theater's entryway. Tim Cook, Apple's chief executive officer, strode into the room flanked by incoming *CBS Evening News* host Norah O'Donnell. Journalists and photographers backpedaled before him with boom microphones and cameras capturing his every move. The fifty-eight-year-old Cook was trim and muscular, the product of predawn workouts and a lifelong diet of grilled chicken and steamed vegetables. He had been at the helm of the world's largest publicly traded company for nearly a decade, overseeing a period of tremendous revenue growth that had lifted its valuation to $1 trillion. His ascent to that corporate pinnacle was a remarkable journey for the product of a small Alabama town, where a future managing a Denny's would have been more probable than a rise to become one of the world's most admired CEOs.

In many ways, Cook was Ive's opposite. He had risen through the ranks from the supply side of the company. His gift was not the creation of new products. Instead, he had invented countless ways to maximize margins, squeezing some suppliers and persuading others to build factories the size of cities to churn out more units. He considered inventory evil. He knew how to make subordinates sweat with withering questions. Though he had started as a wizard of spreadsheets, he was rapidly distinguishing himself as a master politician who had forged global alliances with the presidents of both the United States and the People's Republic of China. A single sentence from his mouth could send the world's stock markets into free fall.

The clicks of cameras that saluted him were deafening. Ive stepped into the commotion and greeted Cook. Then they both turned toward the computer to play their parts in a set piece of contrived spontaneity.

Ive acted as though he were showing his CEO something he had never seen before. Cook feigned earnest curiosity as though he were unaware that this was all a ritual of marketing. The staginess of it left some in the audience smirking.

The moment was so awkward that Ive could hardly stand it. He lingered under the lights for only a few minutes, until he'd finished his lines, and then stepped away as the cameras zoomed in on Cook. Almost no one noticed as Ive glided through the crowd and slipped out a side door, disappearing.

The truth was, Ive had been slipping out of focus for years. Apple was no longer his beautiful creation. He was no longer the star of the show. The cameras no longer clicked for him, and news anchors no longer invited him to wax poetically about design. The outside world wanted to know what the company was going to do about tariffs, immigration, and privacy. They wanted Cook. The creative soul of Apple had been eclipsed by the machine.

Chapter 1

One More Thing

Jony Ive steeled himself outside the stately two-story home in Palo Alto. It was early Tuesday morning, October 4, 2011, and a storm system shrouded Silicon Valley's usually sunny flatlands with heavy clouds. In better circumstances, Ive would have been arriving in Cupertino. Apple was hosting a special event there that day to introduce a new iPhone that he had designed. Instead, he was skipping the show to see his boss, friend, and spiritual partner, Steve Jobs.

Ive entered a house that had become a hospital. Doctors and nurses shuffled around inside where Jobs, sick with pancreatic cancer, was confined to bed. In the study-turned-bedroom where he lay, a TV was wired to provide the world's only video stream of Apple's product event, a private screening for Apple's longtime showman.

Visiting the home weighed heavily on Ive. Since Jobs had gone on medical leave at the start of the year, he had continued to summon Ive and other leaders from Apple's design, software, hardware, and marketing teams to the house. Inside, the passing of time could be measured by the CEO's weight loss and diminished movement. His face had grown gaunt, and his legs had atrophied into rigid branches. He seldom left the bed where he was surrounded by photographs of family, prescription bottles, stacks of paper, monitors, and machines. Still he refused to stop working.

"Apple's not a job for me," he would tell them. "It's part of my life. I love this stuff."

On that day, as Ive headed for the room where Jobs lay, he passed an image by the photographer Harold Edgerton, the man who froze time. The photograph showed a red apple suspended against a space-blue background the instant after it had been shot by a bullet, exploding its core.

ABOUT FIFTEEN MILES AWAY, Tim Cook pulled into the asphalt parking lot outside Apple's headquarters at 1 Infinite Loop. The company's thirty-two-acre campus, a ring of six off-white buildings, was located in Cupertino, just off Interstate 280 and behind BJ's Restaurant and Brewhouse, a national chain and an unassuming base of operation for a company racking up yearly profits of nearly $26 billion.

It was perhaps the most important Tuesday of Cook's career. Two months earlier, Jobs had elevated him from chief operating officer to chief executive officer. The timing of the promotion had caught the world by surprise. Apple and Jobs had concealed the severity of Jobs's illness and deprived staff, investors, and the media of insight into his deteriorating health. When he ceded power to his longtime lieutenant, he assured employees and investors that he would remain involved in product development and corporate strategy but said that Cook would lead the business, a position that thrust the company's back-office manager to the forefront of its product release events.

In a few hours, about three hundred journalists and special guests would descend on Apple's campus for Cook's first keynote address. Such events were usually held in large auditoriums or convention centers in San Francisco, but this one was being hosted in a cramped lecture room known as Town Hall on the back side of its campus.

Apple's showrunners had chosen the small venue, close to home, intentionally. Jobs was the ultimate showman. Cook was not. Apple's cofounder had turned corporate presentations into product theater, winding up the audience with carefully crafted stories that framed the purpose of a new device with a simplicity that excited potential customers and fueled sales. He had hyped the iPhone as being three things: a phone, a music player, and an internet communication device. He had turned the slender MacBook Air into a silver rabbit so thin it could be pulled out of a brown office envelope. And he had persuaded the world that iPods weren't about playing songs but about changing the way people discovered and enjoyed music. Cook was more comfortable evaluating supply chain logistics than standing in front of an audience. When he had taken the stage at previous events, it had been in a secondary role to detail the number of computers sold or stores opened. But with Apple's leading man ill, the time had come for Cook, the understudy, to step into a starring role.

Holding the meeting in Town Hall reduced the risks of Cook's debut. The venue was the equivalent of Off Broadway, with fewer seats for journalists and critics who might pen bad reviews. It was also on campus, so Cook had been able to walk over to the venue from his office throughout the week for repeated rehearsals. He had spent hours going through his scripted remarks in a bid to become desensitized to potential stage fright. The size of the venue meant that there were fewer cameras, a smaller crew, and less noise. The familiarity of the surroundings focused his attention on the most important task: delivering his lines.

STAFF NICKNAMED THE PHONE coming out that day "For Steve."

Over the past three decades, Jobs had cemented his place as a man of such original vision that he had drawn comparisons to both Leonardo da Vinci and Thomas Edison. Working from his parents' ranch home in Los Altos, California, he and his friend Steve Wozniak, a self-taught engineer, developed one of the first computers for the masses, a gray box with a keyboard and power supply that could display graphics. In 1977, their company became formally incorporated as Apple Computer Inc., a name inspired by Jobs's favorite band, the Beatles, and their record label, Apple Records. Jobs's brazen salesmanship of their computers was dismissed by some as all pitch, no substance, but the Apple II computer became one of the first commercially successful PCs, earning the company $117 million in annual sales before it went public in 1980. It made Jobs and Wozniak millionaires and secured their place in Silicon Valley mythology with a rags-to-riches tale that had begun in a garage.

A masterful marketer with an eye for design, Jobs redefined the PC category in 1984 with the Macintosh, a computer for the masses that could be controlled by the click of a mouse rather than by pecking on a keyboard. He pitched it as a machine that would democratize technology and dethrone the largest computer maker, IBM. Working with the advertising agency Chiat/Day, he developed an Orwellian Super Bowl spot titled "1984" that cast the Macintosh and Apple as a sledgehammer-wielding Olympic sprinter who shatters a giant screen projecting Big Brother. He unveiled the computer a week later at one of Apple's first signature events, captivating the audience in a darkened auditorium in Cupertino by turning the computer on and allowing it to speak for itself, saying, "Hello, I'm Macintosh. It sure is great to get out of that bag."

But a sales slump in 1985 led the board to oust Jobs in favor of John Sculley, a former PepsiCo executive. Sculley pushed Apple to new sales heights until Microsoft's Windows software began whittling away at Apple's market share. The company was a late entrant to the laptop market, and infighting led to Sculley's ouster. His successor, Michael Spindler, who joined the company in 1993, flooded the market with Apple computers, a strategy that only deepened its woes. The company lost nearly $2 billion in two years and was on the cusp of bankruptcy in 1996 when it struck a deal to buy a desktop computer company called NeXT that Jobs had launched while in exile.

Jobs returned to Apple and ignited one of the most remarkable business comebacks in history. He culled its product lineup, used NeXT's operating system as the foundation of OS X, a faster, more modern software system, and spearheaded the development of a translucent, candy-colored desktop called the iMac that returned the company to sales growth. He then pushed Apple beyond computers into consumer electronics with the iPod, which came out in 2001 and put thousands of ninety-nine-cent songs into people's pockets. The iPhone followed in 2007, introducing a touch-screen system that changed communication and became one of the best-selling products in history. Its successor, the iPad, which came out in 2010, redefined tablet computing. The string of product successes turned Jobs into a cult hero.

Apple's most ardent customers were as fervent about and protective of the company as members of a religious cult. Some tattooed its corporate logo or advertising phrases onto their wrists. As CEO, Jobs assumed an almost messianic hold over them, and his daily uniform—a black turtleneck, Levi's 501 jeans, and New Balance sneakers—added to his ecclesiastical bearing. He could distort reality. He refused to accept limits in engineering or manufacturing that might impede one of his ideas, and he could persuade his team of designers and engineers that they could achieve what seemed impossible. He was so convincing that some believed he might even outlive death.

THOUGH JOBS HADN'T ATTENDED rehearsals ahead of that day's event, some of Apple's leadership arrived at Town Hall that morning wondering: *Will he show up?*

Staff saved an aisle seat at the front of the lecture room for him, draping a black piece of cloth with the word *Reserved* in white over the back of a tan-colored chair. Apple's general counsel, Bruce Sewell, who sat in the adjacent seat, knew that the odds were against Jobs filling it. Jobs's health had worsened in recent days, but he had surprised everyone before and even some of his closest advisers had not given up hope that the empty seat would be filled by the time the event began.

The lights were low when Tim Cook slipped into the front of the room from behind a dark screen with a white Apple logo. His thin lips formed a flat grin as a few people applauded politely. In a Brooks Brothers spin on Jobs's casual and fashionable Issey Miyake turtleneck, Cook wore a black broadcloth button-down shirt and spun a presentation remote in his hands as he paced in front of the crowd.

"Good morning," he said. "This is my first product launch since being named CEO. I'm sure you didn't know that."

He smirked, hoping that his dry humor would take some tension out of the room. A strained chuckle rippled through the audience. Though the joke hadn't landed, Cook pushed forward. "I love Apple," he said. "I consider it the privilege of a lifetime to have worked here for almost fourteen years, and I am very excited about this new role."

His voice grew confident as he shifted his focus to Apple's growing retail business. The company had just opened two amazing stores in China, he said. The one in Shanghai had set a record by welcoming a hundred thousand visitors in its first weekend, a total that Apple's flagship store in Los Angeles had taken a month to achieve after its debut. Cook transitioned to business highlights from Apple's Mac, iPod, iPhone, and iPad products, complete with line graphs and pie charts. "I'm pleased to tell you this morning that we have passed the quarter of a billion unit sales mark," he said. "Today, we're taking it to the next level!"

Cook ceded the stage to Jobs's other top lieutenants. Mobile software chief Scott Forstall detailed new messaging capabilities, services head Eddy Cue followed with a demonstration of iCloud, and marketing chief Phil Schiller revealed the iPhone 4S, which featured longer battery life and a better camera but looked like its predecessor. The event culminated with Forstall doing a live demonstration of Apple's new virtual assistant, Siri, that with the push of a button and a vocal question pulled up the weather, showed stock prices, and listed nearby Greek restaurants.

"It's pretty incredible, isn't it?" Cook asked as he reclaimed the stage. "Only Apple could make such amazing hardware, software, and services and bring them together in such a powerful yet integrated experience."

His enthusiasm failed to win over the hardened tech press. The journalists and technology analysts in the audience were unimpressed. One technology analyst told the *Wall Street Journal* that the presentation had been "underwhelming." Another expressed disappointment that Apple had stuck with a 3.5-inch iPhone screen instead of pushing it to 4 inches. Fans grumbled on Twitter. Investors dumped shares, sending Apple's stock price down as much as 5 percent and erasing billions of dollars in market value. It was a box-office rejection.

Cook and the rest of Apple's leadership had no time to process the public reaction. As the event ended, Jobs's wife, Laurene, texted some of his top lieutenants—Cook, Phil Schiller, Eddy Cue, and Katie Cotton, the vice president of worldwide communications—to say they should come to the house. As the group huddled, they were gripped by fear: either Jobs didn't like the event and wanted to chew them out, or his health had worsened.

They sped to the house, some fifteen minutes away, hopeful that Jobs's wrath awaited. It was easier to imagine him angry than process their unspoken despair that his health had prevented him from attending.

When they arrived at the Tudor-style home, Jony Ive had already departed after spending time alone with Jobs that morning. Laurene told the parade of business executives that Jobs was doing poorly and wanted to speak with each of them individually. He wouldn't be reprimanding them. He wanted to say goodbye.

THE FOLLOWING AFTERNOON, October 5, 2011, a symphony of notification dings rang across Infinite Loop. An alert appeared atop Apple employees' iPhones, delivering the news "Steven P. Jobs, Apple cofounder, dead at 56." It was among the first times in history that a founder-led company's employees had learned about their longtime chief executive's death on the revolutionary product he had created and they had brought to life.

A group of two dozen software engineers were in the middle of a meeting about product plans when the notification hit manager Henri Lamiraux's iPhone. He stopped the meeting to share the news and watched as fellow programmers pulled out their phones to confirm what few wanted to believe. Without speaking, they shuffled out of the room in silence.

Less than fifteen miles away, Ive sat in the garden outside Jobs's home. The October sky above was hazy that day and his shoes were too tight. Cook joined him and they sat together for a long time. Ive felt numb as he recalled the last words Jobs told him: *I will miss our talks together.*

Farther up the peninsula, General Counsel Bruce Sewell, who had left immediately after the show for a business trip, was stuck in the cabin of a plane that had just landed at San Francisco International Airport. People's phones began to buzz, and hushed gasps reverberated around him. Nearby, someone whispered, "Did you see that?" Sewell hadn't turned his phone on yet but knew immediately that his boss had died. Though the people in the cabin around him didn't have a personal relationship with Jobs, they felt a connection to him through the Apple devices they held in their hands. Now they were wrestling with a question he'd been anticipating the entire flight: What would Jobs's death mean for Apple and the world?

Obituaries of Jobs dominated the front pages of the *New York Times* and the *Wall Street Journal*. He was credited with having transformed the fruit orchards of the San Francisco peninsula into a global innovation hub. Though he wasn't a hardware engineer or software programmer, he had defined Apple's product goals, assembled its talent, and prodded its teams to deliver what many initially considered to be impossible. He had enabled all of that through his charismatic leadership style and willingness to take big risks that had inspired loyalty even in the face of his occasionally caustic demeanor. "Apple has lost a visionary and creative genius, and the world has lost an amazing human being," Cook said in a letter to employees. "We will honor his memory by dedicating ourselves to continuing the work he loved so much." He reassured them that Apple was not going to change.

JOBS HAD ANTICIPATED the pitfalls ahead. He had been troubled by how the Walt Disney Company had floundered after its cofounder's death; he had lectured Polaroid's leadership after it had forced its founder, Edwin Land, out of the company; and he had become alarmed as Sony had lost its way without the direction of Akio Morita, the marketing virtuoso behind the Walkman. He believed that once-great companies often declined after they became monopolies, innovation slowed, and the products they made became an afterthought. Eventually, they put salespeople in charge and prioritized how much they sold instead of what they sold. Companies such as Intel and Hewlett-Packard were different. "They created a company to last, not just to make money. That's what I want Apple to be," he told his biographer, Walter Isaacson. (In 2015, Hewlett-Packard was split up after a seventy-five-year run. By 2020, Intel was falling behind rivals in manufacturing more compact and powerful silicon chips.)

Much like Jobs, Walt Disney had built an empire through a combination of vision, ambition, and chance. He had grown up on a Missouri farm and dreamed of becoming a cartoonist. In 1923, he had moved to Hollywood and founded Disney Brothers Studio with his older brother Roy. Disney had concentrated on storytelling and helped come up with the brothers' first hit character, Oswald the Lucky Rabbit. The distribution contract gave the rights to Oswald to the distributor, Universal Pictures, so Disney recast the rabbit as a big-eared character named Mickey Mouse. The character had taken off when Disney had added sound, a novelty that had made it a global sensation. He had hired animators and developed new characters such as Goofy and Donald Duck before pushing into feature-length films with *Snow White and the Seven Dwarfs*.

Disney had structured his company much as Jobs structured Apple. It had been flat, staff members hadn't had titles, and everyone had been called by their first name. "If you're important to the company," Disney said, "you'll know it."

The philosophy at Apple was the same. The company had only three C-suite titles before Jobs's death: chief executive officer, chief operating officer, and chief financial officer. Another seven people served as senior vice presidents on the executive team. There were about ninety vice presidents, who developed and managed the products the company sold. Below them, there were senior directors and directors. On paper, everyone reported to the finance chief. The structure eliminated bureaucracy, a feature Jobs disdained and disregarded by communicating directly with some of Apple's most talented employees.

Walt Disney had created a similar informality by building a company where everything flowed through him. After he had died of lung cancer in 1966, the company's output flatlined as people began asking "What would Walt do?" rather than taking their own creative leaps. By the 1980s, the company's share of the film market had dropped to 4 percent. It hadn't regained its box-office or financial footing until Michael Eisner had become CEO in 1984 and backed a string of film hits.

Polaroid was another Jobs obsession. He considered its co-founder, Edwin Land, to be one of America's greatest inventors. Land had been defined by many of the traits later associated with Jobs, such as vision, drive, and salesmanship. He had preceded Jobs in championing the idea of a company that sat at the intersection of technology and the liberal arts. He had started Polaroid after creating a process for coating products with polarized film to reduce glare, including sunglasses. He had later invented a process to create instant photographs. From the introduction of its first camera in 1948 until Kodak developed a similar product in 1976, Polaroid had been the world's preeminent camera maker. Land's next invention, an instant home movie camera, had flopped, and he had been pushed out of the company. After his departure, Polaroid had refined its existing products rather than introducing new ones, leading Jobs to upbraid its management during a visit to the company around 1983 that it had become irrelevant.

Sony was the company Jobs knew best of all. In the 1980s, he had visited its headquarters in Japan and met with its cofounder Akio Morita. Like Jobs, Morita and cofounder Masaru Ibuka had relied on their instincts when making product decisions. The Walkman had been born out of Ibuka's request for a portable music player to take along on international flights. Morita had tested an early prototype and brought it to market four months later with clever print ads that showed the device beneath the banner "Why man learned to walk." The product had exploded in popularity, and over the next decade, Sony had developed eighty models. Under Morita's direction, Sony had acquired record labels and movie studios in the belief that the company would benefit from controlling the songs and films that played on its music players and TVs. Morita had ceded his chairmanship in 1994. The company had tapped one of its star marketers as chief executive officer, and he had aimed to make Sony behave more like a traditional corporate titan than a company that acted on its founders' intuition. Instead, the company's electronics business had languished, and it had failed to deliver another hit.

Three great companies, led by three creative founders; none was the same without them.

JOBS WANTED APPLE to defy the fate of Disney, Polaroid, and Sony. In 2008, he hired Joel Podolny, the Yale School of Management dean, to create Apple University. He wanted a curriculum that would teach Apple newcomers what differentiated the company from its peers. When Podolny asked during the interview process how many classes should be offered and how large the faculty could be, Jobs scoffed. "If I knew the answers to those questions, I wouldn't need to hire someone like you," he said. Podolny persevered in creating a curriculum with classes such as "Communicating at Apple," which emphasized clarity and simplicity in products and presentations. There were also case studies on important decisions such as Apple's to make the iPod and iTunes compatible with Microsoft Windows.

But it would take more than codifying Jobs's thinking to ensure that Apple succeeded. The CEO wasn't preoccupied with Harvard Business School concepts of organizational behavior. The company he had built operated like a starfish. He sat at the intersection of legs that focused on excellence in marketing, design, engineering, and supply-chain management. He would crawl out to the end of a leg when he wished to and get personally involved, directing each division as he saw fit.

Before his death, Jobs pressed to keep the legs of Apple's starfish together. He approached members of the executive team individually and pushed them to commit to remaining several more years at the company. "Tim's going to need you," he told them. He urged the board of directors to give the executive team retention stock grants, a request it fulfilled in an emergency meeting days after his death. Each executive received 150,000 restricted stock units with half becoming available in 2013 and half in 2016. Valued at the time at about $60 million each, the grants went to Sewell, Forstall, Schiller, finance chief Peter Oppenheimer, hardware chief Bob Mansfield, and supply-chain guru Jeff Williams. Cue, who had recently joined the executive team on a trial basis, received a smaller grant. Ive was believed to have been granted more than $60 million, but his award was not disclosed because he had arranged to avoid being classified as an officer of the company to keep his compensation secret. The largest allocation went to Cook, who received 1 million shares valued at $375 million, not just elevating Cook in the eyes of Wall Street but also putting him above the rest by giving him the kind of wealth typically reserved for one of Silicon Valley's deified corporate founders.

THE MORNING AFTER JOBS DIED, Cook brought the executive team together in the company board room on the fourth floor of 1 Infinite Loop. Jobs's chair, which was second from the end, remained empty. Cook sat to its right and Schiller to its left, as they always had. They would keep the empty chair between them at meetings for some time, a visual reminder that he was always present.

Cook encouraged everyone to share memories of Jobs. For many of them, losing Jobs was like losing a parent. Jobs had approved almost every business decision they had made for more than a decade. They shared stories and personal memories about him. Cook had assured the group that he wanted to preserve the heart and soul of the company Jobs had created and signaled that he had no intention of making any immediate changes. As they told stories, there was a collective sense that the greatest mark of respect they could give Jobs was not only to keep the company alive but to keep it at the forefront of technology by making great products.

TWO WEEKS LATER, thousands of Apple employees filled the grass courtyard inside Infinite Loop for a memorial service celebrating Jobs's life. Stores around the empire were closed for an official day of mourning so that employees worldwide could tune in for a livestream of the event. The crowd on campus roared as Cook came onto a stage flanked by billboard-size black-and-white portraits of Jobs. Cook described Jobs as a visionary, a nonconformist, an original, the greatest CEO and most outstanding innovator of all time. He didn't tell any personal stories about their relationship. Though Jobs had entrusted the company to Cook, he had considered his longtime chief operating officer to be an enigma. To those who worked alongside them, their connection was their corporate devotion. Cook's dispassionate words reflected their businesslike bond.

"I know Steve, and Steve would have wanted this cloud to lift for Apple and our focus to return to the work that he loved so much," Cook said. He put his hand over his heart before ticking off the principles that Jobs had made central to the company's identity: the conviction that a team, not individuals, achieve great things in business; the imperative that staff refuse to accept work that was "good enough" and always push to deliver the "insanely great"; and the commitment that every product they create be beautiful.

"He thought about Apple until his last day," Cook said, "and among his last advice for me and for all of you was to never ask what he would do. 'Just do what's right,' he said."

Jobs's guidance gave Cook stature. It showed that Cook was among the last to have spoken with the visionary and reminded staff that the late CEO had chosen Cook to lead them. Cook's speech warned them that the company's future would be connected with but not chained to its past. Without Jobs, its identity would have to change.

Ive followed Cook to the lectern and tucked his sunglasses into the neck of his black T-shirt. He laid his notes down and looked out at the mourners gathered before him, a terrifying sight for someone who loathed public speaking.

The crowd looking up at Ive had seen him regularly eating lunch with Jobs on the edge of the very courtyard where they stood. They knew that Jobs considered Ive the second most important person at Apple after only the CEO himself. They recalled that when Jobs hadn't been in his office, he could often be found in the nearby design studio, where Ive's insular team of about twenty industrial designers had sketched the Apple products that reignited the company's business and earned Ive more operational power than anyone else at the company. In recent days, Ive had agonized over finding the right words to capture their deep working relationship and longtime friendship.

"You know Steve used to say to me a lot, 'Hey, Jony, here's a dopey idea,'" he began. "And sometimes they were really dopey."

The crowd laughed.

"Sometimes they were truly dreadful," he continued. He paused. The silver watch on his left wrist sparkled in the sun as he moved his index finger to the next line. "But sometimes they took the air from the room and they left us both completely silent. Bold, crazy, magnificent ideas. Or quiet, simple ones which in their subtlety, their detail, they were utterly profound."

Entranced, the crowd fell silent as Ive explained that Jobs had treated the creative process with reverence and appreciated that ideas were fragile, susceptible to being squashed before they took flight. The two had often traveled together, and Jobs's demand for excellence had been so great that Ive said he had never unpacked his own bags after checking into a hotel. Instead, he would sit on the bed and wait for Jobs to call and say, "Hey, Jony. This hotel sucks. Let's go!"

The crowd laughed, then grew quiet again as they listened to Ive describe what it had been like to develop a new creation with Jobs, how he would push for months to do what many said was impossible.

"He constantly questioned: Is this good enough? Is this right?" Ive said. "And despite all his successes, all his achievements, he never presumed, he never assumed that we would get there in the end."

As Ive scooped up his notes, he told the crowd that Apple had arranged for a special performance in memory of Jobs. "Will you please help me welcome our friends Coldplay," he said. He turned from the lectern and stalked back to his seat beneath a nearby white tent as the British band that had been featured in one of Apple's iPod commercials began to play their first hit, "Yellow."

As lead singer Chris Martin wailed into the microphone, Ive and the executive team looked on, their blank faces concealing their grief and all the anxieties roiling within them. The single most important figure in their professional life was gone. How could Apple go forward without him?

The answer would depend largely on Cook and Ive.

children, who told him that they felt stigmatized by how the hearing aids looked. He told friends that an improved design would both help children hear and alleviate the stress of being different. He expected their focus at school to improve, their grades to rise, and their parents' worries to recede. The product he designed featured a rectangular white receiver with a belt clip and headphones to enable students to hear what a teacher said into a microphone.

Computers weren't central to the design curriculum at Newcastle, but before graduating, Ive was introduced to his first Macintosh. His father, still a school inspector, had become a champion of Apple's computers, encouraging schools to buy them because they were easier to use for students learning design than the more widely available BBC Micro computers. Ive quickly understood why. The Macintosh, with its point-and-click mouse, was more intuitive than any other computer he'd come across. He felt a connection with it that led him to research the company. He loved the rebelliousness of its "1984" ad and saw that spirit in the Mac. "I had a sense of the values of the people who made it," Ive told *The New Yorker* years later.

IN ORDER TO GRADUATE, Newcastle students had to take a familiar concept and reimagine it. The school called it the "blue sky" project, a name that encouraged students to imagine futuristic products that weren't constrained by current production capabilities.

Ive had become interested in the rise of credit cards and meditated on the disconnect between the cheap plastic material and the amount of money people spent with a swipe. He also puzzled over how a merchant could track a purchase instantly, even though a card user had to wait for a mailed paper statement. He imagined a world in which people carried circular medallions the size of a pebble that they would place on a minicomputer at checkout. The glossy black medallion would charge on a device the size of a pocket calculator questions. He obsessed over a sketch Grinyer had pinned above his desk that showed the angled vents of a computer housing. His eyes widened when Grinyer told him that he'd made them while working in California.

"What's it like?" Ive asked.

"It's not like here," Grinyer said. In the United Kingdom, designers often developed concepts that clients spurned. In San Francisco, Grinyer said, designers' ideas were funded and made as envisioned. To Ive, it sounded like "the hot place at that time," Grinyer recalled years later. "He was just amazed and loved hearing about that attitude there, the idea that there was an environment where design could take flight."

Ive's enchantment with San Francisco contrasted with the reality of his experience at Roberts Weaver. His impulse to break boundaries with his designs made some clients uncomfortable and led them to reject his ideas or demand changes. The former was wounding; the latter, vexing. For a toolmaker who wanted a new jackhammer, Ive sketched a sleek, futuristic handheld drill with an artful diamond-shaped handle. Its maroon-and-purple color scheme gave it a prettiness that was hard to imagine in the hands of a heavyset man in a hard hat. The client dismissed it as too elegant. "He just had ambitious ideas and would draw things beautifully," Weaver said. "You'd see the form, but you wouldn't know how to make it."

When he returned to Newcastle, Ive went to greater lengths to define what products should be before embarking on a design. He considered the children's hearing aids of that era stigmatizing. The National Health Service provided a product that coupled a radio receiver that children clipped to their lapel that was connected by a wire to a bulky pink earpiece. He wanted to make a device that looked as modern and cool as the Sony Walkman people carried everywhere. He visited a primary school and spoke with hearing-impaired

ing and lifted it toward the light. His hands trembled with panic as his eyes glimpsed small deviations from his design specifications. He abruptly rose and left the group, upset.

Worried that the part was flawed, Forlenza grabbed a red Sharpie and handed it to Ive. "Circle the things that are wrong," he said. "I'll get them to fix them."

Ive glared at him. "I've got a different idea," he said. "Get me a bucket of red paint. I'll dip this in it and wipe off the things that are right."

Ive held an aluminum sheet above his head and rolled it beneath the overhead lights, showing Forlenza how the reflection revealed almost imperceptible blemishes. He wanted them eliminated. Forlenza explained the problem to the supplier, and when the group returned two weeks later to review the part again, the blemishes were gone.

Ive ratcheted up scrutiny of the supply chain as Apple's product line expanded. When SARS broke out in 2003, the company was preparing to produce its first desktop made of aluminum, the Power Mac G5. The tower computer was the width and height of a paper grocery bag with smooth aluminum sides framed by front and rear panels that featured tiny holes like those in a citrus zester.

Ive wanted to be there as it rolled off the assembly line, so he and members of the operations team flew to Hong Kong on some of the first post-SARS flights. They then headed to Shenzhen, where Ive spent the next forty days sleeping in the factory dormitory and walking the manufacturing floor. He could be intense as he surveyed an assembly line. During the assembly process, he would grab colleagues and point at a factory worker who was crudely handling parts.

"I don't want him touching our products," he would say. "Look at how's he's touching the side of it!"

During the production of the Power Mac, Ive halted manufacturing because he decided that a plastic air vent looked as though it had been spray-painted. He considered paint an impure way to cover up poorly made parts and used it as seldom as possible. Though the vent was on the rear of the unit, he worked with the factory to develop a metal-plating process to give the part a nickel finish. It was time-consuming and expensive, but he refused to compromise.

One day, Ive walked by Forlenza and handed him a piece of paper with a sketch in black ink of an L-shaped stand with a hinge connected to a computer display. "Do you think you can make that?" he asked. Forlenza looked at the drawing with disbelief. He was focused on finishing the desktop in front of them while Ive imagined a future product. They discussed the difficulty of bending a piece of aluminum as the sketch illustrated and agreed to investigate it.

When Ive and Forlenza went to the Hong Kong airport to return home days later, it was still almost empty because of the epidemic. They grabbed seats at an empty bar in the airport lounge and ordered coffee. As Ive sipped on a cappuccino, he stared down the stainless-steel bar and quietly said, "I can see every seam in this bar."

Forlenza followed Ive's gaze down the bar. He saw nothing but thirty feet of smooth silver metal. He decided that Ive, who had a glum look on his face, must have X-ray vision.

"Your life must be fucking miserable," he said.

OVER THE FOLLOWING YEAR, Forlenza and his team went to Chicago and worked with an auto parts company to design a mechanical process to make Ive's L-shaped computer display stand. The design created two manufacturing challenges: aluminum that thick would have dark lines, and bending it could crinkle its exterior like an orange peel. Ive wanted to minimize both imperfections. Forlenza's team, with assistance from the vendor, met his expectations partly by analyzing the grain refiner in the aluminum and determining that elements of boron had contributed to the dark lines, which could be reduced by adjusting the grain. When Forlenza told Ive what they had learned by going deeper into the supply chain, Ive smiled. "We never had this kind of control," he said. "Now that we know how it's made, we can control how it looks."

Control became part of Ive's ethos. He began extending his influence beyond sketches and models to materials. He co-opted Forlenza, turning the operations executive into an extension of the design studio. Ive disliked the name of Forlenza's operations team: Enclosures. One day in the studio with Forlenza, Ive went to a whiteboard with a red pen and scribbled a dozen alternatives on the board, finally underlining the term "manufacturing design," combining it with the names of the teams he worked most closely with: industrial design and product design. The result was a triangle of design that would shape Apple's products for years. The industrial designers defined how the product looked; the product designers determined how the components worked; and the manufacturing designers oversaw the way everything was assembled. Although product design reported to hardware and manufacturing design reported to operations, the leaders of both groups embraced the opportunity to work closely with the studio. They spent much of their time with Ive, who was quietly reshaping the company to put the design studio at the forefront of operations. With time, he would teach the group to think of themselves as craftsmen, an extension of him.

Working together, the trio of design teams would improve the quality of products, reduce the number of defects, and increase output. Everyone at Apple was happier, especially as the demand for products swelled.

AFTER HIS RETURN, Jobs pressed to create a portable music player. The nascent MP3 market sparked dreams of a next-generation Sony Walkman. The project took flight after Jon Rubinstein, the head of hardware engineering, discovered that Toshiba's semiconductor unit had created a miniature disk drive that would hold a thousand songs. He pushed to buy the rights to every disk Toshiba made. To run the project, Jobs hired Tony Fadell, a hardware engineer who had worked on General Magic's personal digital assistant. Rubinstein and Fadell assembled the components, while Apple's head of marketing, Phil Schiller, contributed the idea of creating a wheel to scroll through songs, a concept inspired by a Bang & Olufsen phone. They handed the ingredients to Ive to package.

The design concept struck Ive during his daily commute between San Francisco and Cupertino. While meditating on how to give the brick of components aesthetic appeal, he imagined an MP3 player in pure white with a back side of polished steel. The metal would feel significant, providing a weight that would convey the amount of work artists had put into the thousands of songs the device held, while the white player and headphones would make the device look simultaneously bold and inconspicuous, planting it between the original black Sony Walkman and its brilliant yellow successors.

The design faced resistance internally. Colleagues questioned the stainless-steel case and molded body, and challenged Ive's vision for engraving Apple's logo on its rear rather than on its front. They also expressed doubts about the idea of white rather than the more commonplace black headphones. Despite those competing views, Jobs supported the proposals by Ive and the design team.

From its shape to its color, the device was a subtle extension of the Walkman-inspired hearing aid Ive had made at Newcastle. Apple was already working with white for its computers. The studio favored it because the designers believed that color could alienate people, especially in mass production. White was fresh, light, and acceptable, allowing them to make a single model and forgo needing to make a rainbow of colors to appeal to everyone. For the iPod, Ive wanted a new white. Satzger, who led color materials, worked with his colleagues to create a saturated white they called Moon Gray.

After Apple released the iPod in October 2001, its ad agency, TBWA\Media Arts Lab, considered the white case its most unique feature in a crowded market that included about fifty other portable MP3 players. James Vincent, a Brit at the agency, proposed showing black silhouettes of people wearing white headphones dancing against colorful backgrounds. The spots, which debuted in 2003, were put to songs such as Jet's "Are You Gonna Be My Girl?" The combination of the ads and the arrival of iTunes, a digital store offering songs for 99 cents, helped the iPod explode in popularity. Apple went from selling 1 million in 2003 to 25.5 million in 2005. Its annual revenue soared by 68 percent, to $14 billion, as the iPod transformed the beleaguered computer company into a consumer electronics giant.

Despite the triumph, Ive was disappointed. His design team had been less central to the product's development than it had been to the iMac. He wanted a bigger voice in conceptualizing what Apple made. At the time, he reported to Rubinstein with a dotted line to Jobs. His insistence on specific design features provoked clashes over details ranging from the polish he wanted on a computer to the specially designed screws he insisted it include. Rubinstein, who brought together all aspects of a product from chips to firmware to design, rejected some of Ive's ideas as too expensive. Ive bristled. He disliked conflict and disdained design compromises, so he went around Rubinstein to Jobs. Colleagues compared them to two little kids fighting for Jobs's attention. Jobs's advisers urged him to stop enabling Ive. Eventually, Rubinstein left and later became chief executive officer of a rival device maker, Palm. Afterward, Jobs streamlined Apple's reporting structure so that Ive reported directly to him, ensuring that the designer was the second most powerful figure after the CEO.

"There's no one who can tell him what to do, or to butt out," Jobs told Isaacson. "That's the way I set it up."

Ive's tangles with Rubinstein showed his willingness to play internal politics. Some colleagues considered him to be the most politically shrewd member of Apple's leadership team. He earned a reputation on campus as a British gentleman, often opening and holding doors for colleagues, even years after he had become one of the company's top executives. He had a generous spirit and endeared himself to colleagues by arranging the delivery of flowers or champagne to their hotel rooms during family vacations. A by-product of that kindness and generosity was greater loyalty across a company that worked to bring his designs to life.

But he also could be harsh. If he didn't like an interaction with a colleague, he sometimes pushed to oust that person from Apple, leading people in human resources to recall hiding staff from him. Within the design team, he demanded mutual respect and collaboration. He had little tolerance for ego and wanted to make sure all voices were heard, including people who were soft-spoken. Designers said Thomas Meyerhoffer, the Swedish designer whose translucent laptop had influenced the iMac, fell out of favor with Ive after asserting himself in meetings and dismissing his colleagues' work. Eventually, Meyerhoffer left to set up his own agency. Valuing his talent, some team members tried to persuade him to stay, but Ive and the group found they worked harmoniously without him.

After Ive brought in a new color specialist, Satzger, who had a prominent role, found himself similarly closed out of the group. Each member of the roughly fifteen-person team interviewed potential new hires and discussed the candidate before extending an offer. During one such process, Satzger was preparing to interview a candidate when he was told that the schedule had been rearranged. Ive gathered the group to discuss their interviews and fired the first question at Satzger.

"What did you think?" Ive asked about the candidate.

"I didn't interview him," Satzger said.

"Why not?" Ive asked.

Confused, Satzger explained that the schedule had been rear-ranged, which he thought Ive had known. Satzger left Apple soon thereafter. Ive told people that he fired Satzger. Looking back years later, he thought that Ive had rearranged the schedule intentionally and asked him the first question to lessen Satzger's credibility with his colleagues.

"Being a British gentleman can be a tool," said Parsey, a Brit whom Ive had displaced as the number two of the design team. "It opens doors like you wouldn't believe. That doesn't mean that's who you are. It's a tool. Look at all the British classics, how they charm their way in, these guys are fucking pirates stealing countries."

IVE'S NEWFOUND CLOUT reshaped Apple. At most product compa-nies, engineering defined a product, as it had at Apple with the iPod, and design packaged it. Jobs's elevation of Ive meant that the studio led product development and engineers worked to fulfill its demands. Designers defined how a product would look and had an outsize voice in its functions. Staff began summarizing their power in a single phrase: "Don't disappoint the gods."

Ive solidified the studio's position by carefully maintaining the environment where its members worked and controlling who came and went. He wanted the workplace to be quiet during meetings and focused on making the most aesthetically and functionally pure products possible. If staff from engineering or operations didn't bring reverence to the discussion, if they talked too loudly, or even worse, if they brought up costs, they would later discover that their badge no longer accessed the studio; their admittance had been silently re-voked.

The unspoken judgments fueled a company adage: Don't talk to Jony unless spoken to.

Ive leaned on his colleagues to maintain order. When an operations staffer spoke about the manufacturing challenges of a proposed design, Ive would later pull his manager aside and say that the staffer's comments had disturbed the design process. He wanted people in the studio to understand what the designers wanted and find ways to do it—not erect barriers by mentioning costs or production limitations. "I'm not going to let someone drive the bus just because their fucking legs are long enough to reach the pedals," he would say.

His attention to detail permeated product development. If a cost-conscious supplier chose to use low-priced, reground plastic, Ive could detect with a glance that they had violated Apple's requirement to use virgin material. The higher-priced plastic was critical to its consistently high-quality computers. His perceptiveness spurred the operations team to ensure suppliers met his demands because he would catch any miscues. In a reference to the movie *The Sixth Sense*, the team jokingly said of Ive, "He sees dead people."

The designers enjoyed more perks than their peers did. They went to off-site retreats at five-star resorts in California's wine country. They stayed at luxury hotels in Asia, while colleagues in operations and engineering stayed at more traditional three- or four-star places. In Hong Kong, Ive's hotel of choice was the Peninsula, a five-star colonial building that served afternoon tea as a string quartet played. Apple pampered them.

Inside a company of nerdy engineers, they embodied art school cool. They dressed casually in T-shirts, hoodies, and designer jeans. They drove expensive cars, among the priciest being Ive's Aston Martin DB9, which had cost about $250,000. They obsessed over their hobbies: De Iuliis perpetually searched for the world's best coffee; Julian Hönig, an avid surfer, shaped his own boards; and Eugene

Whang created a record label, Public Release, and DJ'd under his nickname, Eug, at clubs.

They lived like rock stars. After product events, they loaded limousines with bottles of Bollinger champagne and went out for dinner and late-night drinks. They became regulars at the Redwood Room, a historic bar in downtown San Francisco that served artisanal cocktails, and sometimes traveled to Los Angeles for parties with Apple's advertising agency. Drugs from quaaludes to cocaine, which a designer kept in a bullet-shaped snifter, could be available. It was all part of a work-hard, play-hard culture in a group of Renaissance men devoted to art and invention.

THE DESIGN TEAM'S POWER made it a safe harbor for engineers exploring new ideas. When an engineer named Brian Huppi wanted to develop ways to control a computer without a mouse, he approached designer Duncan Kerr about investigating how to do so. Ive loved the idea. With his blessing, Kerr and a team of software engineers, including Bas Ording, Imran Chaudhri, and Greg Christie, started an R-and-D project to find a way to control a device with the touch of a finger. They soon discovered a Delaware-based company that made a touch pad to control a computer. The team bought one of the pads and reworked it, projecting the image of a Mac onto a table to see what it was like to navigate the screen by finger. They wrote code to zoom in on maps, drag files, and rotate images. When Kerr shared the technology with the design team, the group was gobsmacked.

When Jobs visited for a private demonstration, he was less enthusiastic. He dismissed the idea. It seemed clumsy and—as it was still the size of a large table—impractically big. But Ive persisted, with a nudge as subtle as the one he'd given Jobs years earlier on the iMac design. "Imagine the back of a digital camera," he said. "Why would it have to have a small screen and all of these buttons? Why couldn't it be all display?"

Jobs warmed to the idea, and multitouch, as they called the technology, became the foundation of the iPhone. Interest in making a phone had been simmering for several years at Apple. The company's leaders found existing mobile phones clunky and cumbersome. They also feared that a rival might make the iPod redundant by combining an MP3 player and a phone into a single device. To avoid that fate, Jobs set what became known as Project Purple into motion.

From about 2005 to 2007, engineers and designers slaved to create the new device. Ive imagined the phone's touch screen being like an infinity pool, a glowing window that would transport people to a broader world of music, maps, and the internet. The design team worked through several concepts before settling on a style inspired by Sony's product minimalism. A matte black face and a brushed aluminum frame contained an expansive screen. Fadell's hardware team brought it to life with components, and Scott Forstall, a top software executive, led the creation of its transformational software.

On December 19, 2006, Ive and Forlenza arrived in Shenzhen, exhausted after the two-year grind. They entered a factory's poorly lit conference room, where a hundred of the first iPhones ever made lay on a table. They were supposed to pick thirty of the best-made models to feature at a product unveiling event. A group of forty factory workers lined the room as Ive and Forlenza surveyed their work.

"Don't say anything," Ive leaned over and whispered to Forlenza. "We could ship any of these." He was used to only a fraction of the first models being free of defects, but the products before him looked as refined as any camera made by Canon, the gold standard in mass-produced electronics. The manufacturer gave Ive the confidence that Apple could make millions of phones with the kind of handcrafted precision of something on display in a museum. Ive grabbed Forlenza's shoulder and whispered, "Now we can make anything."

A MONTH LATER, Jobs unveiled the iPhone at San Francisco's Moscone Center, trumpeting it as an iPod, phone, and internet-connected computer rolled into one. The first call he placed was to his closest collaborator.

"If I want to call Jony, all I do is push his phone number," Jobs said, pressing Ive's number on the iPhone in his hand.

"Hello, Steve," Ive said from a flip phone in the audience.

"Hi, Jony, how are you doing?" Jobs asked, beaming.

"I'm good. How are you?"

"It's been two and a half years, and I can't tell you how thrilled I am to make the first public phone call with iPhone," Jobs said.

The call drew comparisons to Alexander Graham Bell connecting by phone with Thomas Watson more than a hundred years earlier. Apple sold 11 million iPhones in its first full year on the market, a tenfold increase from the number of iPods sold after the music player's debut.

Jobs and Ive took more pride in the way the iPhone became a cultural sensation than its unit sales. Over the years, they had discussed how to measure success and agreed that it would not be dictated by share price or sales volume. Under those metrics, rival Microsoft became successful only to eventually stagnate. Instead, they decided it would come down to their subjective opinion: Were they proud of what they designed and built?

The iPhone seemed to be the pinnacle of what they could achieve. But as sales took off, Ive weighed leaving Apple. His twin boys were born in 2004. The forty-year-old was also burned out after fifteen years of eighty-hour weeks. He bought a $3 million, ten-bedroom house with a lake near his parents in Somerset and had his father oversee a renovation in anticipation of spending more time near them. He told his longtime friend Clive Grinyer that he was tired and thinking of retiring and making high-end luxury goods with his friend the designer Marc Newson. But the iPhone's popularity and Jobs's illness changed everything.

IN MAY 2009, Jony Ive arrived at the San Jose airport to welcome Steve Jobs home. The CEO, who had cancer, was returning from Memphis after a liver transplant. Ive and Chief Operating Officer Tim Cook greeted him, and Jobs's wife, Laurene Powell Jobs, uncorked a bottle of sparkling apple cider to celebrate. But all was not well.

Simmering beneath the celebration was Ive's irritation. He was struggling around that time with how Apple's growth had compressed its product development cycle, particularly for the 54 million iPods it sold annually. He had an idea to change the screen of one in a new exciting way, but it arrived after a production deadline. He lamented to colleagues that it would have to wait another year because his creative epiphany was two weeks too late. Then, after Jobs became sick, newspapers began casting doubt on the future of Apple. They reasoned that its cofounder had birthed the company and later resurrected it. Without him, it had withered. If the past was prologue to the future, they suggested, Apple was doomed.

Jobs relied on the team he had assembled at Apple to do the work but largely took credit for himself. He disapproved of staff doing interviews and discouraged talk about Apple's creative process. The strategy preserved product secrecy and lessened the likelihood that top talent would be poached by rivals. It also fed a public perception that every product was the result of individual genius, not teamwork. Though Jobs's first public iPhone call had been to Ive, that subtle gesture hadn't fully conveyed Ive's contributions to it. On the way to Jobs's house from the airport, the designer unloaded.

"I'm really hurt," he said, according to Jobs's biographer Isaacson. (Through a spokesman, Ive disputed this account.) Ive complained about the perception that Apple's innovations flowed from Jobs. The truth was that Ive and many others were critical to their success.

JOBS AND IVE TEAMED UP on one more major product. In many ways, the iPad was the least taxing and most rewarding product they made. They had been toying with the idea of making a tablet before they made the iPhone, and Jobs had resurrected work on it before his transplant. The iPhone would inform the design of the tablet, which would use the same software. The biggest question was: What size should it be?

Ive kick-started the evaluation by making twenty models in various sizes with rounded corners. He invited Jobs to the studio, where he laid them out for review. They went from model to model, evaluating each one's look and feel. They settled on a nine-by-seven-inch rectangle that sat flat on the table like a legal pad. As they worked with it, Jobs decided that it was too severe. Ive understood; the device lacked the kind of rounded edges that had made past products like the iMac approachable. He later rounded the edges slightly so that a finger could slide beneath it and lift it off a table.

In January 2010, Jobs introduced the iPad at the Yerba Buena Center for the Arts in San Francisco. He reclined in an easy chair to demonstrate how simple it was to scroll the internet and read a digital book. The product became an immediate hit, with Apple selling 25 million iPads in a little more than a year.

Jobs's health worsened after the iPad's release. When he became too sick to go to work in 2011, Ive began regularly visiting him at home. The two would discuss ongoing work on the iPhone, the plans for a new Apple campus, and the yacht Jobs was having built to sail around on with his family.

After Jobs died on October 5, 2011, colleagues worried about what would happen to Ive without Jobs. They had watched over the years as Jobs's feedback in the studio had sharpened Ive's work like a talented editor enhancing the storytelling of a gifted writer. Ive assured people that he would be fine. He drove himself, he said. Still, the studio felt listless and unmoored without its longtime patron. "Everyone felt fallible," a designer said.

Ive hired a high-end photographer named Andrew Zuckerman and threw himself into a project developing a book entitled *Designed by Apple in California* full of glossy close-up photographs of Apple's hit products against white backgrounds. He pored over his past work, wallowing in mementos from a lifetime of work with the creative partner he had lost.

TWO MONTHS AFTER Jobs's memorial service, Ive received a knighthood for his work in design. He was nominated by the Design Council, a nonprofit dedicated to supporting industrial design. The honor elevated Ive to Knight Commander of the Most Excellent Order of the British Empire, or KBE, the most senior rank of the British Empire, making him Sir Jonathan Ive.

On a sunny day in London in late May, Ive cinched a powder blue tie around his neck and donned a black tailcoat for the ceremony at Buckingham Palace. The formality of the occasion would have amused Jobs, who had chided Ive about Brits' stuffiness, but it carried tremendous importance for the British designer and son of two teachers. The recognition validated his life's work, and in class-based England, the commendation gave him a formal title. Though similar recognition had been given to famous Americans such as Steven Spielberg and Bill Gates, Ive viewed the distinction with the pride of a native son.

Accompanied by Heather and his twin boys, Ive entered the palace ballroom and took a seat in front of a small stage with two thrones made of gilded wood. A royal official soon summoned him forward to receive the honor. As Bach's Concerto for Two Violins in D minor played, Ive walked in measured steps toward the thrones with his head lowered. He bowed, with a grin, before Princess Anne, the queen's daughter, and kneeled as she tapped his left and right shoulders with a sword that had belonged to her grandfather King George VI.

Later that day, Ive shed the tails and tie for a reception at the Ivy, an exclusive restaurant in the heart of London's West End. The Design Council had rented a private room with stained-glass windows and assembled a guest list of Ive's friends, including the actor Stephen Fry, Duran Duran front man Simon Le Bon, and the influential designers Paul Smith and Terence Conran. Ive's sister, Alison, and his mother and father were there, as well.

As the attendees sipped champagne and nibbled on appetizers, Ive mingled with a smile. The room grew quiet when former prime minister Gordon Brown, who had supported Ive's knighthood, proposed a toast. Ive, who craved attention but felt uncomfortable in the spotlight, put his left hand on his son's shoulder and smiled sheepishly as Brown recalled visiting the Cupertino design studio and seeing the team at work.

Ive's father, Mike, looked on with a smile. Over the years, he had made countless things. He'd built hovercrafts and cabinetry, restored old cars and made wedding bands, fashioned school curriculums and forged teapots, but on display in the room that day was the product he admired the most. He told friends that he considered Ive to be his finest creation.

Chapter 5

Intense Determination

The Honda Accord could be seen before dawn, humming down the dark emptiness of the 101 freeway past the dim shadows of low-slung office buildings and shopping centers. Though Apple had given Tim Cook a base salary of $400,000 and a $500,000 signing bonus, he didn't put a lot of value on what car he drove. It was just four wheels that got him to the gym and the office and back to his Palo Alto apartment at the end of the day.

After relocating from Texas in 1998, Cook rented a 544-square-foot apartment in Palo Alto more fit for a college student than an executive. The tight quarters and location reflected the reality that he would spend most of his time at work. He was only a twenty-minute drive from Apple's Infinite Loop campus. The sleepy, tony suburb of Palo Alto lacked the vibrancy of San Francisco, thirty miles north. It was home to Stanford University and a host of startups that clustered their operations around a tree-lined main street, University Avenue, where restaurants and coffee shops buzzed with talk about the latest dot-com company and the newest venture capital investment. People walked or biked everywhere, which was perfect for Cook, who enjoyed bicycling around town just as he had in Robertsdale.

In his first days at Apple, Cook called an operations meeting. He wanted to know every detail about the company's dreadful supply chain. Apple had spent the previous year trying to escape from the clutches of a cash flow quagmire. Before Jobs's return in 1997, unsold computers had been piling up everywhere. The company was running its own factories in California, Ireland, and Singapore. It had a surplus of computer parts and was holding nineteen days of inventory. CFO Fred Anderson had tried to resolve the balance sheet woes with a program called "Crossing the Canyon" that had aimed to reduce inventory. The operations team, hollowed out by departures after Jobs's return, detailed the headway they had made as Cook peppered them with questions: "Why is that? What do you mean?"

"I saw grown men cry," said Joe O'Sullivan, who was the acting head of operations when Cook arrived. "He went into a level of detail that was phenomenal."

The meeting set the tone for how Cook would lead. He whipped staff forward through interrogation with a precision that reshaped the workplace. Intense, detailed, and exhausting, there was little margin for error. He seemed to absorb and retain all the information his underlings provided and learned the business faster than anyone had expected. Jobs had asked O'Sullivan to spend four months with Cook to teach him Apple's operations; Cook mastered them in four or five days. He peeled away at issues with question after question. Silence followed. His Socratic style created a tense atmosphere that caused staff to squirm.

"Joe, how many units did we produce today?" Cook would ask.

"It was ten thousand," O'Sullivan would answer.

"What was the yield?" he asked, referring to the percentage of units that passed quality assurance before shipment.

"Ninety-eight percent."

Unimpressed by the efficiency, Cook would probe deeper. "So how did the two percent fail?"

O'Sullivan would stare at Cook, thinking, *Fuck, I don't know.*

The operations team learned to scour every aspect of production and prepare answers to any question Cook could imagine. They drilled down into the performance of specific parts and each assembly line's production results. Their boss's appetite for specifics led everyone to become "almost Cook-like," O'Sullivan said.

COOK CALLED INVENTORY "fundamentally evil." Computers and computer parts that sat on shelves were like vegetables; if they lay around long enough, they spoiled and would have to be trashed. Apple's operations team had reduced the number of days of inventory by two-thirds since Jobs's return. Cook wanted more. In 1998, he traveled to the company's offices in Singapore to assess how it could make more improvements. The local operations team prepared for his arrival. They flew in Mountain Dew, his drink of choice, supplied the conference room with bananas and energy bars to power his fourteen-hour workdays, and stocked up on the ingredients for steamed chicken and vegetables because he often ate the combination for both lunch and dinner. Seated at the head of a conference table, Cook rocked back and forth in his chair as he listened to manufacturing staff outline the state of its business. The team considered his rocking as an encouraging sign. As they had adjusted to their expressionless, measured boss, they had discovered that his habit of rocking signaled that he was satisfied with the material being presented. When the motion stopped, it indicated that he had found a problem and planned to fire off a question that would expose a flaw.

That day, the group had put together a presentation on inventory turns, a measure of how often it used and replaced its inventory. The more inventory turned, the less money a company spent on spoiling parts. The operations team detailed how it had increased its turn rate to more than twenty-five times a year, up from eight times, making it second in the industry only to Dell.

As they wrapped up, Cook stopped rocking. He stared at them in silence. "How would you get to a hundred turns?" he asked.

"I knew you'd say that," said O'Sullivan, who had begun anticipating Cook's unquenchable appetite for excellence. "We're nearly there."

O'Sullivan detailed a plan to make further improvements. He felt good as he finished, but Cook stared at him, betraying no appreciation for the extra effort.

"How would you get to a thousand?" Cook asked.

A few presenters laughed at what they considered to be an impossible request, but Cook stared at them icily. He was serious.

"A thousand turns?" O'Sullivan asked with disbelief. "That's like three times a day."

No one spoke. Cook watched as the operations staff looked at him incredulously. The marker he set became their goal.

Within a few years, Apple built computers to order and held almost no inventory on its books. The operations team's pursuit of that goal included painting a yellow line down the middle of factory floors. Components on one side of the yellow line remained on a supplier's books until Apple moved them across the line to be assembled into a new computer. That reduced Apple's costs because, under generally accepted accounting principles, the company wasn't in possession of the inventory, even though it sat in its own warehouse, until the parts moved down an assembly line. The concept, which was groundbreaking at the time, became an industry standard.

APPLE WAS IN the early phases of its revival when Cook arrived. The candy-colored iMac shipped five months after he'd been hired. During production, Apple fell behind schedule and needed to add equipment to catch up. An operations executive recommended adding seven production tools that cost about $1 million each, but with the company still short on cash, O'Sullivan thought it could catch up by adding three tools, enough to lift production from seven thousand units per day to ten thousand. Cook overruled him. "We're going to ship as many as we can as fast as we can," he said. The fourteen tools Cook authorized doubled Apple's capacity, enabling it to meet the surge in demand that followed the iMac's release. Though he could be frugal, Cook proved that he was willing to gamble and spend money when warranted.

About six months after Cook joined, Jobs approached O'Sullivan on campus, energized by Cook's performance. "What do you think?" he asked.

"I don't know," O'Sullivan said.

"What do you mean, you don't know?" Jobs asked.

"There's no magic," O'Sullivan said.

At a company that had thrived under Jobs's charisma, Cook quickly proved himself to be the CEO's foil. He was stoic and reserved, seldom showing emotion. He focused on numbers and gorged on spreadsheets. He worked long hours, hitting the gym before dawn and working into the evening. He rallied his team behind the Lance Armstrong quote "I don't like to lose. I just despise it." His disciplined approach delivered results. In his first year, the company cut its inventory from a month of product to six days. A year later, he slashed it down to just two days' worth and watched as the savings flowed to Apple's bottom line.

"He's like a relentless holding midfielder," said O'Sullivan, describing his new boss in soccer terms. He compared Cook to Roy Keane, a legendary defensive midfielder who had anchored Manchester United's defense through a series of title-winning seasons. "He's not the center forward, scoring goals and getting photographed at the nightclub," O'Sullivan said. "He's the quiet guy who does it all and goes home."

As Apple got its inventory under control, Cook began remaking its manufacturing operations. In 2000, he started shutting down its factories in favor of outsourcing production to contract manufacturers. Apple had employed the practice for years, but Cook pushed it into overdrive. While at Compaq, he had gotten to know Terry Gou, the founder of Foxconn Technology Group. The Taiwanese entrepreneur had built one of the world's most reliable assemblers of electronics. With $2,500 borrowed from his mother in 1974, he had set up a factory in Taipei to make plastic knobs to change TV channels before expanding into PCs in the 1980s. He had transformed computer manufacturing by setting up factories in China, where land and labor were cheap. Production contracts with Dell, Compaq, and others had increased the company's workforce to thirty thousand and its sales to $3 billion. Known to work sixteen-hour days, he demanded that products be released on schedule and in line with customers' specifications. He would personally step in to solve production problems. His demanding, hands-on style appealed to Cook, who brought the same approach to Apple.

In 2002, Cook tapped Gou to make Apple's next major computer release. The company planned to replace its gumdrop-shaped iMacs with a flat-panel version. At Apple's request, Foxconn built a supply chain to make 1,500 units a day, but Jobs decided to change the computer's target market from high-end professional customers to general consumers. Cook needed Foxconn to expand its capacity tenfold, and he needed it done fast. Apple's quarterly results would hinge on the success of the new computer. Cook flew to China with some of Apple's top engineers to oversee the expansion. He stayed there through Thanksgiving and Christmas, working on the factory floor to identify problems as computers came off the assembly line and escalate issues to Gou when needed. It was a high-stress environment, but Cook kept his composure throughout, radiating the same calm he had shown during challenges at IBM and Intelligent Electronics. It was emblematic of his situational management style that he remained emotionally indifferent and defused situations by adapting to changing circumstances. As December came to an end, Apple loaded the computers onto planes to ship them out of China, allowing the company to book the new units as sales for the September-to-December period and fulfill Wall Street's quarterly sales expectations.

Over time, Cook upended Apple's approach to suppliers of the components inside its computers. The procurement people tasked with buying parts had abided by the maxim that both parties needed to win for a relationship to work. Cook advocated a different approach: be relentless and uncompromising. In negotiations, he would take an unapologetically pro-Apple position on points such as price, delivery timetables, and more. He wouldn't yield an inch. Instead, he would identify and acquiesce to things the other party wanted that weren't priorities for Apple. He would often stay silent during talks to make suppliers uncomfortable. Sometimes he would go for long stretches without speaking, then lean forward. Here's what I want to do, he would say. Everyone at the table would hang on to what he said because it was often the first time he'd spoken in a meeting. Suppliers compared the tactics to military psychology techniques. After nearly finalizing terms with a chip supplier in the mid-2000s, Cook called the supplier and said he'd reconsidered. "I think you mistreated us, and don't think we'll negotiate anymore," he said. He then went silent for days, as the supplier worried it had lost the deal. "What he's hoping for is that you'll come back with a huge give in the eleventh hour," recalled the supplier, who eventually got the deal done. "Smart guys would have to say, 'Keep the faith.' It was old-school negotiations."

Cook's frugality extended to his personal life. Though his compensation topped $400,000 a year plus stock grants, he continued to live in the cramped apartment he was renting in Palo Alto for several years after joining the company. Colleagues joked that the only thing in the place was a single set of utensils, a plate, a bowl, and a cup. Rumors spread that there were termites in the place. Jobs and Jon Rubinstein, Apple's hardware engineering chief, eventually stopped by the apartment to stage an intervention.

"Dude, you've got to buy a house," Rubinstein said.

Nearly a decade after moving to Palo Alto, one of the nation's priciest housing markets, Cook finally purchased a relatively modest house, spending $1.97 million for the 2,300-square-foot home. Jobs lived a mile away in a stately home that was more than twice the size.

IN A COMPANY rife with drama, the operations team became notable for its lack of it. Cook didn't tolerate politics, and he expected everyone to collaborate. At the end of every quarter, he held a meeting to review where operations had fallen short of goals. A dozen of his lieutenants would write what they thought had gone wrong on Post-it notes and stick them onto a whiteboard. A problem could be as simple as forecasting sales of 100,000 iMacs and coming up 3,000 short. The notes were grouped together, ranked, and discussed. The meetings fostered a culture of accountability. No one wanted to throw a colleague under the bus. "It was like a confessional," O'Sullivan recalled.

The process helped Cook identify top performers. Jeff Williams, whom he had brought over from IBM, became his number two. Williams, who had grown up in Raleigh, North Carolina, and was as stoic as his boss, had earned a mechanical engineering degree from North Carolina State University and an MBA from Duke. Then there was Deirdre O'Brien, who had earned a degree in operations management at Michigan State University and an MBA at San Jose State.

She assumed the oversight for forecasting demand. And there was Sabih Khan, an economics and mechanical engineering major from Tufts University who had distinguished himself during a major manufacturing problem. "This is really bad," Cook had said. "Someone should go to China." He had later looked at Khan and asked, "Why are you still here?" Khan had stood up, driven to San Francisco International Airport, and flown to China without a change of clothes.

Cook's demanding stoicism created fear. Middle managers screened staff before allowing them to present to Cook to ensure they were deeply knowledgeable about the issues involved. They dreaded wasting Cook's time. If he sensed that someone was insufficiently prepared, he could lose patience and say "Next" as he flipped the meeting agenda page. Some people exited meetings in tears.

The result was a team largely fashioned in Cook's image: engineers and MBAs with an exhaustive work ethic. Even so, Cook pursued a diversity of perspectives from other areas in the company. When he went to fill a vacant position in human resources, the woman in the role on an interim basis encouraged him to interview other people. She thought she was too right brained and emotional to work with someone so rigidly left brained and analytical. He urged her to apply.

"I want to work with someone who thinks different," he said.

COOK'S DEVOTION TO WORK troubled his boss. Jobs pestered him to have more of a social life, a nudge that came from the personal enrichment the CEO had experienced by starting a family. On occasion, Cook would get an invitation to Jobs's home for dinner and arrive to find Jobs, his wife, Laurene, and other guests Jobs wanted him to meet. Cook once even got a call from his mother, Geraldine, after Jobs had called her and talked with her about Cook. "He knew how important family was in his life, and he wanted it for me, too," Cook recalled.

Though Cook had a strong connection with Jobs through work, he had only a distant relationship with Jony Ive. The inherent tension between operations and design sometimes put the two at odds. Cook's duty was to control costs by manufacturing as many products as possible with a minimum number discarded because of defects. Meanwhile, Ive's team was scrutinizing products coming off the assembly line to make sure that they were a close approximation of their sketches and models. When Ive identified an imperfection, it could disrupt production. It ate up time and added costs.

But Apple's supply chain began to be reshaped by the mixture of Cook's demand for operational excellence, Ive's insistence on superior design, and Jobs's willingness to spend what was necessary to make amazing products.

The iPod became a tipping point of that successful combination. Between 2002 and 2005, the number of iPods the company sold swelled from a half million to 22.5 million a year. During the development of the iPod Nano, which would become the company's best-selling product, Ive pushed to improve the tools that formed the MP3 player's colorful aluminum shell by asking that factory workers polish the tool that formed its casing. The production step departed from the more common practice of polishing a product's casing after it was forged, a step that Ive thought lessened product quality. Foxconn worked with Apple's operations team to fulfill Ive's request. Having learned to do so, Foxconn could then market its newfound expertise to other consumer products companies.

Chinese suppliers would clamor to work with Apple because its production demands and the number of products it sold could help manufacturers build their businesses. Cook's operations team would use their demand to Apple's advantage, hammering suppliers for lower prices than they offered anyone else in the market. The suppliers often agreed to the onerous terms, knowing that they could learn cutting-edge manufacturing techniques from Apple's engineers and then market those capabilities to other consumer electronics companies eager to catch up to Apple in product design. The dynamic deepened Apple's dependency on China and China's dependency on Apple.

THERE WASN'T MUCH INK spilled over Cook in those days. What he did was boring. The company's sex appeal was in its creative endeavors: the curvaceous backside of the iPod Nano, the striking ads featuring dancing black silhouettes wearing white earbuds, the oaken tables of the Genius Bars in its stores. The growing number of Apple fans didn't give much thought to how iPods were assembled, boxed, and sent to stores. They didn't care who was tallying the sales or getting the company's online store up and running. But companies in Silicon Valley had taken notice of Cook's work in those areas of the business.

When Hewlett-Packard began searching for a new chief executive officer in 2005, Cook's name topped the recruiters' list. His ability to turn around the mess at Apple had earned him a reputation among its competitors as one of the industry's top operations executives. Advisers to Jobs feared that they would lose Cook. They urged Jobs to promote him to chief operations officer and ease restrictions on executives serving as directors at other companies. It was not an easy ask. No one outside of finance at Apple carried the title of chief other than Jobs, and no one but Jobs, who also chaired Pixar, was on another company's board. Jobs was reluctant to empower his lieutenants. When he had previously given someone at Apple the title of chief, naming John Sculley CEO in 1983, Sculley had turned around and ousted Jobs from his own company. By that point, Cook was five years into overseeing manufacturing and sales, and he filled in for Jobs when he took a leave in 2004 because of pancreatic cancer. The advisers implored Jobs, "You can't lose him. No one else can do what he did. He reinvented the supply chain." The pressure worked. On a flight to Japan in the fall of 2005, Jobs turned to Cook and said, "I've decided to make you COO."

The promotion cemented Cook's hold on Apple's back office and allowed him to join the board of directors at Nike. His competent control of manufacturing, sales, and logistics freed Jobs to focus on the company's creative core: design, engineering, and marketing. A yin and yang formed, with Jobs's volatile divisions developing Apple's next transformational product, the iPhone, and Cook's detached operations team bringing the complex product to life on Foxconn's factory floor. One group created demand; the other fulfilled it. One group thrived on magic and invention; the other dominated through method and process.

COOK'S PROMOTION coincided with a masterstroke.

The most critical component of the iPod was its flash memory. Jobs expected its forthcoming Nano model, with its colorful, light-weight aluminum cover, to trigger a surge in demand. To meet it, he asked Cook's team to negotiate with the leading memory suppliers to secure a sufficient supply of memory. Eventually, they struck a multiyear deal with Intel, Samsung, and others for an up-front payment of $1.25 billion. Cook's top lieutenant, Jeff Williams, led the negotiations. The company cornered the market, boxing out its competitors, and was able to fulfill demand when it spiked as Jobs predicted.

The team pulled off a similar feat when Ive's design team dreamed up a new way to manufacture laptops. The design called for the laptop's case to be made with a machine that carved it out of a solid piece of aluminum. The technique, which was used by luxury car manufacturers and watchmakers, was unheard-of in computing, but the company expected it to reduce the thickness of the laptop by as much as 30 percent. The complex process would require thirteen unique machining steps, as well as laser drilling. It was ambitious and expensive.

At most companies, the cost would prohibit such an elaborate undertaking, but Jobs ignored such a practical concern. He saw the potential to make a sleeker, lighter laptop. It fell on Cook to make his and Ive's dream a reality.

To guarantee that Apple could make enough laptops at a reasonable price, the operations team struck a deal with a Japanese manufacturer to buy all the machines it could produce for the next three years. The computer-controlled machines, known as CNC machines, cost as much as $1 million each. Apple's purchase of ten thousand of them roiled the manufacturing industry, which was unaccustomed to such a large order from a single customer.

The design and manufacturing breakthrough paved the way for the MacBook Air, a three-pound laptop so thin that Jobs revealed it by sliding it out of an envelope. It also became part of the preferred manufacturing process for the iPhone. Perhaps more important, it transformed the computer and electronics industries. Soon other companies were trying to ape Apple and make similarly minimalist laptops.

The iPhone further tested Cook's operations prowess. The early model that Jobs carried onstage in January 2007 featured a plastic display. When he put it in his pocket, Jobs found that his keys scratched the surface. He decided six months before launch that Apple needed to replace the plastic display with glass. Cook and others worried that a glass display wouldn't be durable enough to survive a customer dropping their phone. They feared Apple stores would be overrun by people with cracked screens demanding a replacement. Cook's lieutenant, Williams, even told Jobs that the technology to make more resilient glass wouldn't be ready for three to four years.

"No, no, no," Jobs said. "When it ships in June, it needs to be glass."

"But we tested all the current glass, and when you drop it, it breaks one hundred percent of the time," Williams said.

"I don't know how we're going to do it, but when it ships in June, it's going to be glass," Jobs said.

Afterward, Jobs called the CEO of the glassmaker Corning and told him its glass sucked. The CEO, Wendell Weeks, went to Cupertino to meet with Jobs and told him about an unproven product called "gorilla glass" that had a protective compression layer. Cook and Williams worked with Corning to transition a Kentucky factory to producing enough glass in six months to fulfill the demand for history's best-selling product.

Each time Cook did the impossible, the company's fortunes rose. His unseen work became Apple's secret weapon.

WHEN JOBS'S CANCER returned in 2009, he took another leave of absence and left Cook in charge. Just as before, Cook kept Apple humming, but he faced more questions from Wall Street and the press than he had five years earlier because Jobs's health had visibly deteriorated. On an earnings call with Wall Street analysts shortly after Jobs went on leave, Cook sought to rebut critics with his version of Apple's creed. "We believe that we are on the face of the earth to make great products, and that's not changing," he said. "We are constantly focusing on innovating. We believe in the simple, not the complex. We believe that we need to own and control the primary technologies behind the products that we make and participate only in markets where we can make a significant contribution. We believe in saying no to thousands of projects so that we can really focus on the few that are truly important and meaningful to us. We believe in deep collaboration and cross-pollination of our groups, which allow us to innovate in a way others cannot. And frankly, we don't settle for anything less than excellence."

Dubbed the "Cook Doctrine," it had the same rallying clarity that defined Jobs's communication. It also showed that Cook's decade at Apple had imbued him with a deep understanding of the company's unique culture. It cemented his position as Jobs's most likely successor. There wasn't a true challenger. Three of Apple's most talented engineers, software developer Avie Tevanian and hardware executives Jon Rubinstein and Tony Fadell, had already left the company. Rising software star Scott Forstall was considered too young, hardware leader Bob Mansfield was regarded as too narrowly focused, and product marketer Phil Schiller was thought of as too divisive. Jony Ive was better at managing a small team than worrying about Apple's sprawling business. Retail chief Ron Johnson had the marketing and operational skills required but hadn't been exposed to many other areas of Apple's business. "He didn't have a choice," said one of Jobs's former advisers. "No one else could have taken that job. At least fifty percent of Apple's value was the supply chain."

On August 11, 2011, Jobs summoned Cook to his house. He said he planned to become chairman and make Cook the company's chief executive officer. They discussed what that would mean.

"You make all the decisions," Jobs said.

"Wait. Let me ask you a question," Cook said, trying to think of something provocative. "If I review an ad and I like it, it should just run without your okay?"

"Well, I hope you'd at least ask me!" Jobs said with a laugh.

Jobs said he had studied what had happened at the Walt Disney Company and how it had been paralyzed after its cofounder Walt Disney had died. Everyone had asked: What would Walt do? What decision would he make?

"Never do that," Jobs said. "Just do what's right."

The selection surprised some outsiders because—as Jobs told his biographer, Walter Isaacson—Cook wasn't a "product person." But insiders understood the choice. Cook ran a division devoid of drama and focused on collaboration. Apple needed a new operating style after losing someone irreplaceable.

Cook's parents were thrilled with his ascent. When he had been preparing to step in for Jobs in 2009, Cook's father had driven to the office of their local newspaper, the *Independent,* and offered an interview. Reporter Donna Riley-Lein had met his parents at their house, where they had sat in recliners and bragged about how he called home every Sunday regardless of where he was in the world. "He's always been real smart," his mother, Geraldine, said. "When he left home, I just about left with him."

Knowing that Cook was a bachelor, Riley-Lein asked them if Cook had a lady in his life, "so the ladies in Robertsdale would know and could stop grooming their daughters." The Cooks grew quiet. Riley-Lein quickly realized that she'd stepped into a sensitive subject and thought, *I'm not going there. I'll be asked to leave.*

The article was destined for the *Independent's* front page until Apple's media relations team called and pleaded with the paper's editor not to run it. The editor compromised by running it inside the paper. The tussle highlighted what would become a long-standing push to polish Cook's image and protect his privacy. Cook's upbringing in the NASCAR-loving South was out of step with the California-cool image of Apple that Jobs had marketed around the world. Cook also remained deeply private.

JOBS'S DEATH HIT COOK HARD. During one of his first public appearances after taking the job, Cook took the stage opposite the *Wall*

Street Journal's Walt Mossberg and Kara Swisher for an interview at the 2012 D: All Things Digital conference. The trio took seats in red leather chairs in a hotel conference room in southern California.

Cook was confident and funny, noting dryly that the company, whose reported sales had risen by 65 percent the prior year to $108 billion, was in the midst of a string of a "few decent quarters." He recounted how the iPad had taken off because of its popularity with customers, educators, and businesses.

"It's been unbelievable," he said. "Just a knockout, and I think we're in the first inning."

Mossberg gradually shifted the conversation to how Apple was functioning under the new CEO. The tech reviewer had known Jobs well and had attended his private memorial service. He knew better than anyone else how different Cook was from his predecessor.

"Obviously, Apple has undergone a tremendous change, a big loss, with the death of Steve Jobs," Mossberg said. "What did you learn from Steve as CEO and how are you changing things?"

"I learned a lot from Steve," Cook said. Then he shook his head and closed his eyes. He gulped. Several seconds passed. With his eyes still closed, he continued, "It was absolutely the saddest day of my life when he passed." As the room went silent, he stared out at the audience before him, lost.

"Maybe as much," he said. He stopped again. "As you should see or predict that," he said. "I really didn't. At some point late last year, somebody kind of shook me and said, 'It's time to get on.' And so that sadness was replaced by this intense determination to continue the journey."

COOK QUICKLY BROADENED Apple's attitude toward social causes. Less than a month after Jobs's death, he introduced a corporate match program for charitable gifts, paving the way to direct contributions by the company to the Anti-Defamation League and others. The step contrasted with Jobs's long-standing opposition to matching and preference to return cash to shareholders, who could donate as they saw fit. But it was in line with Cook's own track record of volunteering at a local soup kitchen and funding scholarships at his alma mater, Auburn. The change immediately generated goodwill among staff, and his company-wide emails that began "Team" augured a more inclusive, communicative style than his predecessor's.

Not everyone was reassured. Silicon Valley leaders predicted that Apple would falter. Loyal customers worried about future innovations. And Wall Street fretted about the road ahead.

Cook ignored the noise and followed Jobs's advice: "Don't ask what I would do. Do what's right." He continued waking up each morning before 4:00 A.M. and reviewing sales data. He drilled down into small details, discovering through questions that one model of iPhone was outselling another in a small city in Georgia because the AT&T stores there were running different promotions from those being run in the rest of the state. He held a Friday meeting with operations and finance staff, which team members called "date night with Tim" because it would stretch for hours into the evening, when Cook seemed to have nowhere else to be. For the most part, he focused on business and operations and avoided meddling in the creative areas of the business that Jobs had led, such as design and marketing. He declined invitations to join meetings with the software design team, and he seldom swung by the place where Jobs could be found daily, Apple's design studio.

"I knew what I needed to do was not to mimic him," Cook later said of that period. "I would fail miserably at that, and I think this is largely the case for many people who take a baton from someone larger than life. You have to chart your own course. You have to be the best version of yourself."

Doubts simmered at Apple about his approach. Soon after becoming CEO, Cook planned to announce that employees who had spent ten years at Apple would receive a commemorative gift: a crystal cube with a recessed etching of the Apple logo. It had been created by Ive's design team, and—just like every other product Apple made—it came in a custom box and unique packaging. Usually, the design team watched with eagerness as Jobs, who appreciated every aspect of what they did, enthusiastically unboxed their latest creation like a birthday present. The showmanship infused an extra dose of magic into what they made. They hoped that Cook would do the same.

After staff members packed into Town Hall, Cook told everyone about the award. Then he matter-of-factly held the cube up for everyone to see. It was less like a magic show and more like an elementary school show-and-tell. The designers looked at their new leader with horror and wondered, *Does he get it?*

That was when they knew things were going to be different.

Chapter 6
Fragile Ideas

Jony Ive arrived at Infinite Loop feeling energized. It was January 2012, and for the first time since Jobs's death, he felt a sense of purpose.

For months, he had been wrestling with finding a way to honor his boss, creative partner, and friend while also proving to a doubting world that Apple could carry on without its visionary leader. He wanted to develop what he called a new platform, a product that could be a vessel in the years to come, capable of adding new features and utilities that would transform the way people lived just as the iPhone had.

In recent years, engineers and designers across Apple had been turning over the question: What comes next? They were researching an array of possibilities, and one that kept bubbling up was health. In 2010, weary after a year of blood testing that had made him feel like a human pincushion, Jobs had engineered the acquisition of a startup called Rare Light, which claimed it could use lasers to detect glucose levels in blood, a potential life changer for diabetics. Its technology wasn't ready, but its existence became the seed of conversations about how to create a health device.

As Ive passed the courtyard, the site of his recent memorial speech, he was invigorated by a possible answer. He had been thinking a lot about how to make technology wearable. It was an idea that was percolating around Silicon Valley, courtesy of a gadget maker called FitBit that made a clip-on waistband pedometer to count steps. Ive wanted to take the concept of wearable technology further.

He gathered some of the design team inside the studio for a brainstorming session. They carried their sketchbooks and prepared to trade ideas about future products. As they settled in, Ive walked to a whiteboard with a dry-erase marker in his hand. He began to write a series of cramped lowercase letters. Then he turned around to face the designers. On the whiteboard behind him was a single word: *smartwatch*.

THE WORLD OUTSIDE INFINITE LOOP was turning on Apple. In the months after Jobs died, impatient investors and customers wanted to know what product was coming next. Jobs's practice of casting himself as the sole creator of the iPod, iPhone, and iPad had fueled doubt about what Apple could achieve without him. Oracle founder Larry Ellison, one of Jobs's close friends, predicted that the company was destined for mediocrity and positioned for the type of long decline that had occurred after Jobs had left the company in the 1980s. "He's irreplaceable," he told Charlie Rose during a CBS interview. "They will not be nearly so successful, because he's gone."

As Jobs's longtime collaborator, Ive felt pressure to silence the doubters. The idea of a smartwatch alleviated some of that stress, but when he raised it with Apple's leadership team, he immediately faced skepticism.

Software chief Scott Forstall, another of Jobs's favorites, raised concerns. The engineer behind the iPhone's operating system worried that strapping a miniature computer to people's wrists would distract them from everyday life. He feared that it would amplify an unintended consequence of the iPhone, a device so engrossing that it consumed attention, disrupted conversation, and endangered drivers. He fretted that a watch would worsen the interruptions in everyday life by moving notifications from people's pockets and purses to their wrists. Though he didn't rule out a watch, he said it should have capabilities beyond those already available on an iPhone. He preached caution.

Forstall's doubt irritated Ive. The designer believed that ideas were fragile, tentative things that came at unexpected times from unknown places. Rising up out of the ether, they initially seem obvious and brilliant but can quickly be deemed impossible, squashed by the recognition of some insurmountable hurdle that could prevent them from becoming a reality. He and Jobs shared a belief that ideas should be nurtured, not crushed. Now one of his most important ideas in months was being battered by a colleague's misgivings.

Instead of putting his complete support behind the watch, Forstall championed a project that Jobs had favored: reinventing TV.

Before his death, Jobs had told his biographer, Walter Isaacson, that he had dreamed up a way to reinvent television with a solution that would end people's mindless scrolling through channels to find what they wanted. "It will have the simplest user interface you could imagine," he had said. "I finally cracked it." But whatever the idea was, Jobs hadn't shared it widely.

After he had died, Apple's executive team had asked some of its top engineers to make a presentation on what the company could do in TV. Without a road map from Jobs, the software and hardware teams felt like antiquity scholars assigned to decode an ancient script. They delivered a wide range of ideas, including revamping the company's streaming video device, Apple TV, with a new remote control, home screen, and search system. It was all conjecture about what Jobs had imagined but failed to share.

Forstall, whose staff was involved in the presentation, championed the idea of creating a system that would pull TV channels into a single place so that people could search for shows with their voice. The system would also surface shows that people watched regularly and offer up related programs that they might enjoy. But for it to work, Apple needed TV networks to buy into it, a lengthy process that would be beyond its control.

With the external pressure mounting, it fell to Tim Cook to decide on Apple's next move: Ive's watch project or Forstall's TV effort. It was a choice that deepened a long-standing unspoken rivalry between two of Jobs's geniuses, one whom he had considered his creative partner and another whom he had emboldened by saying that hardware was a vessel for software, which he considered a product's true soul.

IF JONY IVE WAS Jobs's prodigy of industrial design, Scott Forstall was his prodigy of software design.

Born in 1969, Forstall was raised alongside two brothers in Kitsap County, Washington, across Puget Sound from Seattle. His mother was a nurse and his father a mechanical engineer. At school, he excelled at math and qualified for a gifted group with access to a classroom full of Apple IIe's. Coding came naturally to him, and he enjoyed writing words that made a machine perform tasks. He developed a local reputation as a computer whiz. In high school, he worked at the nearby Naval Undersea War Engineering Station writing software for nuclear submarines. He had the world's most powerful computers at his fingertips and marines with attack dogs behind him.

An overachiever, Forstall played soccer and performed in school theater. Drama became his favorite extracurricular activity because everyone worked toward a single goal and the audience supported the work. He played the lead in a production of *Sweeney Todd*, committing so deeply to the role that he hyperventilated backstage to make himself look crazed during a murder scene. He graduated from high school as co-valedictorian, sharing the achievement with his high school sweetheart and future wife, Molly Brown. Both attended Stanford University, where Forstall earned a degree in symbolic systems, a combination of philosophy, psychology, linguistics, and computer science. His studies put him at what Jobs had called the intersection of technology and humanities.

After graduating, he was asked to interview with Jobs's post-Apple computer company, NeXT. Its innovative operating system, called NeXTSTEP, had been developed to power a sleek black computer aimed at college and university researchers. When sales of the computers had faltered, Jobs had focused the company exclusively on software and sought more programmers. The hiring process was more like joining a club than a company. A dozen software engineers interviewed every applicant, putting equal weight on their technical skills and personal hobbies. Afterward, the group voted on each candidate. The emphasis on extracurriculars, a practice later adopted by Apple, created a company of expert programmers who were also part-time musicians, avid skiers, and die-hard surfers. Forstall's lifelong interest in drama—he would eventually become a Broadway producer—satisfied the hiring team, who wanted engineers with outside interests because it made the office a more interesting place to work and resulted in more thoughtfully made products.

Ten minutes into Forstall's interview, Jobs burst into the room, escorted the NeXT engineer out, and took over the evaluation. He fired a series of questions at Forstall and then grew silent. "I don't care what anyone else says the rest of the day, we're giving you an offer, and you're going to accept it," he said. "Pretend you care in the interviews, though."

Forstall worked on NeXT's software tools for applications and tried to stay on Jobs's radar. Each quarter, the Apple cofounder hosted an all-hands meeting with the company's four hundred employees. Forstall would spend hours the night before scribbling down questions he hoped would impress the CEO. The following day, he would lob his most challenging and imaginative question at Jobs. His colleagues viewed each calculated exchange as a testament to his eagerness and ambition.

The ties between Jobs and Forstall deepened after the Apple acquisition in 1997. Forstall moved into a management role leading future releases of the NeXT-based operating system, MacOS, and earned the loyalty of Jobs by delivering product on time and fostering an environment of creativity. He allowed staff to take a month after a software release to work on any project of their choosing. The policy helped birth new products for Apple, including the software design that became the impetus for a streaming video device, the Apple TV.

In 2004, Forstall had a severe reaction to a stomach virus, lost thirty-plus pounds, and ended up in Stanford University's hospital unable to eat without vomiting. Jobs called daily and eventually told Forstall that he was sending his personal acupuncturist to the hospital. "They're probably not going to like me bringing in an outside practitioner, so if I get stopped, I'm just going to dedicate a wing," said Jobs, whose reputation for being a jerk could overshadow his generosity to friends and colleagues. The acupuncturist stuck needles into Forstall's back, arms, and head. Afterward, he could hold down food. He later told people that Jobs had saved his life, adding to the CEO's messianic reputation.

Back at Apple, Jobs tapped Forstall to lead the company's nascent phone effort, code-named Project Purple. Forstall assembled a team of software designers and engineers, many with NeXT backgrounds, who worked to cram the Mac's powerful operating system into a phone and designed the features that allowed users to navigate a screen with the flick of a finger. Jobs met with Forstall and his top software designers weekly to vet everything from the way the home screen looked to the way users would be able to pinch to zoom photos.

When the iPhone debuted in 2007, its iOS software transformed computer interaction and ignited the smartphone revolution. Its success strengthened Forstall's relationship with Jobs. They would have lunch together regularly in the Apple cafeteria, where workers swiped their badges at the checkout register and the price of their meals was deducted from their paychecks. Jobs insisted on swiping his badge to pay for lunch. "This is great," said Jobs, who wasn't taking a salary. "I only get paid a dollar a year. I don't know who's paying."

The first iPhone didn't permit app downloads because Jobs didn't want a developer's software to infect phones with a virus. He favored building and selling only Apple-made apps. Forstall advocated opening the iPhone up to app makers and had his team start building protections for the software so that Apple could safely allow app downloads. Later, partly due to pressure from Apple's board of directors, Jobs agreed to create the App Store. At an event to tout its creation, he turned the stage over to Forstall to introduce the tools developers would use to build the multibillion-dollar app economy that would give the world Uber, Spotify, and Instagram.

THE HIGHER FORSTALL ROSE at Apple, the more he aped Jobs's style. *BusinessWeek* called him "the Sorcerer's Apprentice." Like his mentor, he favored black shirts and jeans and demanded excellence from staff. He obsessed over minor details such as speeding up the rate the iPhone's screen refreshed. At the time, some images on the screen updated as slowly as thirty times a second. He wanted faster updates so that scrolling to the bottom of a contact list, for example, would be seamless and match the speed of a finger swiping up the screen. "If we lose a frame, we break the illusion of the device. People will just see it as a computer," he told his software team. Engineers told him that they couldn't refresh the screen that often because the graphic processing chips were too slow, but Forstall insisted. Eventually, they found a way to refresh the screen sixty frames a second, a software leap that made it harder for rivals to copy the iPhone's performance.

Forstall's ascent came at a cost. To create the iPhone, Jobs pitted Forstall's software team against a hardware team led by Tony Fadell, a godfather of the iPod. Forstall and Fadell vied against each other for talent and clashed over Forstall's strict secrecy around the software team's work. After Forstall's design ideas triumphed, the animosity between the teams deepened because the hardware engineers believed that Forstall had blocked new features, such as a better camera, by discouraging his software engineers from prioritizing them. Forstall also rankled services chief Eddy Cue by insisting that the iTunes system for the iPhone be developed by the software team instead of Cue's staff, which had been managing the music service on computers for years. "Scott was very controlling on the iPhone," said Henri Lamiraux, a top lieutenant. "It was his world, and he didn't want anyone else in. He thought if people took pieces of the iPhone software, it was going to collapse."

The most problematic clash occurred with Ive. In 2010, Apple was in the final stages of producing the iPhone 4. A prototype issued to Forstall repeatedly dropped calls while he was on the phone. He feared that the problem was software related and called on staff to figure out what was wrong. After his team found no coding issues, Forstall discovered that the problem was occurring because of the phone's design. Ive had wanted a slimmer, lighter iPhone, which had been achieved by wrapping its metal antenna around the edges of the device. Forstall was apoplectic. He blasted the flawed design in conversations with Jobs and complained that it had been hidden from his

software team. Ive bristled at the criticism. After the phone was released, customer complaints mushroomed, triggering a crisis dubbed
"Antennagate." Jobs held a press conference to address the problem
during which he declined to apologize. Instead, he defiantly said
that the issue had been blown out of proportion. He acknowledged
that calls could drop if people touched the phone's lower left corner.
"There is no Antennagate," he said, mocking the media criticism. He
offered users a free case that he said would address the problem.

The Antennagate conflict heightened one of Ive's other frustrations with Forstall. Working with Jobs, Forstall had designed a software system that Ive considered to be out of step with the phone's
industrial design. Ive's design team had obsessed over the rounded
corners of the phone and become advocates of Bézier curves, a concept from computer modeling used to eliminate the transition breaks
between straight and curved surfaces. The Bézier geometry gave the
iPhone rounded corners that arched like a sculpture. A standard
rounded corner consists of a single-radius arch or a quarter circle,
whereas their curves were mapped through a dozen points, creating a
more gradual and natural transition. Meanwhile, Forstall used a standard three-point curve in the corners of iPhone apps. Each time Ive
opened his iPhone, he could see the difference between the phone's
carefully crafted corners and the software's clunky corners. He was
powerless to change those features because Jobs excluded him from
software design meetings. He could only look at them and fume.

FORSTALL'S OBSESSIVE CONTROL over iOS and voracious ambition
irritated colleagues. He took pride in having the most patents at the
company, a total that eventually swelled to 288, and he could be aggressive in increasing the tally. In 2012, Mac software engineer Terry
Blanchard developed a system of giving selected contacts VIP status
so that their emails would land in a separate inbox for important

messages. He presented the concept to his boss, Craig Federighi, and Forstall. Initially, Forstall challenged it, asking why people had to select VIPs manually rather than Apple doing it for them with an algorithm. Then, after the project was approved, Forstall scheduled a meeting with Apple's patent attorney to put himself down as the chief inventor. A colleague told Blanchard, who rushed to meet with the patent attorney and protect his place on the patent, which he wound up sharing with Forstall.

Such behavior made Forstall a divisive figure inside the company. He maintained the loyalty of the iOS engineers he led but incurred the disdain of staff across divisions with whom he clashed.

After Jobs died in 2011, Forstall's peers on the executive team mused that Forstall thought he should be CEO instead of Cook. They expected his ego and history of conflicts to be among the biggest challenges Cook faced within the executive ranks. His own lieutenants worried that his history of political clashes could spell trouble. Even his most loyal lieutenants recognized the predicament. The head of software design, Greg Christie, told Henri Lamiraux, the vice president of iPhone software, that Forstall wouldn't last long without Jobs to protect him.

"Are you crazy?" Lamiraux demanded. "Scott *is* the iPhone."

"He's not going to survive," Christie said. "Nobody likes him."

FORSTALL'S MOST PRESSING PROJECT was the development of the company's first mapping system. Apple's leadership of the smartphone market depended on it.

Since its launch in 2007, the iPhone had relied on Google's map and search services for navigating the real and digital worlds. But when Google had launched its own operating system, Android, it had gone from friendly partner to rival. To help Android gain market share, Google planned to give it sophisticated mapping features, such as turn-by-turn navigation, before it gave those features to the iPhone. It was an advantage that had the potential to accelerate its push to dethrone Apple as smartphone king.

Forstall's software team proposed retaliating with a simple plan called "Maps 2012." Though modestly named, it was ambitious in scope. It called for Apple to create a mapping system that would be global and dynamic, so that users could zoom in on images in real time and get turn-by-turn directions to a destination. The job unnerved some engineers. It required acquiring a database of all the world's information: every street, every address, every business. It also meant that images of every place on the planet would need to show up in high resolution on everyone's phone. In a few years' time, Apple wanted to not just beat what Google had spent a decade doing but surpass it.

Apple spared no expense as it pursued its own version of maps. Forstall's lieutenant Richard Williamson, who led the project, was granted approval to spend up to $5 million at a time without having to get clearance from finance for each purchase. The large sum allowed him to write checks for building out data centers and hiring staff. Apple acquired several smaller companies with experience in mapping and built features Google didn't have, including a flyover view that would display circling, three-dimensional images of sky-scrapers from cities around the world. To collect the images, they hired Cessna airplanes with cameras to survey cities such as Cincinnati line-by-line like a lawn mower. But the company was tightfisted in its negotiations for digital mapping data.

Apple's marketing team, headed by Phil Schiller, led talks with the leading provider of mapping information in the market, a Dutch company called TomTom. It provided the GPS systems for most cars and typically collected a fee of about $5 for every car that relied on its data. It wanted a comparable licensing fee paid for each iPhone that Apple sold, a proposal Schiller deemed unacceptable because the system for a car cost thousands of dollars more than an iPhone's. He wanted different terms and pressured TomTom's leadership to lower its fee. TomTom resisted. Talks grew heated until the parties reached an agreement for Apple to pay less for a smaller package of data that excluded information such as some freeway data. The outcome disappointed some software team members, who wanted a more robust data package and worried that the strained negotiations had created an adversarial relationship with a critical supplier.

The data TomTom provided were rudimentary and out of step with Forstall's aspirations for high-quality cartography. He spent hours in meetings with software designers obsessing over the fonts, colors, and images. They brought a Japanese artist on board to sketch how freeways could look and devoted entire sessions to debating what color of blue they should make the ocean and what font they should use for road signs. They envisioned freeways with realistic curves, but the mapping information from TomTom didn't include detail about the width of roads. A divide developed between what the software designers wanted and what the software engineers could deliver. Most everyone agreed that the data were to blame.

Williamson approached Forstall and Schiller to tell them that the project wouldn't be ready on time. The deadline was tight, the goals were ambitious, and the team was struggling with the data. He worried that customers were so dependent on maps that a single flaw would lessen their loyalty to iPhones. He proposed keeping Google Maps on the iPhone and offering a prerelease of Apple Maps, known as a beta version, so users would know that it was being improved.

"We don't ship betas," Schiller said.

IN APRIL 2012, charter buses rolled into the parking lot at Infinite Loop to ferry Apple's leaders to an annual corporate off-site meeting at Carmel Valley Ranch, a resort south of Monterey. The event, known as the Top 100, was an exclusive gathering of the company's top decision makers and most talented staffers. For years, Jobs had finalized the list of attendees and required everyone to ride in buses to the gathering—a tradition that Cook continued.

As staff arrived at the five-hundred-acre resort, Williamson took in the stunning surroundings. The Santa Lucia Mountains framed rolling green hills with a golf course and a vineyard. Wild turkeys roamed the grounds, where activities for guests included beekeeping and falconry. It was an elite vacation destination, but Williamson knew that he would have little time to enjoy it. He had been asked to give a presentation on Apple Maps.

The Top 100 leaders gathered in a large conference room with floor-to-ceiling windows shaded by tall oak trees. Williamson gazed out at the crowd and started talking through his team's work. On a screen, he pulled up a simulation of what a trip would look like. He showed a car moving through the streets of San Francisco and demonstrated his team's work to allow users to zoom in on streets without the image having to pause to refresh. He turned on a three-dimensional view of San Francisco's Market Street and drove toward the Financial District. A few people began clapping. Williamson could see that Forstall and Schiller were excited. He knew that his idea of delaying a release of Apple Maps was dead.

THE SOFTWARE TEAM TOOK an only-in-my-backyard approach to testing its new mapping system. Forstall used Apple Maps on his drives around the Bay Area and found that it worked perfectly. Other members of the team tested it across California. The remainder of the world, eighty-one countries, was vetted by about eight quality assurance staffers. Afterward, they said it was ready to ship.

In June, Forstall stepped onstage for Apple's annual Worldwide

Developers Conference at Moscone Center in downtown San Francisco, eager to show off his team's work. The crowd of five thousand programmers hooted and cheered as he strolled before an enormous black screen. In the five years since the launch of the iPhone, he had become the event's star and the company's face of software. He smirked at the crowd and provided a series of software updates. Then he paused before launching into his final feature.

"Next," he said, "is Maps."

The screen behind him showed an image of Lake Tahoe, with its foot-shaped outline rendered in a sky blue color and framed by emerald green mountain ranges. Forstall clicked forward to show how business listings appeared and demonstrated how buildings were rendered in three dimensions. Then he tapped on a button labeled "Flyover" that caused the map to pan out and reveal a video image of San Francisco's Transamerica Pyramid, which rotated as though being seen through a helicopter window. A few people in the crowd gasped.

"Beautiful," he said. "Just beautiful."

He was confident that Apple had outdone Google.

APPLE MAPS IMMEDIATELY ran into trouble. Within hours of its release, Apple customers were reporting that their maps showed that London had fallen into the Atlantic Ocean and Paddington Station had disappeared. In Dublin, users discovered an airfield that didn't exist, leading a local pilots' association to issue a warning not to attempt an emergency landing there. Even the flyover feature had failings. In New York, the Brooklyn Bridge melted off the screen as though an earthquake had annihilated it.

The fiasco became Cook's first full-blown crisis.

As the negative press mounted, he called a meeting in the company's boardroom with some of Apple's executive team, as well as Williamson. Forstall called in from New York, where he'd gone to spend a long weekend with his wife. The atmosphere was tense. Just over a year after succeeding Jobs, Cook cared deeply about the media coverage of the company. For customers, negative press could turn Apple from extraordinary to ordinary overnight.

"We need to issue an apology," he told Forstall. "I want you to sign it."

The request caught Forstall off guard. Amid all the criticism of Maps, he hadn't anticipated that Cook would want to issue an apology. It was a word that Jobs had never wanted to utter publicly, even during Antennagate.

"Why are we going to put out an apology?" Forstall asked. "What is the goal?"

In Cupertino, Cook shifted slightly and stared at the speakerphone in the center of the boardroom table. The executives around him interpreted the question as a challenge of their new CEO's leadership. A few of them leaned forward. No one spoke.

Across the continent, Forstall tried to channel his mentor. Instead of an apology, Forstall suggested that Apple tout that the usage of Maps was strong despite the issues and commit to improving the app. He made his case against issuing an apology. One of his concerns was that if Apple apologized for the failures of Maps, it would discourage staff from taking on hard projects. Why would anyone develop a difficult product if they would later be publicly shamed for shortcomings?

Cook was unmoved. It was clear to everyone in the room that he had made up his mind; Apple would issue an apology.

Working with Apple's communications team, Cook drafted a letter saying that Maps had fallen short of the company's commitment to making "world-class products." He tried to sidestep the uncomfortable truth that the company had forced more than 100 million iPhone customers to download the dysfunctional app in a recent software update. While Apple worked to make Maps better, Cook suggested that people download the rival offerings from Google and Microsoft. It was one of the first times the company's CEO had directed customers to use competitors' products, a painful concession that underscored just how badly Cook thought Forstall and the software team had failed.

"Everything we do at Apple is aimed at making our products the best in the world," Cook wrote. "We know that you expect that from us, and we will keep working non-stop until Maps lives up to the same incredibly high standard."

NEARLY A MONTH LATER, Cook summoned Forstall to his home in Palo Alto. It was a Sunday. Forstall had been to Cook's house before for work meetings, but those meetings had typically been planned in advance. This house call was unexpected.

The headaches with Maps floated in the back of Forstall's mind as he drove down the tree-lined streets near Cook's four-bedroom home. In recent weeks, Cook had appointed Jeff Williams, his top lieutenant, to assist Forstall in turning Maps around. No matter how hard Forstall pushed his team, it was clear that the problems couldn't be fixed quickly. TomTom typically updated its data every quarter and reissued it to automakers. It wasn't prepared to make improvements overnight.

To Cook, it was clear that Forstall had screwed up. Maps shouldn't have been released in the condition it was in, but the bigger problem was Forstall's resistance to acknowledging the mistake.

Cook met Forstall at the door and showed him inside. His living room floor was stripped down to the foundation, and the concrete was polished into a sleek iron gray. It was the same sterile floor that was common in many offices.

Cook had a severance package for Forstall to sign. As Forstall reviewed it, Cook let him know that Apple would begin informing

Forstall's team that he was being let go later that day and would put out a simple press release the next day about his departure. It was devastating news for Forstall, who a year earlier had lost Jobs, his mentor, and now was losing his job at the company that his mentor had built.

"Is this because of Maps?" Forstall asked.

"No," Cook said. "It's not."

"What is it?"

Cook demurred.

WHEN COOK WORKED on the press release announcing the firing, he pressed to portray it as addition by subtraction. Colleagues thought that the episode exposed one of Cook's biggest fears as CEO: that the all-star team Jobs had assembled would abandon him. They appreciated his worry that investors might interpret an executive exodus as a sign Cook wasn't the right leader. It was part of the reason the board had locked in the executive team with lucrative stock packages after Jobs's death. And it was why Cook wanted the press release to emphasize who was staying at Apple, even if newspaper headlines would focus on Forstall's ouster.

Cook suggested that the release highlight the new roles and responsibilities that Ive, Bob Mansfield, Eddy Cue, and Craig Federighi would be taking. He also wanted to assure investors that he would lead a search to find a new head of retail to replace John Browett, who would also be leaving. The release, which came out on October 29, aimed to fulfill his wish. Headlined "Apple Announces Changes to Increase Collaboration Across Hardware, Software & Services," it mentioned collaboration twice and Maps once.

Staff at Apple interpreted it as a sign of a new age of management. Jobs had favored rivalries between executives, encouraging people with egos to advance ideas that he could pick from to make great products. He could keep those dueling personalities in check. Whereas he might have torched Forstall and fired a subordinate directly responsible for Maps, Cook used the fiasco to eliminate disharmony on his leadership team and send a signal to the company that he wanted everyone to work together more than they had in the past. Without Jobs there to connect all the different areas of their businesses, they would have to make those connections themselves.

In the process, Cook had eliminated one of the only perceived rivals to his leadership. He also rewarded Apple's most important employee, Ive, with a responsibility the designer had long wanted, influence over the way Apple's software looked.

Cook had taken an action that would earn him loyalty, but his fear of being abandoned by his top lieutenants was something he would have to face again and again.

Chapter 7
Possibilities

Jony Ive kept his eye on his newest creation. From his office, he constantly tracked his team's progress, watching through a twelve-by-twelve-foot wall of glass that looked out over the studio's waist-high tables, one of which was topped by the earliest models of Apple's watch.

His office was a portrait in minimalism, with an oblong sapele desk designed by Marc Newson and dozens of hard-covered sketchbooks arrayed on shelves. The sketchbooks had yellow spines that formed long, pure lines of color that testified to a lifetime devoted to the control of simplicity and beauty. Leaned against the wall behind the desk was a framed print of Banksy's *Monkey Queen,* showing the iconic bust of Queen Elizabeth II, complete with crown and a necklace of glittering jewels, but with the face of a chimpanzee. It was an especially cheeky image for a man who, earlier that year, had been knighted with a sword from that very queen's family. There were only 150 signed prints of the image, valued at nearly $100,000, all made by a pseudonymous street artist who was believed by fans to be Robert Del Naja of the band Massive Attack, a friend of Ive.

Beside the print sat a poster from Good Fucking Design Advice that read:

> Believe in your fucking self. Stay up all
> fucking night. Work outside your fucking
> habits. Know when to fucking speak up.
> Fucking collaborate. Don't fucking
> procrastinate. Get over your fucking self.
> Keep fucking learning. Form follows fucking
> function. A computer is a Lite-Brite for
> bad fucking ideas. Find fucking inspiration
> everywhere. Fucking network. Educate
> your fucking client. Trust your fucking gut.
> Ask for fucking help. Make it fucking
> sustainable. Question fucking everything.
> Have a fucking concept. Learn to take some
> fucking criticism. Make me fucking care.
> Use spell check. Do your fucking research.
> Sketch more fucking ideas. The problem
> contains the fucking solution.
> Think about all the fucking possibilities.

On November 2, 2012, Ive arrived at the studio early. It had been just four days since the announcement of Scott Forstall's ouster. The news had left some of his engineers shell-shocked, while others in different divisions had celebrated with champagne. Ive, who excelled in corporate combat, prepared to take over the responsibilities of his onetime nemesis.

After firing Forstall, Tim Cook decided to reorganize the software division and carve up its responsibilities. He handed Ive oversight for software design, which Apple called "human interface," and gave Craig Federighi management over software engineering. The decision sharp-

ened Apple's structure. When Jobs had returned in 1997, he had put
the entire company under a single profit-and-loss statement and created an organization whose senior vice presidents managed the various
areas of the business. That so-called functional structure meant that the
heads of hardware, software, marketing, operations, finance, and legal
reported directly to him. He had then created special project teams
to build the iPod and the iPhone. As those products succeeded, the
company's structure flexed to suit Jobs. Software development was split
by products, with Forstall leading iOS and Federighi leading MacOS.
Cook wanted to adhere more rigidly to the MBA ideals of a functional
structure, anticipating that it would help the company during a period
of unprecedented growth. The company had added twelve thousand
employees in the previous year and increased its workforce by 20 percent to seventy-three thousand. Giving Federighi duties for software
engineering meant that he would oversee building what Ive and a team
of software designers imagined creating. The process would mirror
what Apple had done in hardware for years. But some of the human
resources leaders worried that the order Cook wanted to impose would
sow chaos. After all, Jobs had limited Ive's responsibility to supervising
his platoon of designers, believing he was better at creating than shoul-
dering the bureaucracy of management.

Ive embraced the new duties. He was ready to bring an end to
his frustrations with graphics in the iPhone's operating system, iOS.
Steve Jobs had championed a style known as skeumorphism. Ive
thought it made software look as dated and unrefined as the clunky
word used to describe it. The concept's origins could be traced back
to the dawn of the PC era, when software engineers had begun making computer icons that resembled real-world objects such as trash
bins and filing folders. Jobs favored the style because he considered
it intuitive, but Ive detested it. Three decades into the digital age,
he thought that people recognized computer folders and no longer

needed buttons on a calculator to be shaded for depth. He imagined a sleeker software design that was as clean and distilled as the iPhone that carried it. If Jobs was right in saying that hardware was the vessel and software the soul of the device, the time had come for Ive to redefine the soul of Apple's best-selling product.

That day, Ive welcomed Apple's top product marketers, Phil Schiller and Greg Joswiak, and its head of software interaction, Greg Christie, into his design studio. The ground-floor space had returned to life in the year since Jobs's death. Its designers spent most days analyzing new models or drawing in hardbound sketchbooks. The earthy smell of ground coffee from a buzzing espresso machine floated in the air. Ive and the group gathered around an uncluttered table.

Before then, discussions about the look and feel of iPhone software had taken place exclusively in a securely enclosed software section of the building. Forstall had played tastemaker, shaping the designs advanced by the small team Christie led. Jobs had met with them weekly to approve or reject their work. The CEO edited each iteration until the software sparkled with innovation. The meeting location that November made clear that Jobs's role had been assumed by Ive. The designer wanted the team to think big. The iPhone had been around for more than four years. As apps had popped up, the phone had become increasingly cluttered, with icons from Maps to iTunes jockeying for space alongside Facebook and Angry Birds. The first question the group bandied about was: What should the home screen be like? Or as Ive, a man of refined tastes but coarse language, might say: What are the fucking possibilities?

TIME SPED UP FOR IVE after Forstall's departure. In late 2012, he summoned Apple's top brass to the St. Regis hotel in San Francisco. Apple's heads of operations, software, hardware, and marketing marched through the lobby and past the Canadian artist Andrew Morrow's mural *War*. The painting depicted a sword-wielding warrior on horseback in pursuit of a runner. It was a symbolic allusion to mankind's race against time, which seemed somehow appropriate for a group that was trying to outrun doubters and critics who questioned their ability to create another great product without Jobs. Apple's security team prepared for their arrival by locking down a floor of the 260-room hotel across the street from the San Francisco Museum of Modern Art, ensuring that it was clear of recording devices and the curtains over its windows were drawn. Off-site meetings were a longstanding practice. They helped prevent speculation from building among rank-and-file staff who might catch a glimpse of top executives gathering and begin wondering what was afoot. Apple's need-to-know culture, fostered by Jobs, aimed to preserve secrecy, prevent leaks, and encourage mystery.

The meeting that day was a major step toward answering the question that had dogged Apple since Steve Jobs's death: What's next? With a push from Ive, a team of engineers had spent about six months exploring what Apple could do if it developed a smartwatch. They had worked with sensors to track a wearer's heart rate, explored how to relay notifications to the wrist using Bluetooth, and researched other capabilities, including measuring people's emotions. The effort had run its course, and Ive wanted the engineers to present findings so compelling that his colleagues would be persuaded to make the watch Apple's next big bet.

As the engineers prepared to present, they watched the company's captains settle into chairs behind tables arranged in a horseshoe. The room was filled out by a few of their top lieutenants, including three of Ive's designers, Richard Howarth, Rico Zorkendorfer, and Julian Hönig.

Only one major figure was absent: Tim Cook.

After more than a decade of having product development led by its CEO, Apple was embarking on a mission without the participation of its top leader. Cook sent a message that he had no intention of even trying to step into the shoes of his predecessor. The successor whom Jobs said wasn't a product person didn't plan to become one. Instead, he would stay out of the experts' way.

Ive played master of ceremonies. He took a seat near the center of the table, not far from his team of designers. Before him sat a bottle of green juice, which was all he planned to consume that day. He was in the midst of a food cleanse, the latest in a series of diet adjustments after the stress and grief of Jobs's death. The juice alarmed the engineers, who had spent weeks creating a deck of more than 150 slides full of industrial design renderings, details about the size of the device, analysis of what the display could be made of, and insights into how it could alert users to notifications by tapping a wrist. They expected the presentation to take six-plus hours and assumed that the group would have a barrage of questions that could make the meeting take all day. If the meeting was a dog-and-pony show for Ive, as they assumed, the last thing they wanted was for his cleanse to leave him hungry and irritable.

During the presentation, Ive remained mostly quiet as he processed what the engineers presented. His relative silence was typical of his leadership. He seldom spoke in meetings, and when he did, he often strung together several ideas being discussed and raised a possibility that no one else had imagined. Eventually, a few electrical engineers opened a black Pelican case that Apple used to conceal and securely transport unreleased devices. The lid of the case rose to reveal a series of square iPod Nanos with black wrist straps. Engineers removed the iPods from the case and clasped them onto the executives' wrists. Ive held out his left hand and watched as an engineer strapped the iPod tightly around his wrist. He shook his head. "I like to wear a watch loose on my wrist," he said before relaxing the band to create a few millimeters of space.

As Ive admired the makeshift watch, the engineers glanced fearfully at one another. To measure heart rate, the sensors installed on the iPod Nanos needed to have close contact with the skin. They hoped that Ive's decision to loosen the wristband wouldn't ruin their demonstration. The iPod combined the rear sensor with another on its side to chart the heart's electrical activity, a rudimentary electrocardiogram. Ive listened as an engineer explained how to test the EKG and then watched as his iPod display filled with a red line that spiked just like the jagged lines of a hospital's patient monitor. He nodded with approval. In a span of six months, the engineering team had transformed one of Apple's iPods from a music player into a health product.

In the wake of the presentation, Ive could tell that his peers saw the kind of health possibilities he envisioned bringing to people's wrists. The challenges ahead, of course, would be immense. The iPod Nano was way too big. It would need to be miniaturized, waterproofed, and outfitted with a curved display. But everyone left that day motivated to make the watch central to Apple's future.

AFTER IVE RETURNED to Cupertino, he turned his attention back to the future of iOS. He had been insistent that the software designers and engineers pursue a complete visual overhaul of the software.

Ive wanted to bring the eye of a trained artist to the project and turned to Alan Dye, a graduate of Syracuse University's College of Visual and Performing Arts. Like Ive, he was the son of two teachers who had grown up being taught by his father, a woodworker, how to make furniture and craft homemade toys. He had begun sketching letters and words as a kid and had developed a lifelong obsession with typography that had led to a career designing labels and print ads for companies such as Molson beer and Kate Spade. He had been considered a rising star when Apple had hired him in 2006 to work on its product packaging, website, and advertising. His obsession with graphics had resonated with Ive, who made Dye the foundation of a team that would swell into the hundreds.

Working with Dye, Ive began refashioning the icons to harmonize the phone's digital graphics with its physical shape. They quickly addressed the detail that had always irritated Ive, getting the curve of the icons of Apple's apps to match the physical curve of the iPhone. The corners of iOS 7 were redesigned using Bézier principles, creating a smoother, more organic bend that was used for the corner of every app. Ive was so proud of the change that in future work with architects and designers, he used it as an example of how to create the perfect curve, showing them how the number of points in the corners of an app had increased from iOS 6 to iOS 7. He also began looking for a thinner sans-serif typeface for every app and exploring ways to enliven the home screen with brighter colors. The subtle changes signaled a major departure from the past.

A few months after taking over, Ive held a meeting in a large conference room for some of Apple's engineers to share the new direction with them. He told the group that the goal was to strip out all of the dated iconography that Apple had been using for applications such as Photos and replace it with more modern representations. He pulled up an image of Apple's Voice Memos app as an example. Under Forstall, its icon had been a pill-shaped ribbon microphone like the one radio broadcasters had used in the 1950s. "The metaphor is lost on me," he said. "I don't understand what I'm looking at." It was an anachronism that he couldn't imagine users understanding. In its place, he planned to introduce a new icon with the spiky image of a voice as it was rendered in audio recording software. He showed similar changes for the Calendar app and Safari web browser. They were brighter, more dynamic, and more vibrant.

Ive's focus on visual styling vexed the software design team. Though they obsessed over colors and shapes, they prioritized how people interacted with the phone and often built demonstrations of the software they planned to introduce so they could experience how intuitive it would be for users and adjust as needed. Many of them believed that design was how the software behaved and thought that Ive was myopically focused on how it looked. Tensions emerged between the conflicting philosophies as he pushed to eliminate the dark borders around the buttons inside apps. Some members of the team thought those lines helped users quickly identify what button to press when they clicked through a screen. Without them, they worried, the buttons would blur indistinctly into the background, forcing users to search for where to click. At Ive's direction, they shifted from demonstrating how an app worked to making paper printouts that showed how an app looked. They became more like graphic designers than software savants.

One of Ive's big ideas was to bring translucency to the software, so that a layer of text and icons could exist on top of the home page like a frosted window. The concept was central to an advance called "Control Center" that would allow people to swipe their finger up from the bottom of the screen and pull up a translucent page with one-click access to Wi-Fi, Bluetooth, and more. When Ive pressed for the layer to work even over video, the reaction from engineers was almost uniform: We can't do that. The graphics processors of iPhones aren't fast enough, they said. But Ive insisted, and the engineers eventually came up with a system to work around the hardware limitations and deliver the impossible.

The changes were central to a complete overhaul of the entire operating system that Ive demanded be finished in a few months. The engineers called it a death march.

THE KEY TO A PRODUCT'S SUCCESS is purpose. The iPod dominated music because it put a thousand songs into people's pockets. The iPhone flourished because it combined a music player, phone, and computer into a single device. Not every gadget starts with such transformative goals in mind, but every one that succeeds springs from a well of deep thought and consideration.

As the watch developed, Ive was thinking hard about what it should do. One of the first things he wanted to evaluate was the existing marketplace. The smartwatch industry was nascent with only about a half-dozen companies making gadgets that they claimed worked like the two-way wrist radio in Dick Tracy comics. Ive wanted to know about them all.

One day, he welcomed an engineering team to the studio that had gathered information on competitive offerings and printed summaries on eleven-by-seventeen-inch sheets of paper with details about their features and their dimensions. He gathered the group around one of the studio's tables and flipped through the sheaves of paper explaining each watch. He turned past images of a square Sony smartwatch the width of a shirt sleeve and an Italian-made device as thick as a Zippo lighter. As he scrutinized the pages, he grimaced. "These products lack humanity," he huffed.

Ive stared at the bulky gadgets with disgust as some in the group around him nodded in agreement. The devices had only one thing in common with traditional watches: they told time. Ive's team hadn't finalized Apple's design, but Ive knew that Apple's watch would look very different from anything else on the market. It would need to be informed by the past to prosper in the future. The very fact that people would wear it made the way it looked matter more than for something they might slip into their pocket, drop into a bag, or lay on a table. It would be intimate, constantly resting against the skin, and perpetually visible, a battery-and-processor-powered extension of them. A computer that looked like jewelry. A product with soul.

The vision led the design team on a journey through time. They wound back to the beginnings of the clock and sprang forward to explore how modern timepieces were made. They learned how the British had miniaturized towering grandfather clocks to power the rise of the empire with chronometers that enabled sailors to plot their location at sea. They studied how pocket-size watches had become wristwatches to assist armies in timing troop advancements during war. They heard from horologists about how those wristwatches had become fashion pieces in the early 1900s after Louis Cartier had developed the iconic Tank watch with its rectangular case and Roman numerals. And they explored how Swiss watchmakers had made complicated gears to tick off minutes, a craft upended in the 1970s by the arrival of quartz crystals and a battery-powered revolution.

They synchronized the history lesson with purchases of some of the world's finest watches. The orders were routed through a shell company called Avolonte Health, a startup Apple had created in a nearby medical office complex where its engineers were secretly at work on the noninvasive blood glucose–monitoring technology from Rare Light. Such covert businesses were sprinkled around the peninsula, enabling Apple to do research and development without tipping off its rivals. Avolonte's staff trucked packages to Infinite Loop, where the designers opened them to reveal the world's most expensive wristwatches from Patek Philippe, Jaeger-LeCoultre, and more.

As the days of research blurred together, the group carved out moments to brainstorm about what the watch should do. They agreed that it would have to tell time more accurately than any watch on the market. Other ideas followed. It could be a stopwatch and a timer, an alarm clock and a world clock. It could layer in health capabilities that engineers had been working on, such as heart and glucose monitoring. They discussed it tracking people's emotions and recording their fitness levels. But most of all, they talked about it freeing people from the tyranny of their phones, delivering text messages to their wrists, and allowing them to make calls or listen to music on the go—a leap, they agreed, that would require wireless headphones. And just like that, the watch birthed another wearable product.

The fast tempo of idea development made clear they had a platform that could grow. It was reminiscent of how the iPhone had evolved from its start as a phone, music player, and pocket computer to become an elite camera, flashlight, GPS navigator, gaming console, and TV screen. The capabilities it had added each year had made it an indispensable part of people's lives. With the watch, there was similar potential, a long play—a vessel for features that could carry Apple into a brighter future.

AS THE PRODUCT DEVELOPMENT CLOCK TICKED, Ive grew frustrated. Rico Zorkendorfer and Julian Hönig, two top designers who joined him at the St. Regis, had been tapped to lead the effort. They were collaborating with the rest of the team and had settled on a design that looked like a dog tag bracelet. Its rectangular shape had some similarities with Cartier's famed Tank watch without the same elegance. Ive wanted something more fashionable.

The design was complicated by the heart sensors. The most accurate heart rate readings come from inside the wrist, where nurses take people's pulse. But such a design would create a bulky band that would test people's traditional notions of what a watch should be. The design team agreed that that hardware would need to live on the backside of the watch case.

The pressure to find a solution increased each day. The design studio sat at the pinnacle of the product process, and its decisions about how the product should look would dictate software and hardware, as well as what components and manufacturing tools the company's operations team would need to secure. But Ive was as stuck in the design process as an author with writer's block.

In late March 2013, Ive's friend Marc Newson paid a visit to the studio. He was one of the world's most accomplished and versatile designers, a self-made star who had grown up taking apart watches in Australia before going on to start his own studio. He had designed everything from Qantas airplane interiors to Nike sneakers and Louis Vuitton luggage to silicon vibrators. He had married a British stylist, Charlotte Stockdale, kept a garage full of vintage cars, and had achieved acclaim with a solo show at New York's Gagosian gallery. But perhaps most important, he had two decades of experience designing wristwatches.

Ive took Newson to the table where a half-dozen examples of the team's early work were laid out and asked for his thoughts. Newson's eyes scanned across a circular concept, a rectilinear one, and another that was angular. The quality of the work was stunning, yet none of it seemed appropriate. He started talking with Ive about what was missing and kept coming back to the watchband. It needed to artfully intersect with the watch face.

As they talked, they opened their sketchbooks and began briskly drawing lines with their fountain pens. Their hands hurried across the pages, Ive's in exacting strokes and Newson's in squiggles, to create a square-shaped watch face that looked like a miniature iPhone with rounded corners. Other ideas tumbled forward. They traced a curving indention into the rear of the watch case and imagined a way to connect a separate strap to it. From the beginning, the designers had wanted straps that could slide on and off, and finally, they had the design to realize that. The drawings also included a winding crown, a familiar detail the design team had been working with for months as they tried to make their models look less like a wrist computer by including the miniature knob watchmakers had introduced in 1820 to wind and set time.

Ive would become animated by a revelation. Spinning out of the team's exhaustive research into the history of clocks came an idea that gave the newly sketched timepiece meaning. The crown, he and the team grasped, could be a tool for navigation, a dial to turn the volume up or down, a wheel to spin through miniature apps on the wrist, a button back to the home screen. He realized that the primary instrument of the watch of yesterday could be the signature tool of the watch of tomorrow.

In their burst of creative energy, Ive and Newson felt like songwriters who tapped into the collective unconscious, unearthing a concept that had always existed but needed to be sketched into reality.

As they finished, they rushed their sketches into the computer-assisted design room and asked one of the team's CAD technicians to convert it into a three-dimensional model. They stared over the technician's shoulder as he created an image on the screen. It gave them a chance to see if they could turn the concept into a physical object. Much of the idea worked: the form; the thickness; the mechanism for connecting the watch strap. The idea was alive and potent.

Now came the complication: the difficulty of watchmaking.

Chapter 8

Can't Innovate

Tim Cook found himself adjusting to a life in the spotlight, even in the middle of the night.

Early in his tenure leading the world's most valuable company, he was awakened by the sound of someone pounding on and yelling at the front door of his Palo Alto home. The house, less than fifty yards from the street, was hemmed in by neighbors, and Cook, the ever-unassuming son of the South, hadn't wanted the fuss of a twenty-four-hour security detail. As the banging continued, he reached for a baseball bat.

When he neared the door, he heard a man outside shouting about Apple's share price. The sales growth of iPhones had slowed, and a drop in the company's stock price had followed. Investors were upset. They had grown accustomed to a steady, almost uninterrupted increase in the value of their shares. The reversal had been enough to make the man outside scream. Cook didn't engage with him. Eventually, the yelling stopped, and Cook returned to bed.

Weeks later, ahead of a meeting to discuss personal security protocols, Cook casually raised the incident with his security team. He conceded they should move forward with plans to install cameras at his home. But it was a separate incident that opened Cook's eyes to the value of a security detail. He traveled on commercial airlines after becoming CEO and made himself more inconspicuous at San Francisco International by wearing a baseball cap. On one trip through the airport, someone recognized him and asked for a photograph. Soon others gathered around, asking for pictures of their own. Cook felt a hand on his shoulder and a tug. A sharp pain followed. He'd recently injured his shoulder during a workout, and the yank from behind had aggravated a strain.

The airport crowd, the pull, the pain, made clear that he was no longer an unknown operator from Alabama; he was now the world's most visible CEO.

A YEAR AFTER JOBS'S DEATH, Apple was doing its best to show that the magic was still alive.

In late 2012, it introduced a new iPhone that it hyped as an "absolute jewel," with a glass-and-aluminum case that was 18 percent thinner than its predecessor's. Cook supported the rollout with appearances at the grand opening of Palo Alto's new floor-to-ceiling glass Apple Store. He breezed past enthusiastic shoppers and mingled with retail staff hard at work pushing the new device. He was pleased with how the phone was being received. In the new iPhone's first weekend, Apple sold 5 million, a new record and a total that exceeded the company's initial supply. Yet the market reminded him daily that shareholders were unimpressed.

The new model delivered the weakest year-over-year increase in sales in the iPhone's five-year history, sending Apple's share price to a six-month low and erasing $160 billion of its market value, about as much as the total worth of Coca-Cola that year. It was clear to investors that Apple was confronting its first formidable adversary.

AT A WOLFGANG PUCK RESTAURANT in Los Angeles, some four hundred miles south, Todd Pendleton assembled Samsung's scrappiest marketers to plan an assault on the smartphone king. It was fall, the season of the iPhone, and Samsung's chief marketing officer in the United States had an idea for a disruptive advertising campaign that would turn the precious descriptions Apple lavished on its newest device from commercial rites into irreverent ridicule.

Without the visionary force field of Jobs, Apple was vulnerable. His absence had stoked questions about the company's ability to innovate without him and fostered fears among even its most ardent fanboys that it might become less imaginative.

The timing couldn't have been worse. The gaps between the iPhone and rival smartphones were shrinking. Apple's market share was slipping as competitors improved their hardware and software in a sprint to claim a larger piece of what had become the most important consumer product in generations.

Pendleton and his colleagues saw an opportunity in Apple's moment of weakness to awaken the world to the promise of the Samsung Galaxy. The premium smartphone had a bigger screen, alternative features, and sophisticated cameras that were winning over customers around the world.

In recent years, the battle between Apple and Samsung had become personal. In 2011, Apple had sued Samsung in the U.S. District Court for Northern California, accusing the South Korean company of copying the look, design, and interface of the iPhone. Apple's staff had worked weekends, filed patents, and sparked a product revolution. But Samsung had ripped it all off, Apple had said, in a lawsuit that it had deservedly won. Even Google, Samsung's partner, warned the South Koreans that their software design copied Apple's. Samsung tasked Pendleton with leading a commercial counterattack.

A former Nike executive, Pendleton was schooled in the discipline of ambush marketing. The sneaker company had boosted its sales by declining to sponsor the Olympics and then blanketing billboards in Atlanta with its swoosh. The rebellious act had earned it free news coverage and won over customers. Pendleton saw a similar opportunity with the iPhone launch, tech's equivalent of the Olympics.

In Los Angeles, he circulated around the restaurant turned war room where Samsung's marketers watched TVs with video of Apple's product launch event. Nearby, copywriters from the company's ad agency, 72andSunny, scribbled potential lines for a commercial on a whiteboard. They focused on poking fun at Apple's marketing gobbledygook, its tendency to inflate an everyday feature such as a camera by saying it offered "spatial noise reduction," and its habit of calling its phones "jewelry."

Pendleton urged them to find ways to ridicule the absurdity of people waiting in line outside Apple Stores for hours before a product launch. He wanted to contrast the complicated process of buying an iPhone with the ease with which people purchased a Samsung Galaxy. He thought the difference would make for a smart commercial that inspired people to see through Apple's hype machine and buy a better phone. He pushed the team inside the restaurant to get creative as they brainstormed a satirical script.

"What if we had someone waiting for their mom?" someone proposed.

"Oh, that's not cool," someone else said.

Nearby, a film crew waited outside a faux Apple Store to shoot a TV spot the moment the copywriters finished. Pendleton hurried to the set and watched as dozens of actors read through their lines. Cameras rolled as actors playing customers loitered outside the faux Apple Store, waiting to buy an iPhone and chattering enthusiastically about what they'd heard Apple tout as its latest features.

"The headphone jack is going to be on the *bottom*!" one actor said.

"I heard the connector is all digital," another actor said. "What does that *even mean*?"

Meanwhile, people with Samsung Galaxys walked by, holding phones with bigger screens and exclusive features such as the ability to trade playlists by tapping their phones together. People in line at the faux Apple Store watched with awe as they realized the Samsung phones had more sophisticated features.

The commercial became part of Samsung's "The Next Big Thing Is Already Here" campaign, a blistering satire that portrayed the Samsung Galaxy as the phone of choice for hip, cutting-edge people and the iPhone as the device favored by credulous, navel-gazing dweebs. Samsung complemented the spots with billboards outside Apple Stores that advertised the Galaxy III. They also delivered pizza to people waiting in line for iPhones, a classic ambush-marketing maneuver.

The commercial assault irritated everyone in Cupertino. Heightening their annoyance were the obvious similarities between what Samsung was doing and one of Apple's own marketing tricks. It had popularized cool-versus-nerd advertising with the "Get a Mac" campaign that had cast the comedian John Hodgman as a suit-and-tie-wearing PC, opposite the actor Justin Long, playing an untucked-shirt-and-jeans-wearing Mac. As Hodgman boasted about making spreadsheets, Long talked about making digital movies. The commercials forced viewers to ask: Would I rather own an outdated computer used by dorks or an it-just-works desktop preferred by dudes?

Now Samsung was making people ask similar questions about smartphones: Would I rather stand in line for hours outside an Apple Store alongside a bunch of desperate fanboys or chill and enjoy life with a feature-rich device that's convenient and carefree?

AS SAMSUNG'S AD SWAGGERED across the airwaves, Cook embarked on a rare media offensive, traveling to New York to appear on NBC's news program *Rock Center* with anchor Brian Williams. They strolled through Grand Central Station and ascended the stairs to the Apple Store perched above the train depot's vaulted concourse. After they sat down to talk about the business, Williams brought up Apple's acrimony with Samsung.

The South Korean rival was on the cusp of leapfrogging Apple and becoming the world's smartphone king. Its ad had caught the attention of Williams, who described it as "a frontal attack on a giant that would have been unthinkable not too long ago."

"They are trying to paint their product as cool and yours as not cool," Williams said. "Is this thermonuclear war?"

Cook's mouth tightened. "Is it thermonuclear war?" he asked. "The reality is that we love competition at Apple. We think it makes us all better. But we want people to invent their own stuff." He pounded the table for emphasis.

As Williams spoke, Cook rocked, his eyes focused on the newsman. Williams pressed Cook on the cruel cycle of commerce: "If you're the company that can stay fresh without an expiration date, you'll be the first to buck that trend."

Cook stopped swaying. His eyes narrowed as he leaned forward. "Don't bet against us, Brian," he said. "Don't bet against us."

SAMSUNG'S RISE RANKLED COOK. He viewed its phones as nothing more than forgeries, its ads brash and disrespectful, its empire a mess of washing machines and microwaves. It hadn't devoted days and nights to dreaming up simple solutions to complex problems. It hadn't carefully curated its product lineup. Yet it was somehow supplanting Apple as a media darling.

The South Koreans had created a marketing problem, and the onus fell on Apple's marketing team to resolve it.

Under Jobs, the group that developed commercials, ads, and

packaging had been one of the company's strengths. Marcom, short for marketing and communications, was a select team of Apple executives and the leaders of its bespoke Los Angeles–based ad agency, TBWA\Media Arts Lab. They met every Wednesday for three hours to review and refine ideas until they arrived at a piece of advertising gold such as the iPod silhouette campaign, "Get a Mac," or "Envelope," which showed a slender MacBook Air sliding out of a manila envelope to the sound of the pop song "New Soul." Jobs, who had acted as chief marketing officer, had created such memorable spots by pitting two of Marcom's most opinionated members against each other. He had prodded Phil Schiller, the head of product marketing, for ways to tout a new technology, and James Vincent, the CEO of Media Arts Lab, for creative ideas that would grab customers' attention. The result was some of the world's best advertising.

After Jobs died in 2011, Schiller had assumed the leadership of Marcom, stoking concern both inside and outside the company. A round-faced New Englander with a fondness for sage-colored shirts, he had a reputation for being literal and unimaginative. He so regularly shot down others' ideas that he had been nicknamed "Dr. No." His elevation was difficult for Vincent to accept. Jobs had spoken to the shaggy-haired Brit, described by colleagues as a "creative's creative," about leading the group prior to the CEO's death. Instead, Vincent had wound up reporting to his rival, who had become the arbiter of the company's brand.

Under Schiller, Marcom appeared to wobble. An ad about Siri featuring John Malkovich got slapped with lawsuits for embellishing the voice assistant's capabilities. (The suits were dismissed.) Tech reviewers panned a subsequent campaign called "Genius" that depicted an actor as a "Genius Bar" tech support person resolving customer problems, a concept incompatible with Apple's years of "It just works" salesmanship. In a rarity, Apple pulled the campaign. It was the type of ad that some Marcom people said Jobs would have stopped before it started by simply saying, "That's not good enough." As press criticism of its advertising mounted, Apple began building its own team of copywriters and creatives to offer competing ideas to those proposed by Media Arts Lab. The group's woes and tensions were a microcosm of those roiling the company Jobs had built as it adjusted to the absence of its longtime conductor.

In late January 2013, Schiller reacted with alarm when he saw an article on the front page of the *Wall Street Journal's* business section. The headline—"Has Apple Lost Its Cool to Samsung?"—irked him, but an anecdote in the story was more troubling. It described a thirty-four-year-old Apple customer who had ditched his iPhone in favor of a Galaxy S III, inspired partly by Samsung's ad blitz. "If you see this stuff on TV enough, it gets you thinking," said Will Hernandez, adding that he liked his new phone's larger screen. Schiller forwarded the article by email to Vincent, writing, "We have a lot of work to do to turn this around."

Vincent read the note with dismay. He thought that Apple's problems went far beyond Samsung's campaign. The news coverage of Apple had shifted from the great products it made to questions about innovation, reports about suicides at iPhone factories in China, and the company's lawsuits with Samsung. None of it had anything to do with what had distinguished Apple's brand as the cool, rebellious company that people loved. Instead, it felt like a company that had become a big multinational corporation. He drafted a lengthy response, addressed to Schiller:

we feel it too and it hurts. we understand the totally critical nature of this moment. This perfect storm of factors is driving a chilling negative narrative on apple.

in the last few days we have begun developing some bigger ideas for apple, where advertising can absolutely help to begin to change this narrative to a more positive one.

Vincent raised the issues facing the company—"china/usa workers," "too rich," "maps," "apple brand slipping"—and proposed an emergency meeting similar to what Apple had done after Antennagate. He wanted Apple to consider a brand campaign, the first since its release sixteen years earlier of "Think Different."

we understand this moment is pretty close to 1997 in terms of the need for advertising to help pull apple through.

When Schiller read the email, he was appalled. Vincent's note practically said, "The problem isn't us, it's you." Schiller recalled his most recent meeting with Vincent when they had watched the iPhone 5 launch and listened to a presentation from product marketing about the state of smartphone competition. The meeting had made clear that the iPhone 5 was a better product than the Galaxy. Apple's problem was pure marketing. Schiller conveyed his frustration in a reply:

To come back and suggest that Apple needs to think dramatically different about how we are running our company is a shocking response. . . . This is not 1997. Nothing like it in any way. In 1997 Apple had no products to market. We had a company making so little money that we were 6 months from out of business. We were the dying, beleaguered Apple in needing [sic] of hitting a restart button. . . . Not the world's most successful tech company making the world's best products.

The email filled Vincent with remorse. When he reread what he had written, he could see why Schiller had responded so dismissively.

He knew that his note could be costly. Apple was Media Arts Lab's signature client. He tried to repair the damage.

please accept my apologies. this was absolutely not my intention. in re-reading my email I can see how you can feel this way. . . . i'm sorry.

After giving it some thought, Schiller decided to update Cook on the exchange. He emailed the CEO to raise his concerns about the agency. Media Arts Lab had revived Apple's brand in 1997 and, with Jobs's help, provided it with a string of legendary ads, but Schiller told Cook that it was failing to positively shape the public's perception of the iPhone 5. He typed up a message and pressed "Send":

We may need to search for a new agency. I've tried hard to keep this from being the situation, but we are not getting what we need and haven't been for a while. . . . They don't seem to accept that first and foremost they need to do a better job for us this year.

Cook reviewed Schiller's note. He always kept his emails brief, long enough to show he'd read them, but short and direct enough to spur action. He wrote:

If we need to do this, we should get going.

COOK WAS FACING another, more pressing problem that he could not delegate. In early 2013, congressional investigators summoned Apple's tax experts to a meeting in Washington, D.C. Senator Carl Levin, the chair of the Permanent Subcommittee on Investigations, was in the midst of a multiyear probe into corporate tax dodging and the overseas shell companies that corporations used to avoid U.S. taxes. His staff had sent a bevy of companies a questionnaire about offshore entities. In its reply, Apple had left one section blank, and the investigators wanted to know why.

When Apple's tax team arrived, they were ushered into a meeting room with a long, imposing table surrounded by a dozen mismatched chairs. A red leather sofa sat against a wall, adding a thrift store sensibility to the room. It was a far cry from the sleek white tables that occupied the conference rooms at Infinite Loop. The tax team took their seats and watched as the seasoned congressional investigator Bob Roach, an attorney, began reviewing Apple's response to the questionnaire.

"Why didn't you fill that in?" asked Roach, directing their attention to the blank space deep in the survey asking them to indicate the jurisdiction of an entity called Apple Europe.

Apple's tax team looked at the question and back at the investigators. The room went silent. The downward gaze of Apple's team caused the investigators to wonder if the omission had been more than an accidental oversight.

"We're not really a tax resident anywhere," a member of Apple's tax team said.

Roach resisted the urge to lean forward and kept his face straight. "How could that be?" he asked.

Apple's tax team explained that although the United States determines corporate residency by point of incorporation, in Ireland, where Apple Europe was based, it was determined by where the company was managed and controlled. Since Apple's Irish subsidiary had no employees and was not managed or controlled in Ireland, it wasn't treated as a tax resident of Ireland.

After they spoke, Roach realized that Apple wasn't paying taxes in the United States on profits made in Europe because the money flowed to a subsidiary in Ireland, and it wasn't paying taxes in Ireland on the profits in Europe because its Irish subsidiary was managed from the United States. A circular but clever logic trick was saving it billions of dollars in taxes.

Roach left the meeting and called a few tax experts to assess Apple's practices. They all reacted with surprise. He and his fellow investigators sensed that they had unearthed something unique. They went back to Apple and its accountant for more information and additional interviews. They soon learned that Apple had three Irish subsidiaries with no tax residency anywhere, which had collected $74 billion in profits over a four-year period. A favorable agreement with the Irish government meant that it was paying less than a 2 percent tax rate on those earnings. More important, the investigators found buried in the documents the signature of Tim Cook.

Roach took what they had found to Levin. The Michigan Democrat immediately grasped its significance: the United States' biggest and most successful company was not only avoiding paying taxes in the United States, it was avoiding paying taxes in Europe as well. Levin called it the Holy Grail of tax dodging.

Several weeks later, Cook traveled to Washington to meet directly with the investigative team. He breezed into the room and took a seat at the large wooden table. Dressed in a suit and tie rather than his everyday California casual attire, he listened as the investigators shared their concerns about Apple's tax practices. He politely explained why Apple had set up subsidiaries in Ireland. Generally, he said, he believed that U.S. tax laws were unfair because they required companies to pay the same 35 percent tax rate on overseas earnings as they did at home. He considered the rate unreasonable and had been docking Apple's cash in Ireland rather than bring it home.

As the hour-long meeting ended, he watched as one of the investigators pulled out his iPhone.

"I have one more question for you," the investigator said. "I can't get this app to open."

Cook smiled. "I don't know how to do it, either," he said.

AS THE PLANE BEGAN its descent near Cupertino, the team from Media Arts Lab could feel their anxiety increase. It had been a few weeks since Vincent had sent his charged email to Schiller, and the relations between the ad agency and Apple had never been so strained. The team sensed that the future of their sixteen-year relationship with the world's most valuable company hinged on its latest commission: a brand campaign.

The group filed into an Apple conference room carrying storyboards of their creative concepts and began setting up a presentation for Apple's triumvirate of marketing tastemakers: Schiller, Vice President of Marketing Hiroki Asai, and head of communications Katie Cotton. Despite some initial skepticism, the trio approved Vincent's idea for a brand campaign. Apple's only other brand campaign— "Think Different," with its tribute to "the round pegs in the square holes"—had been considered one of the best commercials of all time. It fell on the team from Media Arts Lab to deliver a worthy successor.

The agency's chief creative officer, Duncan Milner, led the pitch. A mellow graduate of a Toronto art school, he had an ear for creative commercials that expressed what brands wanted. The first spot, he said, had been dreamed up by an award-winning copywriter named Michael Russoff. The agency had named it "The Walk."

To set it up, Milner asked the group to imagine an early-morning walk, a moment to let the magic of life in before the madness of the day begins. "Steve always loved to take people on a walk to have a conversation," he said, referring to Jobs. "We'd like to take every Apple employee and user on a walk."

On the nearby screen, a video showed the hills of northern

California. It was early morning. The sun hadn't yet fully risen, and the breeze was blowing through the grass. The camera moved at a walking pace, as Milner provided a voice-over for the commercial.

"It's sad when a founder dies," he said. "You wonder if you can make it without him. Should you put your brave face on for the world or just be honest? You doubt. In meetings, even if it isn't said, there are people thinking: What would he do? What would he say? You wonder if you had him long enough for the magic to rub off. Is it in you? Or did he take it with him? So the doubt goes on for a while. Then, one day, you're sitting discussing something crucially important. A fork in the road. A biggie. You realize you know exactly what to do. You know it yourself without asking what he would have done. You realize everything he believed is still alive.

"Steve knew that his greatest product wasn't something you could hold or use. It wasn't the iPhone or the Mac. It was something much braver. A fearless company. A borderless country. Apple itself. He didn't just think different. He made everyone around him want to. And now we just can't stop."

The screen faded to Apple's logo with the words "Think different."

Looking up, Milner saw Apple's head of communications, Katie Cotton, weeping. She dabbed tears from her eyes. He'd never seen a client cry in a meeting. He froze.

"We can't run this," Cotton said, trying to compose herself.

"Oh, Katie," Milner said. "I'm sorry."

Schiller and Asai were stunned. They couldn't fathom referring to the company's dead CEO in a commercial. They reminded the Media Arts Lab team that Jobs had never wanted to be featured in an advertisement; he had argued against providing the voice-over for "Think Different" before its release in 1997. More important, Schiller said, it wasn't true. In the final two years of Jobs's life, he had been largely absent from campus. The employees had learned to operate without him.

"We can't do that," Schiller said. "We have to show the world that we're moving forward, not looking back, and that we're more than Steve."

Milner nodded. Other proposals that day also failed to impress Schiller, Asai, and Cotton, who sent the Media Arts Lab team back to Los Angeles to come up with something better.

APPLE'S OUTPOST in Washington, D.C., was intentionally tiny by corporate standards. Jobs had disdained politics and considered lobbying wasteful. The Senate's tax investigation stretched the dozen-person team in ways the company had never anticipated.

As winter ended, Senator Levin scheduled a hearing with Apple and summoned Cook to explain its tax practices under oath. Cook pushed his team in D.C. to arrange individual meetings with senators beforehand in a bid to tell Apple's story to them privately and directly before he did so publicly. Having never testified before Congress, he also made personal calls to former president Bill Clinton and Goldman Sachs CEO Lloyd Blankfein. He wanted to know what he should expect.

Blankfein, whose hearing after the 2008 financial crisis had lasted ten hours, told Cook to not let his lawyers dictate his remarks. They want to protect you from legal jeopardy, he said, but that can limit how well you can shield the company from public criticism. He also arranged for Goldman's and Apple's communications teams to speak, paving the way to a press briefing before the hearing.

One morning that spring, Cook went into the conference room in Apple's Washington office and took a seat at the head of the table for a "murder board." Lobbyists, attorneys, tax advisers, and communications officials filled the room. They spent the day pretending to be senators, interrogating Cook in a simulation of the upcoming hearing. Their goal was to terrify him.

Convinced that Apple was being falsely vilified, Cook intended to advance a larger argument during the hearing: if you have a problem with the U.S. tax law, you should fix the law—not blame Apple. But as Apple's attorneys and lobbyists began to ask him questions about its Irish subsidiaries, he responded with questions of his own about tax laws. He wanted to know exactly what to say if he was asked about any of them. The faux hearing veered into a technical discussion of the arcane rules. Then Cook pulled back from the minutiae with a broad question: "What's the story we're telling here?"

The zooming in and zooming out heightened the apprehension floating around the room. A few lobbyists began to worry that Cook was too focused on the minutiae and feared he might digress into a detailed discussion of tax law in the coming hearing, a detour that might make senators see him as an irritating know-it-all. Cook also bristled at some of the lobbyists' questions, reflecting a company ethos that Apple did good in the world and shouldn't be chided as if it were some greedy corporation. Some grew anxious that his coming testimony might be a disaster.

Days later, Cook stood before a panel of senators and swore that his testimony would be truthful. He then sat down and listened emotionlessly as Levin gave a lengthy introduction, lecturing the room about Apple's "ghost companies . . . exploiting an absurdity" in the law. Cook spent more than an hour being badgered by Levin, with much of it coming back to the cash Apple was stockpiling overseas to avoid U.S. taxes.

"I just want to ask you that one question," Levin said. "Is it true you told our staff you are not bringing the $100 billion home unless we reduce our tax rates? Is that accurate?"

"I do not remember saying that."

"Is it true?" Levin asked.

"I said I do not remember saying it," Cook repeated emotionlessly.

"I am saying is it true that you are not going to bring them home unless we reduce our tax rates?" Levin pressed.

"I have no current plan to bring them back at the current tax rate," Cook conceded.

It looked as though Levin had gotten the confession he wanted, but Cook wasn't finished. After a long pause, he added, "Your comment sounds like it is forever, and I am not projecting what I am going to do forever because I have no idea how the world may change."

The murder board team members seated in the hearing couldn't believe it. Cook had countered Levin's efforts to pin him down. He was calm, attentive, and respectful but firm and unwavering in his defense of Apple's tax practices. And rather than talking about specific tax laws, he stuck to the story the team had agreed to tell: Apple is the United States' largest taxpayer, paying $6 billion in taxes annually, and the office it had set up in Ireland, where it kept its overseas profits, had its roots in decades-old computer manufacturing there. In that moment, they saw him morph from Jobs's default successor into the CEO Apple needed.

IN LOS ANGELES, the scramble to come up with a commercial continued. Media Arts Lab worked furiously to finalize ideas that they could show Tim Cook and Jony Ive. They wanted to finish the campaign in time for a summer launch to reset the public's perceptions of Apple, Samsung, and the future of smartphones.

Early that spring, Cook strode into Marcom's conference room to see the work. He was delighted to find Lee Clow, the chairman of TBWA\Worldwide, there with the agency. Clow had rejuvenated Apple's brand twice before, first with the "1984" ad Cook had studied at Duke and later with "Think Different." Cook was glad to see that he would have a hand in another revival.

The jury of decision makers—Cook, Ive, and the Marcom

team—took seats at the table as a video screen came to life. Against a white background, an animated drawing showed a seed planted in the ground that sprouted from a sapling into a giant apple tree. As it grew, a narrator spoke:

> Leave it better than you found it. Crazy idea. Do a little good. It gets less simple. . . . The congratulations come. You grow bigger. And it's expected that your relationship with the world will change. But it doesn't have to if you stay true to your idea. . . . To leave this place better than you found it and make things that inspire others to do the same.

The animation faded into an Apple logo. As the commercial ended, the team from Apple nodded with approval. Cook appreciated the message. As Apple had expanded, its responsibilities had changed. The scrappy underdog that had stood up as an alternative to Microsoft had become a tech giant and in the process had been targeted by everyone from the *New York Times*, which had won a Pulitzer Prize for writing about Apple's outsourcing, to Greenpeace, which had attacked it for its use of hazardous chemicals. Cook wanted to reverse that image.

"I love it," Cook said.

The agency teed up another commercial. A mellow trickle of piano music filled the room as an animated video showed four black dots moving across a white screen like pencil strokes, drawing squares, octagons, and circles. The music flourished as words appeared and disappeared from the screen, line by line:

> if everyone is busy making everything. how can anyone perfect anything? we start to confuse convenience with joy. abundance with choice. designing something requires . . . focus. the first thing we ask is . . . what do we want people

to feel? delight. surprise. love. connection. then we begin to craft around our intention. it takes time. there are a thousand no's for every yes. we simplify. we perfect. we start over. until everything we touch. enhances each life it touches. only then do we sign our work. Designed by Apple in California.

Ive, who was often unimpressed, spoke first. "I love it," he said. "That's us."

Cook nodded. "I love it, too."

For Vincent and the team from Media Arts Lab, the comments were a deliverance. The commercial titled "Intention" reflected their effort to channel Ive's philosophy, which permeated Apple, into a marketing concept that would include print and TV ads and other promotional elements. They hoped it would remind the world what the company was about and restore the public's faith that even after Jobs's demise, Apple remained true to its identity.

In the discussion about the potential campaigns, Cook made clear that he was pleased that Media Arts Lab had given Apple excellent choices. But there was so much negativity around the company's outsourcing and scrutiny of its environmental footprint that he worried that committing to "Leave It Better Than You Found It" would expose Apple to accusations of hypocrisy. Focusing on design and devices, as "Intention" had, would avert that risk and honor Jobs's belief that Apple's marketing should be about its products.

"I'm going to take 'Leave It Better than You Found It' and use it internally," Cook said. It would become one of his favorite turns of phrase, a way he would challenge staff across his ever-expanding empire.

THE LIGHTS at San Francisco's Moscone Center went down, and a hush fell over the packed auditorium. It was early on a Monday morning in June, and a crowd of five thousand developers filled the convention hall for the annual Worldwide Developers Conference. Cook stood offstage and watched as Media Arts Lab's "Intention" advertisement played on a massive screen. The piano music and shapes that filled the screen soon gave way to a series of inspirational words: *delight. surprise. love. connection.*

Applause, two-fingered whistles, and hoots erupted from the congregation of fanboys. Dressed in an untucked button-down shirt, his short gray hair perfectly parted, Cook hustled on the stage with a grin. "Thank you," he said. He listened as the audience roared louder.

"I'm really glad you liked that," Cook said. "Those words mean a great deal to us, and you'll see that reflected throughout the show."

His stage presence had improved in the twenty months since Jobs's death. He spoke with forcefulness, pridefully ticking off the details about how sixty-plus countries were represented in the audience and two-thirds of the attendees were there for the first time. He held the stage for eighteen minutes, making a case that Apple was in far better shape than its critics and Samsung suggested it was. It had a new store in Berlin, swelling Mac sales, and new software and hardware to share.

For the first time, Cook queued up a video about the company's newest iPhone software, iOS 7, narrated by Ive. The British designer, who was seated up front, had taken on a more prominent role leading Apple's software design but still spurned the company's tradition of making marketing speeches. He had prerecorded a seven-minute video that showed the transparent control center that the engineers had developed, the refined typography, the redesigned icons, and its bold new palette of colors. It began philosophically.

"We have always thought of design as being so much more than the way something looks," said Ive. "It's the whole thing, the way something actually works on so many levels. . . . It's about bringing order to complexity."

Ive's work on software soon gave way to his work on hardware. In the wake of the Samsung ad, Cook, Schiller, and others had tired of being asked about their ability to innovate without Jobs. They sought to rebut critics by breaking with event protocol and revealing a new product before it was ready to ship, something Jobs had avoided.

"We don't usually do this, but you're an important audience," Schiller said after taking the stage. "We'd like to give you a sneak peek of something we're working on."

The words *Mac Pro* appeared on the screen behind him. The computer would be revolutionary and radical, he said. A video played showing a white light shimmering off a dark sphere. A slow-building rumble gave way to heavy guitar riffs and crashing drums as the camera panned across a black cylinder-shaped computer. Schiller nodded his head and bit his lip as the crowd roared.

"Can't innovate anymore, my ass," he muttered like a trash-talking athlete. The crowd cackled as he strutted across the stage, gazing up at an image of the computer on a giant screen behind him, his un-scripted line floating away like an ash of insecurity.

WHEN THE COMPUTER LAUNCHED months later, customer interest fell short of what Apple had hoped. After the initial sales of about twenty thousand units, orders plummeted, and the company wound up slashing production. It became known inside the company as "the failed trash can."

The reviews of iOS 7 were mixed. The *New York Times*' David Pogue praised the design for "de-annoyifying" the iPhone, and Tech-Crunch's reviewer thought it was "easier and more enjoyable" to use. But customer complaints mounted as iPhone users found some of the new typography misaligned and criticized the brighter colors for draining their batteries.

The brand campaign, named "Designed by Apple in California," didn't fare much better. Instead of the more abstract "Intention," Apple released a commercial that showed close-ups of its devices, the kinds of images that marketing man Schiller loved. There was a classroom of Asian students doing schoolwork on iPads and a couple snapping a selfie with an iPhone. "This is what matters," a narrator said. "The experience of a product. How it will make someone feel. Will it make life better? Does it deserve to exist? We spend a lot of time on a few great things until every idea we touch enhances each life it touches. You may rarely look at it, but you will always feel it. This is our signature, and it means everything."

Viewers rated the ad below average for all companies and far below the high marks usually awarded Apple. Slate pilloried it. In an article titled "Designed by Doofuses in California," a Slate critic said the commercial betrayed an "underlying arrogance" that showed the company took itself "way too seriously."

The critique stung, but some of the Marcom team agreed. The commercial would be a great spot for other companies; for Apple, it was a B.

THE SENATE'S SPOTLIGHT on Apple's growing cash hoard attracted Wall Street's sharks. Apple's share price was languishing as Samsung stole smartphone market share. Investors wanted the company to pay dividends. In August 2013, Carl Icahn trumpeted on Twitter that he had bought a large stake in Apple and spoken by phone with Cook. Icahn made his message clear: Apple needed to return capital to increase the price of its depressed shares.

One of the original corporate raiders, Icahn had made a name for himself in the 1980s by amassing stakes in mismanaged companies such as TWA and pushing them to cut costs and sell assets. He had spurned analysis and followed his gut, turning to the press to make his case if management didn't listen. He had built a personal fortune of $18 billion with smarts and bluster.

The activist's pressure created a conundrum for Cook. Jobs hadn't believed in returning cash to shareholders. Scarred by Apple's near bankruptcy in 1996, he had favored building a treasure chest that could help the company in an economic downturn and give it the firepower to reinvest in the business when needed.

Cook was less dogmatic, but he lived in his predecessor's shadow. In his first year as CEO, he had committed to $10 billion in stock buybacks. In 2013, Apple increased that to $60 billion. Icahn, who had bought about $2 billion of shares, demanded that Apple almost triple the commitment to $150 billion.

Icahn's campaign deviated from his usual playbook. He believed that Apple was well managed but undervalued by Wall Street. Buying back shares would increase its earnings per share and lift its stock price by a third, he estimated.

But his position unnerved Cook, who was uncertain how to respond. He turned to Warren Buffett for advice and brought in Goldman Sachs to help navigate the tensions.

Icahn then invited Cook to dinner at his luxury New York City apartment on Fifth Avenue overlooking Central Park. Cook surprised his advisers by accepting the invitation, which most agreed that Jobs would never have considered. As an MBA-trained steward of the company instead of a founder, Cook could see the wisdom of Icahn's proposal.

Cook arrived at Icahn's apartment on the evening of September 30 with Apple CFO Peter Oppenheimer. He followed Icahn outside onto the financial titan's fifty-third-floor terrace. The lights of the Upper West Side sparkled across the darkness of the park. They chatted casually until Icahn ushered Cook inside for a three-hour dinner that culminated with a desert of sugar cookies cut in the shape of Apple's corporate logo. As they ate, Icahn argued his case.

"Tim, the obvious thing is you have all this money sitting around," he said. "You should be buying back stock. The company is selling dirt cheap."

Icahn knew that Apple was holding more than $100 billion overseas to avoid paying U.S. taxes. He suggested that the company borrow against it and use the borrowed money to return cash to investors. Then, when the U.S. tax laws changed, it could return the cash from overseas to the U.S. and pay off the debt.

Cook didn't say much, but he made sure that Icahn had the impression that he was open to the idea by giving the corporate raider ample room to lead the conversation. He listened intently as Icahn spoke and nodded in a way that gave Icahn assurance that he appreciated the campaign for buybacks.

After the dinner, Icahn issued a letter calling on Apple to finance buybacks with debt. The company eventually increased its share repurchase plan from $60 billion to $90 billion. Icahn pressed for more. Meanwhile, Goldman worked on Apple's behalf to secure billions of dollars' worth of bonds that the company could use to buy back its shares.

The capital repurchases lifted Apple's stock price and quieted Icahn, who eventually sold his shares for a profit of $1.83 billion. Cook cannily followed Icahn's advice and drove the company's share price up by doing something his predecessor would never have considered.

AS SUMMER TURNED TO FALL, Cook barreled ahead with an effort to add another collaborator to his executive team. The company's head of retail position had been vacant for a year, and with Ive pushing ahead on a smartwatch, he was eager to find someone who could help launch the new product. His previous hire had been a disaster.

John Browett, who had been let go alongside Forstall, had lasted only a few months. With Cook's support, the former head of the U.K. electronics chain Dixons had embarked on a cost-cutting mission that had aggravated retail staff and sparked a minor rebellion in Apple's stores. Though the push had flowed from Cook's long-standing desire to make the stores more efficient, the CEO had dumped the high-profile hire because Browett wasn't a good fit with the company's culture.

With Ive pushing to make a watch, Cook needed to find someone who was less Best Buy and more Burberry. Apple had recently collaborated with the CEO of the British fashion brand, Angela Ahrendts, to provide prereleased iPhones to photograph its fall runway shoot. It was the type of imaginative idea that combined fashion and technology in ways that would benefit Apple's watch launch.

Ahrendts had tripled Burberry's sales, had a reputation as a good manager, and was partners with Ive's longtime friend and Burberry's lead designer, Christopher Bailey. After rejecting executives from the wireless industry as a poor fit, Cook invited Ahrendts to Infinite Loop.

It took some persuading to convince Ahrendts that she was right for the job. She wasn't especially interested in tech and felt more comfortable seated beside a couturier than an engineer. Cook reassured her that that wouldn't be an issue.

"We have thousands of techies," Cook told her. "I don't think that's what I'm looking for."

When Cook announced the hire in October, Ahrendts was heralded as a talented addition to an all-male executive team. She was perceived as charismatic and outgoing, the perfect person to motivate the forty thousand retail employees spread around the world. She had striking blue eyes, sandy blond hair, and a closet full of chic clothes.

Her reputation and fashionable attire created a mirage. In reality, she was introverted and shy.

On one of her first days, the retail staff at Apple's headquarters created two lines just inside the door of their Cupertino office. They began cheering as she approached the door. It was a custom borrowed from rank-and-file retail employees, who sometimes welcomed customers with rousing applause when they entered Apple's stores. As Ahrendts stepped into the building, she froze.

Rather than walk down the corridor created by her enthusiastic new staff, she turned into a doorway and disappeared.

Chapter 9

The Crown

The projects were piling up. At any given moment in 2013, the attention of Apple's reigning tastemaker could be wrenched in many disparate directions, from watch design to software design to the *Designed by Apple in California* photography book. With the company breaking ground on a new headquarters, Ive also assumed more and more oversight of architectural designs, building materials, and construction plans. Any of those individual undertakings would be a burden, especially for a perfectionist, but assuming all those responsibilities at once and alone, without the help of his creative partner Jobs, proved too much.

Time began to blur.

Ive was the most consistent and predictable member of the laissez-faire design team. He typically arrived at the studio at the same time every morning and departed late, after he felt his work was done. He organized and led three design meetings a week, on Monday, Wednesday, and Friday, pulling the group's work forward incrementally like the moon drawing the tide. But as his responsibilities for the software and the building increased, his calendar became more fluid and his presence less predictable.

The paradox was inescapable: at the very moment when Ive was leading an effort to create the world's most accurate timepiece, his personal relationship with time was being unwound.

AFTER IVE AND NEWSON developed a design for the watch, they turned their attention to a charity auction they had been asked to curate by the musician Bono. The U2 front man, who compared the designers to Donatello and Michelangelo, had brought them together to raise money for (RED), the charity fighting HIV and AIDS in Africa. As is the case with most new assignments, Ive and Newson had come to the project cold, with no idea of what they wanted to do or how to do it. Gradually, they curated a collection of well-crafted products that in the end included a "Red Pops for (RED)" Parlor Grand Model A Steinway piano, a magnum of 1966 Dom Pérignon champagne in a red cooler designed by Ive and Newsom, a custom Airstream 16 Sport Travel Trailer, and more. They complemented the assortment of things they loved with a one-of-a-kind original: a custom-made Leica camera.

There were practical reasons to design it. A single product created by two of the world's top designers—one whose tech products changed society and another who had made industrial design into high art— was bound to raise more money for charity than anything in abundant supply. But it was the oddity of it that appealed to Ive. There was something fun about the incongruity of someone accustomed to developing devices for multitudes putting comparable energy into something for only one person. Leica had distinguished itself by building some of the world's earliest compact cameras. Its signature line, the Digital Rangefinder, featured a black metal case. Ive secured Tim Cook's support for a project to make a new camera that stripped away Leica's traditional black exterior and replaced it with a case that was as sleek and simple as a silver MacBook Air. He treated it just like an Apple project. He assigned two designers to lead it, Miklu Silvanto and Bart Andre, and drafted a product engineer, Jason Keats, to work on the components and assembly. They cleared a table in the design studio for the camera, putting the single object on a par with the tables devoted to iPads, iPhones, and Macs, which sold in the millions.

Ive and Newson developed a design that called for the camera's case to be cut from a single block of aluminum. They proposed using a computer-controlled machine to laser-etch a hexagonal pattern into the exterior of the case, giving it a subtle texture reminiscent of the perforated black leather that had enclosed early Leica cameras.

The project was supposed to take a few weeks, but they soon discovered a problem: the original Leica camera opened like a clamshell and had a sandwich of digital components inside. Since they were planning to create a one-piece case, they would have to reconstruct the camera's internal parts, from circuit board to control switches, so they could lower them all into the body.

When the first prototype was finished, Ive exited his glass office and strolled to the table to review it. He twisted the shimmering silver camera in his hands and brushed his fingers across a toggle button on the rear of the unit that looked like a Nintendo controller. It was there to enable users to scroll through digital photographs on the camera's display. But he didn't like the buttons. They protruded too much. He told the team that he wanted the knobs to be as flush and smooth as the aluminum case itself.

"Make it feel perfect," Ive said.

It was a challenging ask. Keats spent days inserting 100-millimeter sheets of plastic film called Mylar on each side of the rear toggle, trying to raise the buttons the minimal amount necessary to make them discernible while keeping them practically flush with the exterior of the case.

The camera design took more than nine months and required 561 different models before Ive was satisfied. Apple estimated that fifty-five engineers had spent a combined 2,100 hours on it. The company reused some of the manufacturing techniques in future Apple products, including the laser-etching process for MacBook speakers. Keats did the final assembly by hand and traveled to Germany to have Leica's engineers ensure that the camera worked.

After the product was finished, Ive invited Cook to come see it. The distance between Apple's centers of power—the design studio and the executive wing—had widened after Jobs's death. Unlike his predecessor, who had visited the studio daily, Cook seldom ventured from the company's capital of commerce, Valhalla, to its core of creativity, the holy of holies.

Ive ushered Cook to the table and held up the polished aluminum camera. He glowed as he explained the laser etching that dimpled its exterior like a citrus zester. He observed that the only notable color on the unit was the red of a Leica logo on the front and in a few small details such as the *A* for "Auto" on the exposure dial, allusions to the (RED) Auction.

Cook nodded expressionlessly as he peered over Ive's shoulder. To people watching from across the studio, Cook had the look of a semi-interested parent examining a child's finished Lego project. Some would later joke that they had caught his eyes scanning the nearby tables that held the iPhones, iPads, and Macs that actually made the company money.

Ive was still engrossed with the camera when Cook turned to leave. He had stayed only five minutes.

AS IVE FINISHED his charity work, his team turned its focus to watchmaking. The design created by Ive and Newson provided the direction the company needed to move forward, and Ive rewarded Newson for his work on it by encouraging Apple's leadership to bring the Australian onto the team in a more formal way. The responsibility for bringing Newson on fell to Apple's mergers and acquisitions team, which structured a high-dollar deal that valued the independent designer more like a company than an employee.

The solution for how to attach the watchband made it possible for the team to fulfill Ive's ambition to make the watch more personal than any other Apple product. It meant that Apple could mass-produce identical watch faces and create an array of interchangeable watchbands that would allow people to customize the device on their wrist with different colors and materials. In weekly meetings, the designers settled on making the bands of silicon, leather, and metal. Ive assigned a lead designer to each material and tasked recruiters with hiring experts in textiles to provide support.

For the leather watchband, a new team of soft-goods engineers worked to identify tanneries around the world that could provide the ideal material. They imported swatches of brown cowhide from Italy, France, England, Denmark, and the Netherlands and examined them to find sections of the leather without stretch marks or nicks. They cut those flawless pieces out of the swatch like a butcher trimming filet mignon from a steer. Later, they would spread the perfect samples out on the oak design table in the studio for review.

Ive would pick up each cut of brown leather as delicately as he might lift a feather. "This is beautiful!" he would mutter with enthusiasm. Then he would run his index finger up and down the surface to evaluate its smoothness. He would fold the sample to evaluate its suppleness and examine how it creased. Then he would put it under a jeweler's microscope and scan the surface of the grain, scrutinizing it for specks or imperfections. Finally, he would carefully return the leather to the spot he had lifted it from before moving to the next sample and repeating the exercise.

Ive attended dozens of meetings like that over the year, evaluating one set of samples with a pebbled surface and others with the raw look of a vegetable tanning process. Ultimately, he chose a smooth top-grain leather in a tawny brown from a French tannery established in 1803, a pearl-colored leather with a pebbled texture from an Italian tannery, and a black leather with a subtle grain from a Dutch tannery. He insisted that the leather swatches be thin and delicate, with engineers adding a laminate between some swatches for durability.

A similar process played out for what would be named the Milanese loop, a smooth, lithe stainless-steel bracelet that Ive obsessed over until the team found a way to weave the metal together into a watchband that was as delicate as fabric. Comparable care was given to the selection of colors for the silicon bands that had a design reminiscent of the work Newson had done with the watch brand Ikepod, creating a watchband that could be adjusted to sizes by pinning a metal button into a hole.

They were single bands, as fully thought out and obsessively reviewed as the entire watch.

IVE WASN'T DONE. He also pushed the group's thinking to the molecular level.

For years, he had been refining the design team's understanding of the materials it used. The effort began to gather momentum in 2004 with his push to control the faint streaks of black in the iMac's aluminum stands and expanded in the years that followed as he sought to create iPhone volume buttons as refined as jewelry. The former, part of his obsession with aluminum, led Apple deep into the supply chain, where it determined that it could minimize streaking by dictating the percentages of the magnesium, iron, and other elements that went into the aluminum alloy. The latter, part of his infatuation with refinement, involved a trip to the Hong Kong Watch & Clock Fair around 2010 to evaluate equipment exhibitors used to make the miniature polished metal buttons that fit flush against the edges of timepieces. Over time, the design team increasingly leaned on materials experts such as Masashige Tatebe, who would stand at a whiteboard and draw pictures of the molecules inside plastics and metals as designers posed questions about how the materials' properties influenced color and reflected light.

Ive proposed applying all those years of learning to create watch cases in different materials, increasing the options for people who wanted to personalize what they wore. To support and broaden its use of materials, Apple acquired part of a Chicago-area firm, QuesTek Innovations, that had pioneered the use of computers to design custom alloys. It had patented the steel used in race cars and rockets, and Ive wanted the firm's help in developing his own line of gold.

Almost from the beginning, Ive was adamant that Apple make a gold watch, which he imagined as a bejeweled halo for the entire product line. He proposed making it in rose gold and traditional gold. The concept frightened Apple's product design engineers, who were responsible for combining the materials, hardware components, and software into a manufacturable product. They knew that gold was a dense but soft metal that could easily be dinged and scratched, a prospect that filled their imagination with customers seeking costly refunds on lightly damaged $10,000 timepieces. To eliminate that risk, they embarked on an effort to design a sturdier, more durable gold.

The work fell to the team from QuesTek, who used computers to conceptualize a sturdier metal. Standard 18-karat gold is 75 percent gold and 25 percent other metals such as zinc and nickel. The proportion of those other metals determines the strength of the gold. Apple's engineers came up with combinations for the rose gold and traditional alloys that included copper, silver, and palladium. The luxurious metal could be cast into blocks of gold and chiseled away to produce a single-body watch case. But what delighted Ive was the engineers' assurance that it was twice as strong as traditional gold, luxurious yet durable.

Ive's ambitions for the watch challenged one of Jobs's principles. Upon his return to Apple in 1997, Jobs had eliminated 70 percent of the products the company was making and drawn a four-square chart on a whiteboard. He had written "Consumer," "Pro," "Desktop," and "Portable" in the squares and said the company needed to make one great product for each quadrant—four products in all. It reflected his philosophy that deciding what not to do was as important as deciding what to do. With the watch, Ive was testing those boundaries, pushing three cases in two different sizes, available in several colors, plus an array of complex bands. His pursuit of personalization would end up with fifty-four different configurations. Instead of a narrow focus, he pushed a broad undertaking that required more decisions and more bureaucracy.

While some managers find empowerment from adding head count under their umbrella, Ive viewed it as an annoyance of corporate bloat that could create obstacles for his creative ideas. More and more engineers and operations staff filtered into the studio to manage all the elements of the watch and a push by Apple's product marketing arm to diversify the iPhone with a lower-priced model in five colors. The newcomers brought Cook's back-office concerns about operations and costs into the sacrosanct studio. Ive's unwritten rules began to be broken.

During a meeting on the watch case in 2013, the lead designers, Julian Hönig and Rico Zorkendorfer, found themselves hemmed in by operations staff, whose job was to deliver the product on time and under budget. They were seated around an oak table near the industrial design studio's kitchen and passing around a stack of printed presentation slides. As Hönig and Zorkendorfer flipped through the papers, they were surprised to find a detailed proposal to reduce costs by using a less expensive manufacturing technique for the watch crown. The design team wanted to cut each crown with a computer-controlled machine, a CNC tool, that would have unrivaled precision resulting in a more beautiful and realistic crown. Yet, the operations staff proposed a low-priced laser-cutting process that would save millions of dollars.

"That's not Apple," Zorkendorfer said.

"That's something Samsung would do," Hönig added.

The product designers, who were also at the table, tried to hide their alarm. They knew how unwelcome Ive would have found the remark. In his absence, a realization sank in: money changers had found a place in the temple of design.

THE TWO-MILE DRIVE from Infinite Loop to the old Hewlett-Packard campus took less than ten minutes. Apple had bought the 100-acre office park in 2010 for $300 million. At the time, it was a sprawling swath of low-slung buildings surrounded by a moat of asphalt parking lots. The plan was to replace it with rolling hills, towering oak trees, and a showcase headquarters built by the world's leading architecture firm, Foster + Partners.

In late 2013, Ive arrived at the former HP campus for a product review. He had been making trips there regularly since the groundbreaking a year earlier. The landscape around him was a wasteland of dirt and crumbled asphalt. Standing not far from the rubble was a single-story prototype of one section of what would be Apple's future headquarters: a three-million-square-foot circle of curved glass that had the futuristic look of a flying saucer. The steel-and-cement prototype, shaped like a wedge of pie, had become a hub of activity, a place where Ive reviewed and vetted every element of the campus with the same care and concern he devoted to future products.

Ive felt a special responsibility for the project. In 2004, he had gone for a walk with Jobs through London's Hyde Park and fantasized along with the CEO about building a new campus for Apple that would be as open and communal as a university quad. Jobs had approved the final design before his death, settling on a four-story circle with an interior courtyard as long and wide as three football fields. It fell to Ive to ensure that the project lived up to his late boss's demanding taste.

For months, Apple's retail team had been searching the world for samples of glass for Ive to review. Getting clear glass for an office building might not seem like a complicated task; corporate real estate developers don't give it much bother, so long as it's transparent. But Ive insisted on inspecting every one for sufficient clarity.

Apple shipped glass samples from Europe and Asia to Cupertino, and Ive came over to inspect them. He wanted to find a piece of glass that was so transparent it would fill the company's offices with natural light that he believed would increase employees' happiness and boost their productivity. Ultimately, he chose a glass that was incredibly clear and relatively thin, two 12-millimeter-thick layers of glass with the ability to minimize noise and control the building's internal temperature.

Now he was visiting to review the options for a tinted glass canopy that would jut off each floor of the building like a hat brim, shading the interior from the California sun. He had requested sample after sample until he got a sheet of glass that had the lowest amount of iron content of any glass made in the world. As he examined it, he found its hue of green off-putting. He preferred cooler colors to warmer ones and asked about the possibility of making the canopy as white as an iPod.

The request ignited a scramble to find a way to take a clear piece of glass and give it the illusion of color. Architects and engineers arrived at a solution that involved applying a silicone finish to the glass canopy that masked its naturally green hue. The result was a building encircled by frosted fins that made the structure look as pure as a wedding cake.

BACK AT INFINITE LOOP, Ive met almost daily with a group of about ten software designers he had assembled to develop the way the watch would work. Just as the product couldn't proceed without its physical design being finalized, it couldn't go to market before it was determined how people would interact with it.

In a symbol of Ive's responsibility for the watch project, the ten-person team was housed inside the design studio, a short walk from his office. They occupied a ten-by-ten-foot space where they erected tack boards to display sketches and illustrations of how the software could look.

Ive would swing by regularly to evaluate their progress and make suggestions. The goal was to create a miniaturized iPhone on the wrist. It needed to be familiar but also original, an extension of the multi-touch technology birthed by the iPhone but adapted to the smaller 1.5-inch screen. He wanted a home screen that could be controlled by the watch's crown.

An early iteration of that concept came from Imran Chaudhri. A British-born human interface whiz with a shaved head and a penchant for black T-shirts and jeans, he had made his mark at Apple as one of the core innovators behind the multitouch technology that had made the iPhone a success. He sketched the idea of a home screen that could be made up of dozens of tiny app icons arranged in a circle, a tribute to the round shape of traditional watch faces. The icons could increase or decrease in size with a spin of the watch's crown.

Ive favored arranging the icons in the shape of a hexagon, which would make it easier to keep the shape balanced as users added and subtracted apps. The team thought it looked great, but some software designers brought on to work on the watch later worried that the icons would be too small for people to use.

With prototypes of the watch still being developed, the group did most of their work on paper and tested interactions on an iPhone rigged with a Velcro strap. They created a screen on it the size of the watch and outfitted it with a digital crown, so that they could evaluate how it would expand and contract the app icons. The miniature screen was too cramped for messaging, so they developed a system called "Quickboard," which suggested basic replies that could be delivered with a single tap.

To alert people to an incoming message, they worked with the product design team to develop a so-called taptic engine that would subtly tap a user's wrist. The concept created an engineering challenge. The taptic engine was composed of a linear resonant actuator, essentially a spring with a weight on the end that bounced on command. In a phone, it got people's attention because it vibrated at the same frequency as a mosquito's buzz, a sound humans are evolutionarily disposed to hearing. But no one wanted mosquitos buzzing on people's wrists, so the engineers worked to virtually eliminate the sound of that vibration until all that remained was the feeling of a gentle touch on the wrist.

WITHOUT JOBS to conduct new-product development, Cook tasked the company's top operations executive, Jeff Williams, with orchestrating the design, software, and hardware teams responsible for bringing the watch to life. He assumed the responsibility for leading a committee with representatives from each division who met regularly to chart the project's development. Ive provided the group with creative direction and Williams with supervision.

The role placed the operations executive in unfamiliar territory and put stewardship of Ive's vision in the unfamiliar hands of someone with decidedly different sensibilities than Jobs. Like his boss, Williams's expertise was on manufacturing products on a large scale, not inventing them. Inside Apple, he was known as "Tim Cook's Tim Cook." The similarities ran from their résumés to their physiques.

Both men were southerners with engineering degrees and MBAs; they were both tall and slender with close-cropped hair and narrow eyes; they were both stoics who listened more than they spoke; and they were both considered frugal. Colleagues considered Cook tight-fisted because he had refused to buy a house for a decade. Williams had earned the label because he continued to drive a Toyota Camry even after his total 2012 compensation had soared to $69 million.

One of the first challenges Williams faced came from hardware engineering. As the industrial design efforts hurtled forward, the hardware engineers struggled to finalize features of the watch. They explored an array of capabilities, including an EKG system to check for heart conditions and an electrical measurement tool for sweat glands called galvanic skin response that would inform people whether they were calm or stressed. Their indecision about which features to include prevented them from identifying the chips and components needed to power the device. Their workspace seemed more science lab than product development facility, demoralizing some of the team, including a senior engineer named Eugene Kim, who temporarily defected to Google.

Alarmed, hardware chief Bob Mansfield replaced the group's top manager with one of his trusted lieutenants, Jeff Dauber. The selection created an odd-couple dynamic with Williams. Dauber was charismatic, opinionated, and openly gay. He kept his head as bald as Mr. Clean and sported a curled mustache. A tattoo sleeve covered his left arm. He had joined Apple in 1999 and personified the rebellious spirit of a company that Jobs had once rallied around the slogan "It is better to be a pirate than in the navy."

One of the first things Dauber wanted to do was bring back Kim from Google. At most companies, rehiring a former employee was commonplace, but at Apple, the idea was seditious. Jobs had demanded complete loyalty from employees and had adhered to an unwritten rule that anyone who left the company for a competitor could not return. When Dauber proposed bringing back Kim to Williams, the operations executive initially hesitated. Williams wrestled with whether Jobs's rule should be upheld or abandoned as the company emerged from its longtime leader's shadow. Eventually, he acquiesced.

Almost immediately, Dauber and Kim sought to bring order to the chaotic hardware engineering effort. They killed features they deemed suspect or impossible to deliver on deadline. Among the casualties were the EKG and galvanic skin response. The EKG would have required approval from the Food and Drug Administration, a time-consuming, bureaucratic process that would take years. The galvanic skin response concept seemed unnecessary. Why would people need a device to tell them when they were excited? Wouldn't they just know?

The elimination of those and other features curtailed the health-monitoring capabilities of the watch and evoked questions about the device's purpose. Elbowing into the $7 trillion health care industry had been part of the justification for pursuing a smartwatch. With the iPhone lifting Apple's annual sales above $150 billion, it needed to develop products in enormous industries such as health where it could generate new sales. Including health features also gave the watch an altruistic focus, fitting with Cook's promise that Apple would make products that enriched people's lives. But every health concept Apple explored failed to take flight. Even the noninvasive glucose-monitoring system Apple had acquired from Rare Light had turned out to be a disappointment. Apple's engineers at Avolonte Health had found that Rare Light's technology didn't work as hoped and started to build their own glucose-monitoring system from scratch. They would spend nearly a decade working to develop a glucose-testing system that was the size of a miniature fridge and far from being miniaturized into a wristwatch. Other ideas had also stirred excitement followed by disappointment.

During a committee meeting of the top executives on the project, including Ive and Williams, someone raised the work of a startup that had claimed to have a microchip that could detect cancer. Ive grew excited about the possibility of being able to help people detect such a lethal disease and seek early treatment. Others in the room shared his enthusiasm. Not only had the disease taken Jobs's life; it had also caused the death in early 2013 of the de facto chief technologist, Mike Culbert, one of the company's unsung heroes. But their enthusiasm waned as they considered the legal risks of providing customers with false positives or negatives and the potential brand damage that could come if Apple became a messenger of doom, tapping people's wrist to deliver a bleak notification: you have cancer and may die.

In the face of those setbacks, Dauber narrowed the engineering team's focus to the heart rate sensors on the back of the watch case. Operating off the knowledge that blood is red because it reflects red light and absorbs green light, they developed a system to monitor the amount of blood flowing through a wrist's arteries by flashing green LED lights hundreds of times per second. Each heartbeat increases the blood flow through the arteries. As the blood moves, it absorbs more green light. Between heartbeats, the absorption of green light decreases. Those differences could be calculated in real time by the sensors and algorithms to determine the number of heartbeats per minute. The team developed a separate infrared sensor that shined a light onto the wrist every ten minutes to get a heart rate reading. Following the direction of Apple's designers, they put the lights inside four identical circles in a smooth ceramic case. The result combined complex engineering with sophisticated style.

To power the system, they designed the smallest circuit board the company had ever made. It connected the watch to an iPhone, which would transmit text messages and emails to the wrist via antenna. The radio-frequency components that provided that connection typically

needed to be encased inside small shields to prevent signals from in-
terfering with one another, but the watch was so small that there
wasn't space for the shields used inside an iPhone. The engineers ultimately created a custom coating that could be sprayed onto the circuit board to protect the integrity of signals. The solution paved the
way to a one-by-one-inch board with more than thirty components
and thirty pieces of silicon. It became the watch's brain.

As the hardware engineers made headway, the workload increased for the product designers responsible for building and testing
early prototypes. The process can be plodding because each prototype
requires testing each component to ensure that everything works harmoniously, almost like checking each bulb in a nonfunctioning strand
of Christmas lights to find the one that's out.

Hoping to accelerate the process, Williams urged engineers to
develop prototypes faster. The engineers thought the push flowed
from Cook's press to deliver the watch by the fall of 2014, so that Apple could show investors that it was more than an iPhone company.
In operations, Williams's team could meet ambitious deadlines and
shave days off a product's shipment date by solving problems on a
factory floor. He wanted the product designers to do the same. Some
engineers tried to explain that the trial-and-error nature of their work
made it difficult to fast-track the product. When Williams persisted,
they developed a fake calendar for him and traveled to China weeks
before the parts they needed would arrive. A more realistic calendar
they kept for themselves revealed the truth: there was no way they
could warp time.

WHEN APPLE'S TOP ONE hundred leaders gathered for one of their
annual off-sites south of Cupertino, the mood was largely upbeat.
Samsung was rising in power and continuing to poke at Apple in its
marketing, but the promise of the watch instilled executives with op-

timism that Apple would deliver a full-throated rebuttal of its critics. The presentations filled the luxury resort's meeting room with enthusiasm and confidence about the future.

The watch had galvanized Apple's workforce and given it purpose. It shook its leadership out of their collective grief and challenged them to lead the company forward. Through it, they rediscovered the spirit of teamwork and creativity that had helped them create the iPod, iPhone, and iPad. They also had other products on the way, including a mobile payment system and new iPhones. There were many reasons for the group to be upbeat.

But Ive was brooding.

One morning, he asked his most trusted operations colleague, Nick Forlenza, to join him for breakfast. He arrived at the shaded, outdoor patio that morning with wraparound sunglasses that concealed his eyes but not his anxiety.

One of the operations engineers working on the watch planned to bolt to a competitor. The defection filled Ive with concern that the engineer would provide the rival with a road map for ripping off the designs and engineering concepts that Apple had spent years developing. He worried that his baby was in jeopardy.

"Do you know how bad this would be?" he asked.

Forlenza listened sympathetically as Ive searched for a solution. Ive asked if there was anything they could offer to keep the engineer. He didn't care what it cost; he wanted Apple's intellectual property to be protected. Agitated, he stared into space, searching for a solution.

"I'm not sure you understand," he said.

He explained that the watch's crown was going to be its signature feature. It was a bridge that would connect watches of the past with his computer-powered watch of the future. It was central to the user experience, allowing people to click through the screens and apps on their wrist. He feared that the defecting engineer would share the idea of the crown and Apple's rival might introduce a poor imitation of the concept, polluting the marketplace before Ive's watch arrived. "This is not just them understanding an aspect of the product," he said. "This is the essence of the product."

Forlenza said he would try to get the operations manager back. In the end the rehiring effort would fail, but Forlenza's assurance helped calm Ive and restore his confidence.

THE RETURN FROM THE RETREAT to Cupertino brought new pressures and responsibilities.

Ive had historically provided minimal input into Apple's marketing. To the degree that he was involved, it was largely in regard to product packaging, where he had defined the minimalist white boxes with tolerances so tight that they eased open dramatically. Jobs had spearheaded the rest of the promotional push, from commercials to events. But in Jobs's absence, Cook had implemented a collaborative approach to marketing that required Ive to provide more input about how Apple sold its products to the world. Ive had such a distinct vision for what the watch should be that he embraced that responsibility.

As the timepiece came together, his conviction about its purpose deepened. He constantly characterized it as Apple's most personal device and preached that its success would depend on people's willingness to wear it. Selling the watch would require the endorsement of cultural tastemakers, particularly in the world of fashion, an industry with unseen influence over what people wear.

In a marketing meeting, he told colleagues that the reaction of tastemakers such as *Vogue* editor Anna Wintour and fashion designer Karl Lagerfeld would have more bearing on the watch's success than would the tech reviewers who assessed the latest Macs.

"Our future isn't in the hands of people like Walt Mossberg," he told Apple marketers, referring to the longtime product reviewer at the *Wall Street Journal*. He respected Mossberg but believed the watch needed to transcend tech reviewers to gain acceptance.

The focus on fashion struck some colleagues as being out of step with Apple's historic focus on marketing technology features. Product marketer Phil Schiller wanted to pitch the watch as an accessory of the iPhone or a fitness device, emphasizing its ability to bring messages to wearers' wrists or keep track of their workouts. Members of Apple's Marcom team worried that Ive was using the watch as a wedge to push the company in a direction that personally interested him. They considered his focus on fashion vain and self-serving.

Ive bristled at their resistance. He remained convinced that no one would wear the watch if it was marketed like a computer. People close to him saw his interest in fashion as an extension of Jobs's legacy of marrying technology with culture. In their eyes, the iPod alone hadn't resurrected Apple; its connection to music had. If the watch was going to succeed, Apple would need to forge relationships in the creative world and win over thought leaders in fashion the same way it had won over record labels and musicians.

Jobs would have crushed such internal tensions by issuing an autocratic decision based on his personal preference. The finality of his decisions had spared Ive, who disliked conflict, from the everyday headaches of corporate infighting. But Cook wasn't intimately involved in the product's development and allowed the corporate feud to simmer by trying to give each side some of what they wanted.

To support Ive's vision, Cook endorsed the hiring of Paul Deneve from the Paris-based fashion house Yves Saint Laurent. Deneve was tapped to develop the go-to-market strategy for the watch, reporting to Cook. The role gave him responsibility for sales, distribution, public relations, and the product range. He was in the studio daily, helping Ive with fashion-related design choices, such as a late decision to have a leather strap with a classic watch buckle. They came up with ideas for how the watch would be displayed in Apple Stores, toying with a concept for converting the stores' famed oak tables into miniature jeweler's cases with the timepieces below glass. Though the company had always been averse to using outside consultants, Ive brought on communications advisers with a background in fashion, heightening the tensions between longtime marketers accustomed to doing things the Apple way and outsiders who had ideas of their own.

In the summer of 2014, Ive arrived outside the company's fourth-floor executive wing for a meeting in the boardroom. He and Cook were joined by members of the marketing and communications teams to discuss the plans for unveiling the watch. The company agonized over its fall presentations. The marketing events introduced the new lineup of products ahead of the Christmas shopping season. Millions of people watched the shows and the press documented every new product, delivering the company hundreds of millions of dollars in free advertising. Since the watch would be Apple's first new product category after Jobs's death, the presentation of it needed to be perfect.

Cook prepared to mediate between the differing views of the people surrounding Schiller and Ive. The marketing team, led by Schiller, favored revealing the watch in September at De Anza College's Flint Center for the Arts, the same Cupertino auditorium where Jobs had introduced the first Mac and iMac. But Ive and his team of outsiders fretted about where they would display the watches for media and special guests afterward. He proposed erecting a tent that would be all white, his preferred color, to double as a place where they could see the watches after the show. To pull it off, Apple would have to remove the trees outside the building, erect the tent, and then replant the trees afterward. It would be expensive.

"How much?" Cook asked.

"They want $25 million," someone said.

Cook faced a dilemma. He appreciated the marketers' impulse to return to Apple's roots and host the event at De Anza. He also recognized Ive's concern about the risks of hosting fashion press at a community college in Cupertino. He rocked as the group discussed the tent. Some silently worried that it would look too much like a wedding tent. Others questioned the logistics of moving trees. Some simply tried to process how a company that had flirted with bankruptcy and adopted the mindset of a Depression-era grandmother had arrived at a place where it was considering spending so much money on a tent. When he'd heard enough, Cook stopped rocking.

"We should just do it," he said.

WHEN THE EARLIEST VERSIONS of the Apple Watch were finished, Ive insisted that *Vogue* editor in chief Anna Wintour see it before the public did. The request was unusual at a company that kept its work shrouded in mystery so that it could reveal new products with a magician's flair. It occasionally briefed journalists it knew well before important launches. With the watch, Ive wanted to go beyond that and brief one of the world's most influential figures in media.

Finding a time to bring Ive and Wintour together was difficult. He spent a portion of the summer in the United Kingdom, and she summered at a forty-two-acre estate in the Hamptons. They eventually scheduled a meeting at the Carlyle Hotel on New York City's Upper East Side. It was one of Ive's favorites, known for luxurious touches such as monogramming guests' initials onto pillowcases in gold.

In August, a product security team ferried several models of the Apple Watch to New York on a jet. They carried the watches in black cases into a suite overlooking Central Park where Ive and Wintour were due to meet.

Ive had never met the notorious *Vogue* editor. Known as the Ice Queen, she was the most powerful figure in fashion, admired for her acute business sensibilities and feared for her demanding nature. She could make or break a designer with little more than a look. Her approval could land a collection on the pages of the magazine, validating it for fashion's most elite and influential readers.

The two of them met privately in the suite. Once Wintour was settled, Ive delicately unwrapped the watches from swatches of leather as though he were unwrapping a gift. His process for showing off the watch typically included a lesson on the history of timekeeping, explaining how there was a parallel between the clock in a town square being miniaturized into a wristwatch and mainframe computers being miniaturized into smartphones. With the Apple Watch, the two fields were converging.

As Ive picked up each watch, he explained the design, the alloys, and the straps. He talked about how each item was made and showed her how the crown was a tool to navigate the miniature computer.

Wintour was mesmerized. She had seen countless designers show off their work over a lifetime in fashion, and she knew how to differentiate between those who were deeply involved with their products and those who relied on staff to bring their ideas to life. It was clear to her that Ive knew every millimeter of the watch and had thought through every detail. She felt as though he could build it himself.

Everything Ive showed her, he did in a loving way. She was struck by the way the design was on a par with a piece of art yet the product was functional. The detail of the presentation reminded her of her many meetings with Karl Lagerfeld, the longtime Chanel designer and one of the most influential figures in fashion.

The meeting between the artist and the Ice Queen had been scheduled to last fifteen minutes. It lasted an hour.

BACK IN CALIFORNIA around the same time, Jeff Dauber was growing increasingly concerned about the watch. The engineers had pushed as hard as they could to make its battery last as long as possible, but the green LED lights required tremendous power, challenging their efforts and forcing compromises. Solutions included limiting the length of time the heart sensors ran and lengthening the battery life by having the watch display information only when a wearer tilted his wrist toward his face. The result was a watch that was often just a dead screen that didn't even show the time.

Separately, engineers noticed that the watch's processing speeds could be sluggish. It could take time to transmit messages from the phone to the watch. Other capabilities also lagged. That left some engineers wrestling with a major question late in development: What did the watch really do?

In August, a month before its debut, Dauber went to Williams's office in the operations building near Infinite Loop. At the time there was tremendous pressure from Apple's senior leadership, including Cook, to ship the watch as soon as possible, a push that many working on it interpreted as a bid to silence the company's critics and reassure its investors. But Dauber couldn't shake his doubts.

"Jeff, if you left your phone at home and got to the office without it, would you go back and get it?" he asked.

"Yes," Williams said.

"If you left your watch at home, would you go back to get it?" Dauber asked.

Williams paused and thought. "No," he said. "I'd get it when I got back home."

"That's why we can't ship this," Dauber said. "It's not ready. It's not great."

Chapter 10

Deals

The United Airlines jumbo jet hurtled westward over the Pacific Ocean, carrying Tim Cook and other prospectors hungry to do business in Beijing.

By 2014, China had solidified its place as the world's preeminent growth market. Farmers across the country had uprooted from their tiny rural outposts and migrated to megacities that were exploding with higher-paying jobs and frenzied construction. A half dozen of them were as big as or bigger than New York. Brand-conscious consumers were spending more on Huggies diapers and Penfolds Cabernet Sauvignon than were people in the United States. Every day, planes ferried executives from California and beyond to the threshold of Tiananmen Square, where they eagerly worked the halls of power to extract a sliver of the country's commercial boom.

Cook had made the trip many times. Since the iPhone's debut in 2007, he had advocated that the company focus its geographic expansion on the world's most populous nation. The iPod had given the Apple brand cache with consumers who pined for its new phones, but to be able to distribute the popular device, Cook needed Chinese government approval. The effort required navigating a bureaucracy so

Byzantine that Apple's staff had turned to the U.S. State Department in 2008 for assistance in securing its earliest distribution deal. Each year the company added more distribution, but Cook's focus seldom wavered from the real prize: a deal with China Mobile, the country's largest operator.

Cook's pursuit of that opportunity transformed him into Apple's chief diplomat, traveling to China to meet with officials from the Ministry of Industry and Information Technology. He charmed the ministers with his concentration and resolve, and in a rarity for someone so private, he made the appeal personal, sharing that his sister-in-law, the wife of his younger brother, was Chinese. He was especially close with their son, Andrew, a talented math student who shared his uncle's love for Auburn football. He told people that his sister-in-law's ties to China deepened his interest in the country.

But this trip wasn't about cultural exploration. Its purpose was to finalize a deal six years in the making. As the plane touched down, Cook was eager to stake Apple's new claim.

IN CUPERTINO, Apple's leadership could see the coming cliff. The iPhone, which had transformed the world and electrified Apple's business, was showing signs of fatigue. It posted its lowest holiday sales gains ever in early 2014, setting off alarm bells both inside and outside the company. Apple's marketing team projected a future in which such woeful growth would become the norm as the smartphone saturated the world, leaving fewer potential first-time buyers. Their forecast fed a fear that the company's most valuable asset had become a dangerous liability.

Word spread through the company that Cook wanted its next major device to deliver at least $10 billion in sales, an artificial benchmark to ensure that any project Apple pursued would be more than a rounding error for a company now reporting $170 billion in annual revenue. The financial target spoke to the law of large numbers, a business theory holding that as the sales of a blockbuster product expand, it becomes harder and harder to deliver the rate of growth that investors expect. Publicly, Cook called the theory a "dogma . . . cooked up" to create fear. He assured investors that Jobs had created a culture in which numbers didn't limit thinking. Instead, he said, the focus at Apple was on creating products that would produce the numbers. But inside Infinite Loop, numbers had begun to inform product development and business strategy. Cook believed in Ive's vision for the Apple Watch and its potential to post the kind of sales numbers Apple needed to deliver revenue increases, even as the iPhone matured. But the perpetual backup planner and risk mitigator anticipated that it might take more than one new business to meet investors' expectations, so he searched for opportunities of his own.

AS A CHAUFFEUR DROVE from Beijing's Park Hyatt hotel to China Mobile's headquarters, Cook gazed excitedly out the window. The normally reserved, resolute CEO was in a joyful mood that day in January 2014. The city teemed with energy as people in winter coats hurried across intersections, some clenching mobile phones in their hands. Some of them were among the 760 million subscribers to China Mobile's wireless service. In a few days, those same people would be able to buy and use iPhones for the first time.

The car arrived at a steel-and-glass skyscraper fronted by a drab concrete wall stamped with China Mobile's name. Cook strode inside to greet Chairman Xi Guohua, his new business partner. The two had met a few times over the past year as they had tried to finalize the complex business terms determining how much China Mobile would subsidize each iPhone it sold. Such subsidies had been critical to every wireless deal Apple had cut around the world, enabling it to reduce the price of an iPhone for customers who signed multiyear contracts with carriers. In China, subsidies tended to be less generous. It took time for Cook and Xi to reach an accord on what China Mobile would pay to discount iPhones, but having found an equilibrium, they were ready to reveal their excitement over their new partnership to the world.

A TV crew from CNBC was set up inside a drab corporate room to interview both executives before a floor-to-ceiling poster of an iPhone. Cook hoped that the TV appearance would send a message to Apple's investors that the company had found a way to extend the iPhone's run by striking a deal to reach hundreds of millions of new customers.

As Cook readied for the camera to roll, Apple's head of communications, Katie Cotton, checked to ensure that every aspect of the image that would be televised in the United States would be flawless. She grew worried when Xi declined makeup ahead of the TV interview. She knew that powder could take the gleam of camera lights off a forehead and eliminate a distraction for viewers. She couldn't stand that possibility. Just before Xi sat down, she pulled out her Chanel compact case, popped it open, and grabbed the powder sponge. He froze as she planted herself in front of him and began to powder his face.

When the cameras rolled, the distracting gleam was gone. The focus narrowed to the interview.

"It's a watershed day for Apple," Cook began. "I'm so honored to be doing business with Chairman Xi and China Mobile."

Cook slouched in his chair, relaxed and at ease.

"Mr. Xi, will you now use an iPhone?" CNBC's newscaster asked.

"Good question," Xi said in Mandarin. "Before China Mobile and Apple joined hands, I used the cell phone of another brand. Now I've decided to switch to an iPhone. I'm very thankful to Tim Cook that this morning he gave me one of the first iPhones made for China Mobile, and it's gold."

Cook smiled. The color he had chosen for the gift was symbolic of Apple's future in the world's most populous country.

"We play for the long term," Cook said during the interview, sweeping his hand upward as though it were charting the company's stock performance. "I see this announcement today as being one of those very key milestones in doing great stuff over the long term for our customers and our shareholders and our employees."

Cook and Xi later headed to one of China Mobile's three thousand stores. They strode inside in lockstep, grinning, as a crowd of customers gathered behind them. As Xi spoke about his company's deal with Apple, Cook held his hands above his head with his fists clenched like an Olympic marathoner crossing the finish line. He had spent six years working toward the deal and knew that the agreement could unlock more sales of the company's signature device, allowing it to outpace the law of large numbers. He grabbed a microphone and gazed over the murmuring crowd. "We have long wanted and waited for this day to come," he said, his voice edged with excitement. "Today we're bringing the best smartphone to the fastest network and the largest network in the world."

In front of him stood a stack of five commemorative iPhones. Each box featured his swirling signature in one corner and the sharply drawn characters of Xi's name in the other, the same as the pen strokes that would have finalized their momentous deal. The autographed mementos would soon give way to the purchases of millions of iPhones by Chinese consumers.

AFTER HE RETURNED to the United States, Cook's search for growth took him to southern California. Winter was turning to spring in 2014 when he entered a plush Santa Monica office park. He was joined by Eddy Cue, who had been elevated to the executive team after Jobs's death and charged with leading a bundle of services Apple provided, including iCloud, iTunes, and its troubled Maps. They were there to evaluate an opportunity to artificially boost revenue by acquiring a new business.

Apple's coffers were awash in cash as its slowing iPhone business continued to produce massive profits. The company had amassed a $150 billion treasure chest. Watching its riches grow, Wall Street had begun clamoring for a deal. Pressure was also coming from inside the company. A year earlier, former vice president Al Gore, a member of Apple's board, had encouraged Cook to consider acquiring Tony Fadell's digital thermostat company, Nest Labs. Gore, an investor in Nest, had helped bring the two men together for a meeting during which Fadell outlined how Nest would birth a line of internet-connected devices that would create a smart home with lights that could be dimmed and raised by a voice assistant. Ultimately, Google had swooped in and bought Nest for $3.2 billion. The deal had irked Gore, who had encouraged Cook to find another acquisition target. With headwinds hitting the iPhone and the Apple Watch still under development, a deal for an established brand or product would allow Apple to add the sales of another company to its balance sheet and relieve pressure for growth.

Cook and Cue breezed into the offices of Beats Electronics and were greeted by its cofounder Jimmy Iovine. The son of a Brooklyn longshoreman, he was a serial hustler with unrivaled pop culture sensibilities. He had begun his career as a recording engineer, starting with John Lennon and Bruce Springsteen before going on to work with Tom Petty, Stevie Nicks, and U2. In 1989, he had started a record label, Interscope Records, and signed an eclectic mix of artists that included Andre Young, aka Dr. Dre, and Trent Reznor's Nine Inch Nails. He had helped Dre popularize gangster rap, transforming the music industry and forging a lifelong friendship. In 2006, Dre had told Iovine that a sneaker company had approached him about an endorsement opportunity.

"Fuck sneakers," Iovine said. "You should sell speakers."

Inspired, the hyperkinetic Iovine had set up a company called Beats and pulled together a collection of the best headphones available at the time. He and Dre evaluated them by listening to songs they had produced: Tom Petty's "Refugee" for Iovine and 50 Cent's "In Da Club" for Dre. They tapped Robert Brunner, the former Apple design chief, to develop the product and help with the brand. Introduced in 2008, Beats by Dre headphones became a cultural sensation. Athletes wore them at the Olympics. Artists wore them in music videos. Sales jumped from 27,000 units to 1 million units in a year. Apple moved thousands of the $350 headphones through its stores. As it took off, the record producer turned entrepreneur, who was friendly with Steve Jobs, regularly urged the Apple CEO to buy Beats. Jobs turned him down twenty-five times, Iovine liked to say, before adding, "Apple will come around eventually."

That day, Iovine met with Cook and Cue in a sun-splashed room that looked out onto Santa Monica's streets. As they sat around a conference table, Cook and Cue probed Iovine about the state of Beats' business, particularly its recent expansion from hardware into software with the introduction of the streaming music service Beats Music. Iovine showed off some of Beats' forthcoming products during the discussion, including a portable Bluetooth speaker with a large red *b* in the front. He worried that Apple was losing its beachhead in music and wanted to help.

"Your heart and your roots are in music," he told them. "How are you going to give that up?"

Iovine was in full sales mode, trying to charm the staid Cook by energetically describing the bright future of the Beats business. In the weeks before Beats Music had launched, he had given friends such as Tom Hanks and Sean "Puff Daddy" Combs early access to the service. He encouraged Cook to try it out. Following in the footsteps of Spotify, it provided unlimited access to a catalog of music for $10 a month. The difference between it and rivals' offerings, Iovine said, was that Beats Music had been created with the help of artists, particularly Reznor. It focused on curating songs and helping subscribers discover music they might otherwise miss, almost like a digital version of the record store managers of the 1960s and 1970s. He liked to say that it was Apple-like. And he was confident that Cook would like it.

AS COOK WEIGHED his next move, he made another adjustment to Apple's leadership ranks. In March 2014, the company issued a press release that its longtime chief financial officer, Peter Oppenheimer, would be retiring at the end of the fiscal year. Luca Maestri, the company's controller, would succeed him.

Outside Infinite Loop, the announcement attracted little attention, but inside its halls, longtime Apple employees worried about the change.

Jobs had believed that accountants and lawyers should be largely kept out of decision making, treated more as implementers than as influencers. In his decade as finance chief, Oppenheimer had embodied that philosophy. He had raised questions about spending when warranted but had generally adopted Jobs's view that sometimes a company needed to spend money to make money.

Since 2011, Cook had been introducing a different doctrine of finance. Where possible, the industrial engineer and MBA wanted to find efficiencies and reduce costs. It was reflected in the short-lived effort by John Browett, the head of retail, to cut costs, and it was evident in Cook's continual push to negotiate lower prices on the components of Apple's products.

Maestri, an Italian, shared Cook's penchant for financial discipline. One of the first things he did was start a review of all of the company's contracts with third-party suppliers. The request put pressure on division heads, who occasionally tapped consultants to assist with strategy, recruiting, and exploration of future opportunities. It was just the beginning of a shift in power that would move finance from the end of the line at Apple to the forefront of decision making.

TWO MILES SOUTH of Infinite Loop, work barreled ahead on Apple's new headquarters, and the forecasts of the total cost were astounding. No manufacturer had ever produced anything like the forty-five-foot-tall sheets of slightly curved glass that Apple demanded for the building's exterior. Creating them would require developing new manufacturing processes and building new factories, an expensive proposition projected to cost as much as $1 billion to fulfill perhaps the largest glass order in history.

Pained by the thought of such extraordinary costs, Cook stressed over how to bring the effort's spending under control. He realized he needed someone who could squeeze costs, strip out excesses, and strike world-class bargains in order to save Apple hundreds of millions of dollars. He needed a man known inside Infinite Loop simply as the Blevinator.

A diehard negotiator from the small Blue Ridge Mountains town of Jefferson, North Carolina, Tony Blevins refused to buy almost anything at price. He pridefully wore a cheap puka shell necklace that he had bargained down to $2 from $5 as a reminder to his staff that nothing should fetch full price. He bragged to friends about his personal triumphs, including his purchase of a man's vintage car worth $8,000 for $2,500. When friends reminded him that his stock in Apple had made him a millionaire and he could afford the car's list price, he shrugged. "But I wasn't going to let him get what he wanted," he said. His unyielding passion for winning negotiations had vaulted him to the top of Apple's operations team.

After Ive selected the kind of glass he thought Apple should use, Blevins invited glassmakers from Germany and China to the Grand Hyatt in Hong Kong. He reserved a series of adjoining conference rooms at the hotel and put each bidding company into its own room. He then went from room to room, pressuring the bidders to lower their prices. He told the Germans, who were seeking upward of $500 a square foot, that the Chinese were asking a fraction of that. He told them they had ten minutes to lower their price. "If you don't agree to this number, the guys next door said they would," he said. Moments after issuing the ultimatum, he exited the room, leaving his astounded colleagues to process his bluffs. As the clock ticked, the various bidders scrambled to figure out if they could reduce their prices and still make enough profit to make the project worthwhile.

Meanwhile, Blevins cycled from room to room, intensifying the pressure. "The project is not moving forward," he said. "We have a cost problem. You have fifteen minutes to give me the best possible deal you can offer."

The blend of bluster and whirl of demands worked. By the time the final bid was accepted, Blevins had reduced Apple's glass cost by hundreds of millions of dollars.

Seele, the German manufacturer that won the contract, created an entirely new manufacturing process with custom-made machines capable of bending the glass into a subtle curve. It constructed a giant European manufacturing facility to do the work. Installing the glass required a completely new, $1 million machine with suction cups designed for the express purpose of hoisting the massive glass panels onto the building's exterior. Architects working on the project were astounded at how the lofty demands of a tech company had forced the construction industry to innovate. In the years that followed, they would marvel as other buildings, such as the Los Angeles County Museum of Art, featured curved glass, which might have been impossible without Apple Park.

After some of the earliest forty-five-foot-tall panels were completed, they were loaded onto a chartered Boeing 747 and flown to Cupertino. They were then installed on the exterior of the prototype building constructed near Hewlett-Packard's old campus.

One day, Cook arrived with a small group of Apple executives to inspect the newly installed panels. He was greeted at the site by some of Foster + Partners' lead architects and escorted through an expansive fifteen-foot-wide corridor alongside one of the world's largest curved office windows. Sunlight poured through the transparent wall and painted the white terrazzo floor before him with a yellow glow. Cook glanced everywhere as he walked, assessing each inch of the miniature version of the future headquarters. As his eyes roamed, he abruptly stopped. Everyone around him froze.

Cook stepped toward one of the glass panels and bent down on one knee. Though he was known for his mastery of numbers, his time at Apple had given him an amateur's eye for design. Jobs and Ive's obsession with detail had infused the entire company with some aesthetic sensibilities. Cook stared at the base of the glass, which was separated from the terrazzo floor by a half-inch strip of silicon beneath a one-inch strip of stainless steel. The steel-and-silicon barriers provided a buffer to protect and stabilize the glass, giving it room to wiggle in an earthquake or a windstorm. But something about the steel seemed off to the kneeling Cook.

"Could this be smaller?" he asked.

THE ENGINEERS WERE growing restless. As the watch project hurtled ahead, a few began to think about what big thing Apple could do next. They didn't get a satisfying answer. In the void, a handful of the company's senior engineers decided to depart en masse.

Their abrupt resignation ripped through their divisions and reached the ears of Cook. Some of the engineers were members of the architecture and core operating systems team. They set Apple's road map, developing the chips and internal capabilities that brought its products to life. Many of them were longtime Apple employees with tremendous institutional knowledge. Losing one would be disappointing. Losing them all would be a brain drain.

Cook faced a conundrum. To halt the exodus, he directed his engineering leaders to empower and inspire the mutinous engineers by asking what they would they like to pursue next.

A car, they answered. They wanted Apple to make a car.

At the time, Tesla was in the process of doubling its staff and plowing money into the development of more sophisticated batteries for its electric vehicles. The electric vehicle company was recruiting dozens of Apple engineers, who told former colleagues that the company's founder, Elon Musk, was going to be the next Jobs. In nearby Mountain View, Google had been working on its own self-driving car and trying to partner with established automakers to bring it to roads nationwide in a few years. The entire peninsula buzzed with the possibility of a transportation revolution.

A group of engineers gathered in a conference room to discuss how to proceed. They reviewed marketing analysis drawn up by the consulting firm McKinsey & Company. It showed that Apple already accounted for the majority of the profits in the $500 billion consumer electronics industry and needed to move into other areas to deliver sales increases for shareholders. The largest options were the $2 trillion auto industry and the $7 trillion health care industry. Turning to MBA-style analysis was disorienting for some engineers. Steve Jobs had disdained consultants; he had thought they made recommendations and moved to their next project without working to determine whether their ideas succeeded or failed. But Cook's hunger for numbers and data led the group to turn to traditional sources of business information as they sought and eventually won the CEO's support for the project.

Initially, they focused on developing an electric vehicle that they hoped would disrupt the auto industry in much the same way the iPhone had disrupted the communications industry. It would not be the first, but it would be the best.

They called the new effort "Project Titan."

ONE NIGHT AFTER WORK, Cook was listening to music and doing market research. The world of streaming music services had become frothy with new entrants. The pioneer, Spotify, had been joined by hip-hop artist Jay Z's service, Tidal, and Jimmy Iovine's Beats Music. As each one had arrived, the services had eroded iTunes' decade-old business of selling 99-cent songs. It was becoming clear to Cook that the music industry was moving toward a subscription-based future.

The shift posed a threat to the iTunes business and the way Apple thought about music. The industry was core to its identity and had been one of Jobs's true loves. An avid fan of 1960s rock and folk music, he had taken pride in the way Apple had provided the record industry with a life raft in the early 2000s as free file-sharing services such as Napster had eroded its sales. iTunes had helped the industry survive. As the service had become the dominant sales engine for digital music, Jobs had espoused his belief that people wanted to own songs, not rent them. He had continued to preach that philosophy even as start-ups emerged offering apps with access to entire catalogs of music for a monthly fee. As those apps began to change the music industry, Apple stayed true to Jobs's thinking. But as Cook listened to music that night, he was beginning to reassess his predecessor's wisdom.

Cook toggled among the various streaming services on the market, Spotify, Tidal, and Beats. He compared how they looked and how they made him feel. Underlying the investigation was a natural question: If the services had the same catalog of songs, what could make one better than another? Each time he returned to Beats, he felt something different. But why? Then it dawned on him: it had human curators.

In the days that followed, the restrained CEO buzzed about his discovery. He shifted from evaluating Beats to wanting to acquire it. Colleagues watching him thought he was smitten with Beats and joked that Cook was behaving like a high school nerd who had been invited to a weekend party by one of the cool kids. Like many jokes, there was a bit of truth in it. Jobs, who had dated singer Joan Baez, had pop culture sensibilities that had put Apple's advertising and products at the forefront of society. He had bridged the gap between what engineers could make and what he anticipated people wanted. His favorite music came from the Beatles and Bob Dylan. Cook lacked his predecessor's swagger. Behind his back, the company's marketers and designers poked fun at his music tastes, which included OneRepublic, a Colorado Springs band known for its earnest pop songs. They saw his pursuit of Beats as a way to recapture Apple's cool.

Beats also offered Cook a solution to the company's failure to enter the streaming music business. Recognizing that the market had shifted, Cue's services team was at work creating their own streaming offering that would allow people to blend their iTunes purchases with a full catalog of songs. But the early designs of the service were discouraging. It looked more like an iTunes list than the colorful modern apps of its rivals. Influenced by Clayton Christensen's book *The Innovator's Dilemma: When New Technologies Cause Great Firms to Fail,* Jobs preferred to be a disruptor rather than be disrupted. He had famously killed the iPod Mini, Apple's best-selling product, and replaced it with the iPod Nano, a lighter, slimmer device that had gone on to even greater sales. Whereas Jobs might have guided the development of an industry-leading music app to replace iTunes, Cook looked to buy outside help. He believed that Iovine could provide the sensibilities Apple needed to create a cutting-edge music service. Iovine had already shown with Beats Music that he could create something in the vein of what Apple would do. The combination of software and music editors channeled Jobs's philosophy that the best products lived at the intersection of technology and the liberal arts.

Cook's proposal of an acquisition ran into immediate resistance. Oppenheimer, who hadn't yet retired, raised concerns about whether Beats would fit with Apple's culture. Dre had a history of violence, including the 1991 assault of a TV host whom he had thrown against a wall and punched in the head. Apple had largely shed its rebellious past and become California corporate, its laid-back dress code disguising an intense, detail-oriented workplace full of people striving for perfection.

The leaders at Apple also weighed a natural question: Why can't we build a streaming service ourselves? Cook had considered that and determined that although Apple could build its own service, bringing in the Beats team would infuse an Apple service with the sensibilities of music lovers and artists. The combination of Iovine and Dre would give whatever Apple introduced credibility with customers.

Striking a deal wasn't easy. Iovine had built two businesses in one. He and Dre had much of their equity tied up in Beats Electronics' headphone business. The nascent streaming service, Beats Music, had granted shares to many of the software developers who had built it. Though it would be cheaper for Apple to buy the streaming service and not the headphone business, Iovine insisted that Apple buy them both.

In the wrangling that followed, Apple's finance team eventually recognized an opportunity. Beats' headphone business was generating about $1.3 billion in sales a year and paying its manufacturers a 15 percent margin for production. By comparison, Apple paid its manufacturers 2 to 3 percent. If Apple pressured the manufacturers to slash their margins, as Cook planned to do, the profits from Beats would surge and the acquisition would pay for itself in a few years.

As talks progressed, Apple came up with code names for the two acquisitions: Dylan and Beatles, a nod to Jobs's favorite artists. It dual tracked the acquisitions with teams of attorneys working on a deal for Dylan, the streaming music service, and separately on Beatles, the headphones business.

By May, Apple had agreed to pay $3.5 billion. It was a sum that Iovine and Dre could barely fathom. As the lawyers worked through final details, Iovine summoned the leadership team of Beats to his home near Beverly Hills. He told everyone that they were on the cusp of finalizing a massive deal. The only thing that could spoil it would be for word of the deal to leak.

Iovine described Apple in ways that made the company sound like the Mafia. He said that it was very tight-lipped and expected its business partners to be the same. He warned his team to keep their mouths shut and turn off their phones.

"Whatever you do, don't talk about this," he said.

Heading into the weekend, Iovine reiterated that on the phone during a call with Dre. "Remember that scene in *Goodfellas* where Jimmy tells the guys, 'Don't buy any furs. Don't buy any cars. Don't get showy'?" he said. "Don't move."

"I gotcha," Dre said.

At 2:00 A.M., Iovine got a call from Puff Daddy, who was screaming that Dre and Tyrese, a rapper, were talking about the deal in a Facebook video. Iovine pulled up the video and cringed as he saw Tyrese bragging about being drunk on Heineken in a recording studio. As Dre pointed his finger into the camera, Tyrese rocked his head from side to side with swagger. "Billionaire boys club for real, homie," he said. "They need to update the *Forbes* list. Shit just changed."

"In a big way," Dre said. "The first billionaire in hip-hop, right here from the motherfucking West Coast!"

Iovine panicked. His business partner had just outed a multi-billion-dollar deal with the world's most secretive company before it was official. Suddenly everything was in jeopardy.

When word of the video reached Cook, he summoned Iovine and Dre to Cupertino. He invited them into a conference room for a private conversation. Iovine was anxious and afraid that Cook was going to kill the deal. Instead of the anger and cursing that would have poured out of Jobs in a moment like that, Cook exuded calm. He told the music executives that he was disappointed and wished that Dre's social media outburst hadn't happened but said that the video hadn't shaken his conviction that buying Beats was right for Apple.

A skilled negotiator, Cook used the social media fiasco to demand an adjustment to the terms of the deal. In the days that followed, Apple shaved an estimated $200 million off its offering price. The reduction led staff at Beats to say that Apple had given Dre just enough of a haircut to make sure that he did not become a hip-hop billionaire.

As the company prepared to announce the deal that spring, Cook gathered its leadership in the Infinite Loop boardroom to discuss its final terms and how to handle the press. The deal would make two music icons, Iovine and Dre, employees of Apple. They would get badges to campus and begin attending meetings. But no one had a clue what their titles should be. As they discussed options, one of Apple's executives raised a possibility: Why not call Iovine "chief creative officer"?

The room fell silent as everyone weighed the idea. Iovine had spent a lifetime as a creator of music, working with some of the world's most recognized artists. He had gone on to a second act as the creator of a business, showing a marketing savvy that had turned a headphones company into one of the hottest brands in the world.

To some, the idea made sense, but not to everyone.

"What about the rest of us?" Schiller huffed. "Are we not creative?"

Ultimately, Cook decided to forgo titles. The deal itself answered Schiller's question.

Chapter 11

Blowout

Anxiety welled up inside Jony Ive whenever the time came to release a new creation. It was impossible to feel that any product was finished. In the race against the market's artificial deadlines, there were always accommodations: engineering advances that couldn't be achieved; materials marked by nagging impurities; and limits imposed by the physics of components that could be pushed only so far. The sacrifices on the way to perfection had left him walking through a world full of Apple products, thinking, *I wish that was better.*

When he arrived at De Anza College in Cupertino on the morning of September 9, 2014, Ive looked uneasy. He had spent three years consumed by a project he hoped would honor his late creative partner and silence the echo chamber of doubt about Apple's continued ability to innovate. He had agonized over the Apple Watch's design, toiled to define its interface, and pushed to shape its marketing. Now the time had come for the world to pass its judgment.

Scattered clouds blew across a sunlit blue sky as Ive passed by the $25 million tent abutting the community college's performing arts center. The towering structure was more building than temporary wedding venue. It stood two stories high, featured rigid ninety-degree corners, and was as white as the clouds overhead. It had been designed to look like the facade of the Flint Center auditorium. Inside, staff moved among long white tables where watches, recently arrived from China, had been slipped onto metal pedestals, creating a rainbow of suspended candy-colored silicon bands.

Some three thousand miles away in New York City, people had begun lining up outside Apple's store on Fifth Avenue in anticipation that those very watches would be available for purchase later that day. The company's years of making hit products had instilled its fans with confidence that it was only a matter of time before it would deliver another one.

Ive eventually arrived at a nearby courtyard where friends and special guests had gathered before the company's product showcase. The designer moved through a small crowd that included the media titan Rupert Murdoch and NBA star Kevin Durant. A *New Yorker* writer at work on a profile shadowed Ive as he sipped coffee and chatted with longtime friends, including Coldplay singer Chris Martin and the actor Stephen Fry. As the *New Yorker* writer asked questions, Ive fidgeted with his fingers. It was all so strange, he explained. "You go from something that you feel very protective of, and you feel great ownership of, and suddenly it's not yours anymore, and it's everybody else's," he said.

His philosophical musing disguised his stress about the day ahead. After years of development, he would be able to wrap the product on his bare left wrist, even if he knew that the watch wasn't ready.

IN FRONT OF THE FLINT CENTER, some two thousand guests began to form a line outside the entrance, eager to get seats for the show. There were fashion writers and editors from Europe, tech reporters from San Francisco, TV crews from ABC and CNBC, all there to cover the scripted spectacle and provide Apple with millions of dollars in free advertising.

Ive entered through a separate door from the frenzied masses and took a seat in the front row between Marc Newson and Chris Martin. Like so many previous shows, Ive had no intention of speaking publicly and left the showmanship to colleagues. He watched as the lights dimmed and Cook strolled onto the stage to enthusiastic applause.

The theater had been such an important part of Apple's history. Some thirty years earlier, Steve Jobs had stood in the same spot and revealed the Mac, the company's most enduring product line. The late CEO had returned nearly fifteen years later and ignited Apple's rebirth by unveiling the iMac. Now Cook planted himself there in a symbolic gesture pointing to a new future.

He usually opened events with lengthy business summaries, detailing the number of stores opened or iPhone customers added. But he dispensed with that and summarized the company's performance with a single sentence. "Everything's great," he said.

Ive grinned as the hall filled with laughter, applause, and a few high-pitched whistles. He watched as Cook conducted a two-hour show that began with the introduction of two large-screen iPhones, the 6 and 6 Plus, which were 17 percent and 38 percent bigger than their predecessors, respectively. The phones fulfilled customers' demands that Apple deliver phones with bigger displays for videos, gaming, and photos. They also combated the competitive pressure from Samsung, which had been selling larger phones for months. Afterward, Cook welcomed Eddy Cue onstage to detail a touchless payment system, Apple Pay, that would enable people to pay for things by holding their phones above a store's checkout scanner. The feature, which recalled Ive's Blue Sky project at Newcastle, pushed Apple into the world of finance, where it could take a small fee on each of the millions of transactions worldwide. Returning to the stage, Cook said that it would "forever change the way all of us buy things."

"Now we've gone through enough to call it a day . . . but we're not quite finished yet," he said. He gazed out at the crowd.

"We have one more thing."

THOSE WORDS WERE LOADED with significance. After Jobs had returned in the 1990s, he had made them a special weapon in his arsenal of showmanship. He would direct an hour of product unveilings, each one topping its predecessor, as the audience *ooh*ed and *aah*ed. Then he would say flatly, "We have one more thing," and reveal something totally unexpected such as the tiny iPod Shuffle or the first Apple TV. The three words "one more thing" embodied his marketing magic and hadn't been used onstage since his death.

Upon hearing Cook utter them, people in the crowd erupted. A few rose to their feet. Many clapped, hands above their heads like concertgoers before an encore.

After the auditorium quieted, the room went dark. The guttural bass of rocket thrusters poured out of the speakers and shook Ive's seat as the screen came to life with a camera zooming out from space to reveal Earth at dawn. Then, at the sound of a finger snap, the planet on screen gave way to a chrome-colored edge, a circular crown, and a stage-size image of a watch that arrived like a spaceship.

Close-ups of the watch followed. The video showed the crown zooming in and out on a display of apps and leather straps clicking into place on the watch case. It was three years of work playing out before an audience of 90 million viewers.

When the video ended, Ive watched Cook return to the stage pumping a watch-clad wrist above his head. Cook walked slowly toward Ive with his arms held out in gratitude. Hundreds of Apple employees in the crowd rose in a standing ovation. Ive locked eyes with Cook as the CEO raised his arms above his head in touchdown-style triumph. The raucous reaction reflected the attendees' relief as much as their enthusiasm. After three years of being dogged by skepticism that the company could make anything new without Jobs, it had proved the doubters wrong. Their jubilance convinced them that the market would embrace Apple's latest creation, just as it had its past ones. But Ive knew that commercial success wasn't guaranteed.

Cook needed to sell the watch to the masses. His pitch began as Ive had designed.

The watch was an accurate, personalized timepiece, a communication tool, and a health device. The trio of capabilities echoed the way Jobs had sold the iPhone as a phone, a music player, and an internet communicator. The iPhone had taken off because people had wanted to replace their clunky mobile phones. The challenge facing the Apple Watch was that no one yearned for a better timepiece. In fact, many people had given up wearing a watch because their phones told the time. They needed to be persuaded to strap something onto their wrists again.

The order in which Cook introduced the watch's features betrayed its shortcomings. Though it had been inspired by Jobs's interest in making a health device, the first-generation timepiece could do little more than read the wearer's heart rate. It couldn't track a walk or run with GPS. It couldn't provide an EKG reading. It couldn't be pitched as a health product. This was precisely the problem hardware engineer Jeff Dauber had tried to raise prior to the event: the watch had no compelling purpose. But his concern and desire for time to develop features was brushed aside because employees thought Cook was eager to release a new product that silenced critics and reassured investors. The CEO had favored speed over substance.

Now, without those features, Cook looked to Ive to persuade people that the watch was a fashionable computer for the wrist. He ceded the stage as a prerecorded ten-minute video voiced by Ive filled the room.

"You know, it's driven Apple from the beginning," Ive said. "This compulsion to take incredibly powerful technology and make it accessible, relevant, and ultimately personal."

Ive introduced what he labeled the "digital crown," which spun to magnify app icons and returned users to the home screen. He said that the infrared sensors on the crystal underside of the watch case could track the wearer's pulse. And he detailed the different metals available—aluminum, stainless steel, and gold—as well as the straps in leather, metal, and silicon that made the watch customizable.

"We're now at a compelling beginning, actually designing technology to be worn," he said.

In the video, Ive betrayed none of his personal misgivings about the Apple Watch. He had reserved those concerns for private acknowledgment to friends who recognized how stressed he was that the watch was being launched prematurely. Apple's engineers never resolved the battery-life issues and settled for a compromise that meant that the watch only showed its clock some of the time. His friends would later joke with him, asking, *Who really wants a watch that charges three fucking hours a day?*

Cook all but acknowledged Apple still had work to do when he told the crowd that the watch wouldn't go on sale until the following spring. It was the first time since the iPhone that Apple had introduced a new product category months ahead of its availability. But the show wasn't over.

U2 took the stage to perform. Bono, U2's lead singer, had been close with the company since 2004 when Apple had released a special edition iPod with the band's signatures. He had become a close friend of Ive and had deep ties with Jimmy Iovine, Apple's newest executive, who had produced U2's *Rattle and Hum* album. The band ripped through a song called "The Miracle (of Joey Ramone)" and announced with Cook afterward that their new album, *Songs of Innocence,* would be a free download to everyone with an iPhone, half a billion people. It was the largest album release of all time.

As the event drew to a close, Cook asked everyone at Apple who had worked on the products released that day to stand. He thanked them for their work as their colleagues applauded.

"I'd like to call out, especially, Jony Ive for his incredible contribution on the Apple Watch," Cook said. He also thanked operations chief Jeff Williams and Apple Pay leader Eddy Cue. It was a departure from Jobs. The late CEO typically took credit for developing the product himself with the help of the entire team at Apple rather than single out the contributions of individuals. Ive, who had complained to Jobs about that practice, now found the entire room applauding for him.

As the audience rose to leave, some of the engineers and designers who had worked on the watch were filled with a numbing and unexpected worry. They had made Apple's first new product after Jobs, received a standing ovation, and then had U2 celebrate it all with a short concert. It felt like a career summit. They wondered: *Where do we go from here?*

IVE FILED OUT OF THE VENUE through a separate door and entered the white pop-up building. Photographers and writers buzzed around the white table of watches suspended on pedestals. The timepieces glowed beneath a cool, clear light from above, designed by an Italian fashion lighting specialist who typically lit runway shows for Prada.

When the frenzy died down, he joined his core team—the twenty-one people most central to the watch's creation—for a photograph beneath a black Apple logo on the building's interior white wall. Ive threw his arm over Newson's shoulder and stared at the camera with a slight smile. A white watch with a silicon sport band dangled loosely from his previously bare wrist.

BACKSTAGE, Cook buzzed with adrenaline. Apple's communications team arranged an exclusive interview for him with David Muir of *World News Tonight* on ABC, which members of the team referred to as "Daddy's network" because it was owned by the Walt Disney Co., whose CEO, Bob Iger, was on Apple's board of directors. Cook tried to impress on Muir the meaning of that day. "It shows innovation is alive and well," he said.

He led Muir to the $25 million tent nearby to see the watch and meet Ive. When they arrived, Ive shook Muir's hand firmly and stepped backward as if to exit the spotlight. He conveyed none of Cook's energy or enthusiasm. He remained nervous about how the watch would be received. Muir tried to put Ive at ease by saying that he understood that Ive had focused on making a watch that people would want to wear.

"The bar for that is very high when it's something that you wear and it's something you're going to wear all day every day," Ive said. "And so we worked extremely hard to make an object that would be, one, it would be desirable but it would be personal because we don't all want to wear the same watch."

IN THE SUN-DRENCHED COURTYARD of Infinite Loop, Ive later joined the design team for a sushi lunch. He listened as colleagues shared press coverage of the watch from tech reviewers and fashion writers. *Vogue's* Suzy Menkes, a leading critic, described marveling at the array of digital watch faces that ranged from a butterfly to flowers.

"I still don't know whether the fashion world will embrace this smartest of watches, or whether a new generation that has its phone as a timepiece will find the wristbands compelling," she wrote. A tough critic, she described its aesthetic as neutral, adding, "But I like the idea of setting the visual aspects according to my mood. And perhaps my wardrobe. A bunch of violets to set off my purple outfits? Why not look at my watch—and dream."

The designers found the early reactions gratifying. Most of the reporters had spent only a few minutes with the watch, but their judgments were the first impressions that would shape the world's view of a product that few people would be able to touch for another half year.

That night, the designers gathered at the Slanted Door in San Francisco's Ferry Building, a high-end Vietnamese restaurant with towering glass windows that looked out on the shimmering lights of the nearby Bay Bridge. They were joined by the members of U2 for a special dinner to celebrate their three years of work. The table filled with spring rolls and spare ribs.

Ive sat by his friend Bono, his hand near a bubbling glass of champagne. The anxiety of releasing the watch melted away. For the first time in years, he felt free to celebrate.

A FEW DAYS LATER, Apple released U2's album along with its latest software update. Hundreds of millions of iPhone owners worldwide unexpectedly found *Songs of Innocence* in their iTunes folder. The download sparked a customer revolt, with some customers referring to the unwanted album as the "U2 virus." The attempt at generosity had backfired, exposing instead the power Apple had to put something onto people's phones without their permission.

Eager to quell the unrest, Cook endorsed the development of a software tool that would enable customers to delete the album with a single click. Apple also created a customer support page to help people delete it. The company stopped short of apologizing, but as the unrest spread, Bono told fans that the band was sorry.

"I had this beautiful idea and we got carried away with our-selves," he wrote during a Facebook chat. "Artists are prone to that kind of thing. Drop of megalomania, touch of generosity, dash of self-promotion and deep fear that these songs that we poured our life into over the last few years mightn't be heard."

He hoped that people who didn't want the album would be able to make peace with having received musical junk mail.

WEEKS LATER, in late September, Ive jetted off to Paris Fashion Week, where the watch was scheduled to go on public display for the first time.

Paul Deneve, who was responsible for the product's sales strategy, arranged a pop-up display at the most celebrated store in fashion, Colette. The former Yves Saint Laurent CEO knew the head of the three-story Parisian boutique renowned for curating style and streetwear from brands such as Chanel and Nike. Exhibiting the watch there was the first step in a broader strategy to make it avail-able in the world's most influential stores, an attempt to infuse a product for the masses with the type of exclusivity that was custom-ary for any fashionable accessory.

Early one morning, Ive and his design partner Marc Newson headed to Colette to show the watch off to fashion writers and influ-encers. They arrived before Colette opened to the public and found the store wrapped with images of their latest product. The entryway featured wall-size posters of Apple Watches suspended against a white background. People on the sidewalk watched through floor-to-ceiling windows as Ive and Newson circled Apple Store tables with inlaid watches.

Soon the fashion icons Anna Wintour and Karl Lagerfeld ar-rived and greeted the Apple designers. Since her personal showing of the watch, Wintour had become a proponent of it, pushing *Vogue* to showcase it in one of the magazine's most influential features, "Last Look," on the final page of its October edition. She had encouraged Lagerfeld, whom she considered similar to Ive, to join her.

Lagerfeld's arrival caught Ive by surprise. Though Lagerfeld and Newson were friends, fashion's man in black seldom turned up at commercial events like the one at Colette, even when he received an invitation. But he was a longtime fan of Apple products, often buying dozens of iPods, filling them with songs, and giving them away to friends. Ive led Lagerfeld around the table of watches, explaining the design and recounting its inception.

Nearby, Newson spoke with a reporter from *Women's Wear Daily* about why Apple was at Colette. "Fashion is popular culture," he said. "Technology is popular culture." The reason they were there was that Apple had fused the two, he said, making something that was "not a big dopey, plastic, horrible thing."

THAT EVENING, Ive went from entertaining to being entertained.

Azzedine Alaïa, another of fashion's most admired figures, hosted a star-studded dinner party for Ive and Newson. A perfectionist known for meticulously handcrafted dresses, he had become a modern-day Gertrude Stein, hosting dinners for cultural influencers that could include the hip-hop star Kanye West and the artist Julian Schnabel. His event for Ive was considered to be one of the most coveted invitations at Paris Fashion Week. Rock stars (Lenny Kravitz and Mick Jagger), actors (Salma Hayek), models (Cara Delevingne and Rosie Huntington-Whiteley), and tastemakers (Valentino designer Maria Grazia Chiuri) turned out to celebrate the Apple Watch at an event that doubled for Ive as a fashion debutante ball.

Ive and Newson, who considered Alaïa to be a godfather, couldn't help but be awestruck by the room full of celebrities. They found it humorous that they had spent the morning with Lagerfeld and the night with Alaïa because the two fashion legends famously feuded with each other: Lagerfeld dismissed Alaïa for making "ballet slippers for menopausal fashion victims," while Alaïa said that Lagerfeld had "never touched a pair of scissors in his life."

As everyone dined around circular tables, the Spanish choreographer Blanca Li took the nearby stage to perform a flamenco dance. Afterward, guests tried on Apple Watches and debated which one they would buy.

Ive, who clung to a glass of white wine, could be found late in the evening at the back of room, watching the spectacle. Newson, who stood nearby, spoke with the *New York Times'* fashion critic and did his best to capture how the Apple designer felt. Surveying the room, he said, "It's a long way from Cupertino, you know."

Chapter 12

Pride

For Tim Cook, the numbers were hard to comprehend.

Rising in California around 4:00 A.M. to review sales reports from around the world, the figures he perused each day astonished him. The introduction of the iPhone 6 and 6 Plus had electrified demand for Apple's most popular product, leading to long lines outside Apple Stores as customers waited hours to snap up the new devices. The company's daily sales figures tracked the sale of 74 million iPhones over the holidays, a stunning 46 percent increase from a year earlier. On average, five hundred iPhones were sold each minute, twenty-four hours a day. Apple moved $600 iPhones at the same rate McDonald's moved $5 Big Macs.

China drove the surge, validating Cook's work to land the long-sought distribution agreement with China Mobile. Purchases by its customers positioned Apple to nearly double its sales in the world's largest smartphone market. Having seen its rival Samsung's success selling phones with larger screens in the United States and China, Cook had anticipated that the iPhone 6 would do well, but it shattered his highest expectations. It was enough to make the normally stoic CEO gloat.

"Demand for the new iPhones has been staggering," Cook told analysts during an October call. "I couldn't be happier."

As the doubts about Apple's future after Jobs receded, Cook assumed the confidence he needed to take a rare personal risk.

IN THE FALL OF 2014, Cook prepared to return home to Alabama for induction into the state's Academy of Honor. The recognition was reserved for one hundred distinguished citizens of the state, including prominent figures such as Auburn's Heisman Trophy–winning football star Bo Jackson and former secretary of state Condoleezza Rice. The induction event afforded him the opportunity to address state representatives at the capitol.

As Cook considered his address, he stumbled over a familiar conflict. He was proud of his Alabama heritage but disappointed in the state's legacy on race and equality. It was a familiar point of tension for sons of the South, who struggled to reconcile their love of their birthplace and the region's values with their horror at its history of slavery and racial discrimination. But for Cook, a personal frustration amplified the discord.

In the previous year, antidiscrimination policies had moved to the forefront of politics as the U.S. Senate had weighed a bill to extend protections to workers against intolerance based on sexual orientation and gender identity. Cook had penned an editorial for the *Wall Street Journal* in 2013 supporting the law, writing, "Protections that promote equality and diversity should not be conditional on someone's sexual orientation. For too long, too many people have had to hide that part of their identity in the workplace." The law hadn't advanced out of Congress, leaving a patchwork of state laws to protect workers. Alabama didn't have one.

Cook wanted to call out the state's leaders for failing to pass a law that would protect lesbian, gay, bisexual, and transgender workers from being fired based on their sexual orientation. He knew that the pressure of his words would be more powerful if state leaders knew the truth.

IN LATE OCTOBER, Cook summoned Apple's head of communications, Steve Dowling, to discuss how to tell the world that he was gay.

Cook had been thinking for some time about going public with something he had long kept private. Two years earlier, he had read CNN newsman Anderson Cooper's announcement that he was gay in an email to writer Andrew Sullivan. Cook admired the succinct and direct way that Cooper had addressed something so personal. It struck him as classy. He and two other executives at Apple arranged a lunch with Cooper in New York, where the usually reserved CEO bantered easily with the newsman, leading colleagues to joke that they should have left the table. Cook's admiration of Cooper made him want to come out publicly in a similarly spare, forthright, and inspiring way.

As Cook weighed how to do that, he reached out to Cooper for advice. He told Cooper that he wanted to write something that explained why he hadn't come forward earlier and why he was doing so now. Though he had largely made up his mind about how to proceed, the conversation influenced his next steps.

OVER THE YEARS, Cook had grown to trust one reporter above all others: *Bloomberg Businessweek*'s Josh Tyrangiel. He had sat down with Tyrangiel twice for one-on-one interviews after becoming Apple CEO and considered the former *Time* magazine writer to be smart and principled. He and Dowling discussed reaching out to Tyrangiel and securing space in *Bloomberg Businessweek* for a personal essay about his sexuality.

Cook called Tyrangiel and invited him to California from New York for a meeting. Under Apple's strict secrecy provision, Tyrangiel didn't tell colleagues on the magazine's tech team about the trip. It was clear that Cook wanted an audience alone.

After Tyrangiel arrived at Infinite Loop, Cook confessed that something had been gnawing at him. Every day, he said, he walked into an office where he kept a photograph of Martin Luther King, Jr., on the wall. Some days seeing it was inspiring; other days it was challenging. Lately, it had been more the former than the latter because he felt torn between his privacy, which he guarded so protectively, and his recognition that he was in a powerful position where he could be an inspiration to others.

Cook said that the time had come for him to step out and say that he was gay. He raised the possibility of mimicking Cooper with a personal essay in the magazine. He didn't want it to be on the cover or be marketed aggressively. He envisioned something understated, inside the magazine. He handed Tyrangiel an early draft of the essay.

Tyrangiel read through the personal essay, which began modestly and built toward Cook's coming out. It ultimately began:

> Throughout my professional life, I've tried to maintain a basic level of privacy. . . . At the same time, I believe deeply in the words of Dr. Martin Luther King, who said: "Life's most persistent and urgent question is, 'What are you doing for others?'" I often challenge myself with that question, and I've come to realize that my desire for personal privacy has been holding me back from doing something more important. That's what has led me to today.
>
> For years, I've been open with many people about my sexual orientation. Plenty of colleagues at Apple know I'm gay, and it doesn't seem to make a difference in the way they treat me.

After Tyrangiel finished reading, he assured Cook that he would hold a page for the essay inside an upcoming edition of *Bloomberg Businessweek*.

ACCEPTANCE OF GAY and lesbian relationships in the United States had accelerated in the 2000s, with the majority of Americans for the first time believing that same-sex relationships should be legal. It was a view that had long been held by people in Silicon Valley, where San Francisco's reputation for tolerance and open-mindedness had made it a destination for gay men after World War II.

By 1980, it was estimated that a fifth of San Francisco's population was gay. Men in its Castro neighborhood encouraged one another to be open about their sexuality. The supportive community that arose made the Bay Area a destination for gay, lesbian, bisexual, and transgender Americans, and their arrival coincided with a period of economic transformation as the PC era spilled into the dot-com boom.

Cook joined Apple amid that period of economic growth. The company had long been among the nation's most progressive in accepting and supporting LGBT workers. It had amended its hiring policies in 1990 to prohibit discrimination based on sexual orientation, and two years later, it extended benefits to domestic partners of employees.

As Cook rose up the ranks at Apple, speculation about his sexuality flourished. A 2008 profile in *Fortune* described him as a "lifelong bachelor," a term that the gossip site Gawker latched onto as code for Cook being gay. In a post analyzing the *Fortune* article, Owen Thomas picked apart the description of Cook as an "intensely private" "fitness nut," who spends time outside the office "in the gym, on a hiking trail, or riding his bike."

"What is this—a *Fortune* profile, or a men-seeking-men personals ad in Craigslist?" Thomas wrote. He added, "We'd be remiss in our duties as a gossip if we didn't wonder if Cook was gay."

Until then, many of Cook's colleagues had assumed that he didn't have time to date because he worked so tirelessly. He seldom mentioned hobbies outside of work beyond cycling, hiking, and Auburn football. Some gay employees at Apple said they had seen Cook out at bars, but they never spoke about that broadly on campus. Steve Jobs, who hadn't known for some time, had tried to set him up on dates with women before learning the truth.

The Gawker article turned something that Cook's colleagues might have suspected into an unspoken fact. In 2011, *Out* magazine named him the country's most powerful gay man, cementing the public understanding that he was gay even if he failed to publicly acknowledge it. A Gawker article after he became CEO said that some people in management at Apple worried whether his coming out would hurt the brand. It said that Cook "is into Asian guys" and speculated that he would pair well with a Google executive named Ben Ling, who was described in countless articles as Tim Cook's boyfriend, even though Ling said they had never dated. The articles created a situation in which Cook's sexuality simmered in the background, understood but unconfirmed.

As Cook settled into his role as the company's leader, he gradually moved to change that. In 2014, he approved Apple's participation for the first time in San Francisco's Pride parade and joined four thousand rainbow flag–waving employees in June to march down Market Street behind a white banner with Apple's logo above the word PRIDE.

A FEW MONTHS LATER, on the morning of October 27, 2014, Cook visited Dexter Avenue Baptist Church in Montgomery, Alabama, where Dr. Martin Luther King, Jr., had served as pastor in the late 1940s. Cook wanted to see the site where Dr. King had organized the 1955 Montgomery bus boycott that had helped ignite the civil rights movement. Standing outside the simple brick building topped by a white cupola, he felt moved by the courage Dr. King had shown to stand up for equality in a world that tolerated hate.

Several hours later, he stepped to the lectern inside the statehouse. He placed an iPad on the lectern before him and gazed out at the people who had recognized him as one of the state's one hundred most important living citizens. Then he began to criticize them.

"We're all familiar with the historical struggle of our African American brothers and sisters for equal rights," he said. "I could never understand why some within our state and nation resisted basic principles of human dignity that was so opposite the values I had learned growing up in Robertsdale, Alabama. . . . My parents worked hard so we could have a better life, go to college, and become whatever we wanted. They moved to Alabama because they found friends and neighbors that shared their values. . . . It was a time of great struggle across our state and our nation. It deeply impacted me."

Cook told the legislators about his visit to Dr. King's church that morning and how it had reminded him about the importance of taking a public stand for equality and human rights.

"I have long promised myself to never be silent in my beliefs in regard to these tenets," he said. "Although there has been much progress, our state and our nation still have a long way to go before Dr. King's dream is a reality. As a state, we took too long to take steps toward equality. And once we began, our progress was too slow, too slow on equality for African Americans, too slow on interracial marriage, which was only legalized fourteen years ago, and still too slow on equality for the LGBT community. Under the law, citizens of Alabama can still be fired based on their sexual orientation. We can't change the past, but we can learn from it, and we can create a different future."

Cook's criticism about the lack of rights and opportunity for the LGBT community generated headlines. It caused some across the state to bristle. A prominent conservative news outlet dismissed how Cook lectured legislators rather than thanking them by calling the speech "low class."

BACK IN CALIFORNIA, Cook held meetings with Apple's top leaders to discuss his coming essay in *Bloomberg Businessweek*. He told some of them for the first time that he was gay and sought their help in assessing the risks of going public in markets less tolerant of homosexuality such as the Middle East and Russia. He had similar conversations with Apple's board of directors, which approved the announcement. In those discussions, he acknowledged that part of the reason he planned to go public at that moment was that Apple was on a steady footing after the recently released iPhone 6 and Apple Watch. Failing as the first CEO after Jobs would have been a personal setback, but failing as the first gay CEO after Jobs would have left a legacy that could limit opportunities for other LGBT executives.

In the eyes of Apple's top executives, it was the latest example of Cook thinking at a different level.

Ultimately, Cook made clear that he wanted to speak now so that he could be a role model in the future for young people who were bullied or worried that their families would disapprove of them. Though he would not be the first business executive to go public as gay, his position as CEO of the world's largest company would carry an outsize influence and serve as an example of how far the LGBT community had come since John Browne had resigned seven years earlier as CEO of BP after losing a legal battle when the press outed him. Cook knew that his announcement could show young people that a generational barrier had been broken.

"I kept it to my small circle, and I started thinking 'You know, that is a selfish thing to do at this point,'" he would later explain. "'I need to be bigger than that, I need to do something for them and show them that you can be gay and still go on and do some big jobs in life, that there's a path there.'"

AT *BLOOMBERG BUSINESSWEEK,* almost no one knew about Cook's plan. Tyrangiel held a blank page in that week's edition and, to protect the secrecy of Cook's disclosure, filled it with Cook's essay just before shipping the magazine to the printer.

It posted three days after Cook's speech in Alabama. Headlined "Tim Cook Speaks Up," it immediately became the dominant news story on the business networks CNBC and Bloomberg News, as well as front-page news in the *Wall Street Journal* and the *New York Times.* They all pointed to his words:

> While I have never denied my sexuality, I haven't publicly acknowledged it either, until now. So let me be clear: I'm proud to be gay, and I consider being gay among the greatest gifts God has given me.
>
> Being gay has given me a deeper understanding of what it means to be in the minority and provided a window into the challenges that people in other minority groups deal with every day. It's made me more empathetic, which has led to a richer life. It's been tough and uncomfortable at times, but it has given me the confidence to be myself, to follow my own path, and to rise above adversity and bigotry. It's also given me the skin of a rhinoceros, which comes in handy when you're the CEO of Apple.

With the article, Cook became the first chief executive of a Fortune 500 company to declare he was gay. His declaration carried more weight because he was the CEO of a company that had been at the forefront of culture for more than a decade. His focus on inclusivity and diversity beyond the gay, lesbian, bisexual, and transgender communities appealed to other minority groups eager to be afforded more opportunities in the business world.

People in the gay community lauded Cook's essay for avoiding a trap that had ensnared other public figures. He had managed to make the disclosure less about feeling liberated from some personal burden and instead emphasized why he thought his sexuality was a gift. His message that his life was better for it resonated with them. As a high-profile CEO, he also provided hope that the business world would be more accepting of others as they came out to their work colleagues.

The essay added momentum for gay rights at a time when the community had recently won validation at the highest levels of government. Cook's declaration landed a year after the Supreme Court ruled that parts of the so-called Defense of Marriage Act, which denied federal recognition of same-sex marriages, were unconstitutional. It arrived a year before the Supreme Court ruled that same-sex marriages should be recognized by all states. Those achievements overshadowed his declaration but didn't lessen the effect he had on corporate America.

Though news outlets were eager to hear more about Cook's journey, he let the essay stand for itself. He didn't do a talk-show circuit or sit for interviews about it. He wanted people to see him not as the first gay CEO but as all the other things that he saw himself as: an engineer, an uncle, a nature lover, a fitness nut, and a sports fanatic. He let the end of his essay have the last word.

"We pave the sunlit path toward justice together, brick by brick," he wrote. "This is my brick."

Chapter 13

Out of Fashion

Jony Ive appeared to have reached a new pinnacle. Only a few months earlier, thousands of people had celebrated him with a standing ovation, cheering the Apple Watch and treating him like a star. The fashion luminaries whom he wanted to impress had feted him in Paris and marveled at his latest creation. His hard work had been rewarded and his newest product had rejuvenated the company with confidence and pride.

In late December 2014, he assembled the software design team for a meeting in an Infinite Loop gathering space known as "The Room." Ive had renovated the area after Scott Forstall's ouster, replacing the screen and theater seating used for software demonstrations with a long oak table and benches like the ones in the design studio.

Ive assumed a place at the head of the table as the software team gathered around the room. He commended the group for their work on the Apple Watch and iPhone and thanked them for all that they had done. They had exceeded everyone's expectations, he said. Then he paused and exhaled.

"I've been at Apple for twenty years," he said wearily. "This has been one of the most challenging years I've had."

Ive's comments and body language confused the team, which couldn't reconcile his downbeat demeanor with the recent outpouring of enthusiasm they had witnessed for the Apple Watch. Instead of being emboldened and cheered by that day's exuberance, Ive stood before them, brooding and distant.

Ive could feel his creative spirit dimming. Behind the scenes, he had spent much of the past three years engaged in corporate conflict. He had tussled over whether to develop a watch with former software chief Scott Forstall. He had then battled over which of its features to promote with chief marketer Phil Schiller. Concurrently, he had confronted rising concerns about costs as he selected construction materials for Apple Park. And he had been sapped by the additional responsibility of managing dozens of software designers. He navigated it all without the support and collaboration of Jobs, the creative partner whom he hadn't fully mourned. The entirety of it left him feeling exhausted and lonely.

Not long after the meeting, Ive's private Gulfstream V jet took off for Kauai, where he owned a custom home near the Nā Pali Coast. He spent the next three weeks on vacation, his longest break in years, but he was still wrestling with fatigue as the new year arrived. His pride in the Apple Watch had been diminished by his lingering frustration about how much he felt he had to fight for his vision, particularly with Apple's marketers, which had resisted his push to market the watch through fashion. From the depths of his frustration, the thought of leaving Apple that had gripped him in 2008 began to well up again.

BY 2015, Ive was being chauffeured to Infinite Loop in the backseat of a Bentley Mulsanne, a $300,000 ultraluxury car with extra legroom and a cream leather interior. The cars were sold with Wi-Fi, enabling riders to work on the go, and handcrafted leather luggage designed specifically to fit into the trunk. He could stretch out in the back and stare out the window as his driver navigated traffic on Interstate 280. He sometimes listened to the business network CNBC on the radio during the ride, keenly aware that Apple's position as the world's largest company made it a nonstop topic of conversation on Wall Street.

He followed the network enough to know that Jim Cramer, the animated, voluble host of *Mad Money* and *Squawk on the Street,* loved Apple and couldn't stop raving about its performance. Cramer endeared himself to Apple staff like Ive by ridiculing skeptics who perpetually discounted the company as a one-trick iPhone pony.

"Year after year, they've kept you from owning the greatest stock in the greatest company on Earth," Cramer said of the skeptics.

Cramer praised Cook for "creating such amazing wealth" for shareholders, including Ive. Cook, Jobs's originally discounted successor, had become a Wall Street darling. In less than four years, Apple's market value had doubled to $700 billion, and its staff had swelled from sixty thousand employees to nearly a hundred thousand. The figures made Ive increasingly uncomfortable. He and others at Apple looked back fondly on the days when a core team of a few hundred developed the iPhone. It was no longer that kind of intimate place where its designers could pull aside the CEO to discuss materials for a line of candy-colored computers. The company now made a multitude of iPhones, iPads, and Macs; pension funds and Wall Street traders tracked every dip in its stock; and tens of thousands of employees counted on it to support their families. Ive's influence over the future of Apple's products affected more people than he cared to imagine. The company's exponential growth troubled him, even as he was being chauffeured in that dream machine paid for with all those profits.

THE INVITATION TO APPLE'S next event arrived in people's inboxes headlined by two words: "Spring forward." Clever, cryptic, suggestive, the words in the RSVP email inspired a flurry of press speculation that Apple would finally tell people when they could buy the company's new watches.

Six months after the watch unveiling, enthusiasm had given way to skepticism. Tech and fashion writers alike questioned the purpose of the watch. When press representatives arrived at San Francisco's Yerba Buena Center for the Arts in early March, they wanted to know: What does it do?

The company's promotion of the watch as a fashion accessory opened it to unforeseen criticism of Tim Cook. In an article headlined "This Emperor Needs New Clothes," the *New York Times* fashion writer Vanessa Friedman had posed the question "Is it time for Tim Cook to tuck in his shirt?" She noted that Apple had hosted an event for the watch at Paris Fashion Week and featured it on the wrist of supermodel Liu Wen in a cover shoot for China *Vogue*'s November issue. "Might it not behoove the leader of such a brand to dress the part?" Friedman asked. She criticized Cook's preference for large, slightly wrinkled, untucked button-down shirts, a style she called "the fashion of no fashion."

The article horrified Apple's communications team, which sniped at Friedman for being petty. The team sought counsel from stylish colleagues about what they could do to help Cook dress better. They were advised to hire Cook a stylist, and on the day of that spring's event, a change was evident in Cook's appearance. He arrived that morning wearing a snug navy zip-up sweater over a pressed high-collared shirt and dark jeans. Rather than Saturday-dad style, he'd adopted what could best be described as Mr. Rogers chic.

In contrast, Ive reached the performance hall about a half an hour before the show dressed in an oversize black sweater that concealed some of the weight he had gained amid the stress of the past year. He strode inside and found his usual seat in the front row beside Laurene

Powell Jobs. He slouched into his seat, his shyness sheltering him from the spotlight and scrutiny of the stage.

When the event began, Cook took the stage with less energy than he'd carried months earlier. He recounted the company's continued expansion in China, where it had opened six stores in the previous six weeks. It had twenty-one stores there and planned to increase that number to forty over the next year.

"Now we've got a few more reasons for you to visit those stores," he said.

In reintroducing the watch, Cook emphasized not only how it looked but what it could do. He positioned it as a modern-day Swiss Army knife, a single device that could tell time, track activity, deliver messages, remember appointments, and pay for coffee—all while looking stylish. To highlight its do-it-all potential, he invited the supermodel Christy Turlington Burns onstage. The towering brunette and longtime face of Maybelline makeup had recently worn the Apple Watch during a half marathon in Tanzania to raise awareness for her charity focused on making childbirth safer in the developing world. She said she wore a silicon wristband during runs and a leather wristband for stylish occasions.

For her, the watch straddled fitness and fashion.

As Turlington left the stage, Cook detailed Apple's pricing plans for the watch. The aluminum Apple Watch Sport would cost $399, the stainless-steel Apple Watch would cost $599, and the 18-carat gold Apple Watch Edition would cost $10,000 to $17,000. They would be available on April 24.

After the event, Wall Street analysts hurried to project how many watches Apple would sell. The company's previous new product, the iPad, had recorded 32 million units in its first complete fiscal year. UBS analyst Steve Milunovich expected the watch to top that with 41 million units sold. On CNBC, anchors asked Daniel Ernst of

Hudson Square Research if people would buy it. "Definitively, yes," he said. "It's a beautiful device. It feels like a piece of jewelry. It's not like some cheap plastic gizmo."

It was as though Ive had scripted the remarks himself. He wanted the world to consider the watch as an extension of every timepiece that had come before it. Only then would it be embraced and worn without embarrassment by the masses.

SUCCESS WAS FAR FROM ASSURED. As Apple started production, it almost immediately ran into manufacturing problems.

Across the Pacific, outside Shanghai, Apple's operations team was distressed to find that the manufacturer they had hired to assemble the watch didn't have enough factory workers. It wasn't just short; it was short more than a hundred thousand people.

It was a staggering number that left many inside the operations team aghast, especially because they needed to find the workers in a matter of days. The search for answers unearthed two underlying problems. Apple had tapped Quanta Computer rather than its trusted partner Foxconn to assemble the watch because it wanted to diversify its supply chain and protect the project's secrecy by using a manufacturing location far away from competitors in Shenzhen, China's primary manufacturing hub. It also timed the product for a spring release, putting the heart of its production process after Chinese New Year. Each year, factory workers returned home to rural parts of China for the holiday and many didn't come back. A similar workforce shortfall would have been unimaginable in the United States, but in China, it represented a small fraction of the 3 million workers at Apple suppliers. With Quanta falling short, only one company could rescue Apple's manufacturing plans: Foxconn, the very company it had spurned.

Operations chief Jeff Williams contacted Foxconn Chairman

Terry Gou and told him Apple needed help. Though unhappy that Apple had chosen Quanta over Foxconn for the project, Gou came through, delivering more than a hundred thousand workers on short notice to staff the assembly lines. The speed reflected the Taiwanese businessman's deep ties in China and his ability to consistently pull off what often seemed impossible. He delivered the workers with an unspoken message for Williams: You owe me.

Once staffed, the assembly lines struggled with the complexity of making the three different watches. The gold watch, in particular, created challenges. A machine cut it from a solid block of gold, creating a shimmering shower of gold flecks that Apple manufacturing engineers watched fall into the hair of Chinese factory workers being paid about $2 an hour. Many of the workers made less money in a month than the value of the gold flecks in their hair. Apple set up a surveillance system to watch for people wiping dust from their hair and walking out with it at the end of the day. The engineers watching marveled at the absurdity of the financial imprecision of Ive's precise design.

A defective part presented a much bigger problem. Late in the assembly process, Apple engineers found a flaw in the taptic engine produced by one of its two suppliers. The part provided the tapping sensation on the wrist that would alert people when they received a notification. The taptic engine being made by one supplier would stop working after a period of time. The faulty part limited the number of watches that could be produced, an untimely error that derailed Apple's high-stakes push to deliver millions of watches on time.

FACED WITH LIMITED SUPPLY, Apple leaned into a strategy to restrict the watch's distribution.

Paul Deneve, the former Yves Saint Laurent chief executive, designed a sales and distribution plan that borrowed from luxury brands such as Louis Vuitton and Hermès, which fostered a sense of scarcity and exclusivity to command higher prices and more prestige for their handbags and clothes. Deneve agreed with Ive that the watch could stand the test of time only if it was perceived as a personal accessory, instead of a computer on the wrist. To boost its desirability, Deneve sought to place the timepiece in high-end stores known for selling the world's most coveted goods. He struck distribution deals to bring the watch to London's Selfridges, Tokyo's Isetan, and Paris's Galeries Lafayette. In the weeks before the watch became widely available, the stainless-steel and gold models went on display in those stores along-side brands such as Cartier and Rolex, lending opulence to Apple's everyday image.

At the end of April, Deneve planned to broaden distribution to Apple Stores, which would be remade to sell treasure, not just technology. He wanted to extend the intangible value of the watch by having individual salespeople help customers select the model and straps that suited them best. The idea was to bring a personal touch to Apple's most personal device. He took the concept a step further by suggesting that customers make an appointment to buy a watch at their nearest store. The head of retail, Angela Ahrendts, supported the idea, introducing a program to convert Apple's workforce of forty-six thousand retail employees into style counselors.

ABOUT THREE MONTHS before sales started, longtime Apple Store employee Jaron Neudorf arrived at one of the company's stores in Calgary, Canada, to begin his training. The sales course for the new watch included instructions on how to assess customers' wealth by studying their clothes and checking to see what brand of watch they wore. He and his colleagues were then instructed to point customers to the watch in the highest price range they might be able to afford. For example, a single mom with three kids should be pitched the lowest-priced aluminum model, while a banker in a suit should be encouraged to buy the pricier stainless-steel version. The concept was disorienting for Neudorf and others, who were accustomed to troubleshooting Macs and repairing cracked iPhone screens. Soon, their stores filled with jewelry displays and underwent a redesign that created an area for customers to try on watches. The change was so dramatic that Neudorf wondered if the shops would be renamed Apple Boutiques.

At Infinite Loop, the new strategy stirred debate. For decades, Apple embraced its identity as a technology company, and it was difficult for some to watch it distance itself from that legacy to adopt strategies from fashion. Its sales executives, who oversaw the rollout of Macs, iPhones, and iPads, worried the embrace of luxury tactics would undermine one of Apple's strengths: its identity as an accessible, premium brand. Under Jobs, Apple had commanded the highest prices in technology by combining Ive's sleek designs with easy-to-use software. Cook's operations wizardry had kept their prices affordable. They feared the watch would make the brand less democratic and more exclusive, turning off loyal customers.

A WEEK BEFORE SALES officially began, Apple took its high-end marketing strategy to Italy.

Having flaunted the watch during fashion events in Paris and New York, Jony Ive arrived in the world's other fashion capital, Milan, in mid-April 2015 to show the Italian grandmasters of craft what his combination of California designers and Chinese manufacturers could do. He knotted a black satin tie around the unbuttoned collar of a white shirt and pulled on a dark suit jacket ahead of a special event for Italian influencers attending the city's annual design fair, the Salone del Mobile (Milan Furniture Fair).

Apple rented a palazzo in the city for a dinner to celebrate the

Apple Watch. Invitations went out to designers and tastemakers such as former British national rugby captain Will Carling and the socialite Umberta Gnutti Beretta. More than a hundred guests milled around the palazzo's spacious interior clutching wine-filled glasses and admiring a rainbow-colored row of watch straps Ive had made exclusively for the event. Eventually, everyone took a seat for a multicourse Italian dinner accompanied by sparkling wine. Ive reveled in the who's who of Italian society and design, thrilled to see his work on display during a weeklong event that had been influencing style for decades.

A few days after the event, he traveled to Florence to speak alongside Newson at the inaugural Condé Nast International Luxury Conference. It represented a departure for Ive, who seldom appeared on the conference circuit. The introduction of the watch had sparked fears that Apple would compete with the traditional makers of jewelry and leather goods. Decades of disruption made whole industries shudder each time a tech company launched a new category. Just two days before Apple Watch sales started, some five hundred guests crowded into the city's seven-hundred-year-old town hall to hear directly from Ive and Newson about the threat they posed.

Moments after the designers took their seats onstage, *Vogue International* editor Suzy Menkes launched into the question many in the room wanted answered: "When we get down to the nitty-gritty of this conference, are you in competition—I don't mean personally— but are your products in competition with the handbags we see in the shops outside, with the many things that have traditionally been described as luxury? Are you now part of that game?"

Ive leaned on the left arm of his chair and gazed toward a Renaissance painting by Giorgio Vasari that towered over the expansive hall. It showed a collision of men and horses as Florentine troops attacked the Republic of Siena in 1554 in a bid to bring the breakaway republic back under Florence's control. The artist's depiction of the battle centuries before seemed appropriate for Ive, surrounded as he was by luxury goods executives frightened by Apple's move into one of their signature categories.

"We don't think about what we do in those terms," Ive said, returning his gaze to Menkes. "Our focus has been on trying our very best to develop a product that's useful. When we started work on the iPhone, the motivation there was we pretty much all couldn't stand our phones, and we wanted a better phone. When we worked on the watch, the motivation was completely different. We happen to love our watches. . . . And so it wasn't because we thought we could design a better watch. . . . It was because we saw that the wrist was a fabulous place for technology."

In his answer, Ive showed the tricky path Apple had walked in creating the watch. Its previous products had been solutions to problems. Jobs had launched the iPhone project because he thought mobile phones sucked. He had pursued the iPad because he wanted something to read on the toilet. In his absence, the watch project had started with less clarity of purpose. It aimed to address several advantages: women would no longer need to listen for their phones buzzing in their purses; diabetics might be able to do noninvasive glucose monitoring; and everyone would benefit from FitBit-inspired fitness tracking. The team had woven those disparate threads—notifications, health, fitness—into a platform that didn't seek to disrupt the watch industry so much as try to relocate technology from people's pockets to their wrists. Recognizing its various goals, Menkes pressed Ive for his thoughts on its purpose

"How do you think most people are going to use the new watch?" Menkes asked.

"People will use it for very different reasons," Ive said. "Some people will be particularly interested in the sort of the health and fitness capabilities and the sort of the coaching the watch can give you. Other people I think will be, you know, happy to be able to be more in touch in different ways. Other people have been really intrigued by some of the more intuitive, more personal ways of communicating."

It would be the watch's ultimate test. As Jeff Dauber had cautioned Jeff Williams before the watch was unveiled, the product didn't have the one compelling reason people should buy it as they had with the iPod, iPhone, and iPad before it; its multifaceted purpose meant that it was being released into the world for market testing. It would be the users who would tell Apple the watch's reason for being rather than the other way around.

WHEN THE WATCH ARRIVED a few days later, its availability was extremely limited. The manufacturing challenges and sales strategy had curtailed its rollout. Apple Stores offered appointments to try on the watch but directed customers to place their orders online. Immediate purchases were largely limited to a handful of high-end stores in major cities such as Paris, London, Berlin, and Tokyo.

In the United States, the Maxfield fashion boutique in West Hollywood was one of the nation's only stores where people could enter and exit with one of the new timepieces on their wrist. Deneve had tapped it to sell the watch because it was one of the most influential stores in fashion. He hoped that availability there would create a ripple of interest that would swell into a wave of demand. Instead, the line outside the store reflected a culture clash where Apple fanboys in fanny packs brushed shoulders with fashionistas carrying Burberry bags. The divide between the customers in line mirrored the split playing out inside Apple as Ive's push to focus on fashion collided with the company's historic focus on technology.

In Selfridges' Wonder Room in London, a writer from the

technology outlet the Verge asked to try on the $17,000 Apple Watch
Edition. A security guard brought out a leather box with a recessed
Apple logo. Magnets held the box lid in place over a gold version of
the watch nestled into a suede interior. The writer slipped the watch
onto his wrist and examined it with some disappointment. He tried
a smaller size and had the same reaction: "Neither felt like a luxury
timepiece, they just felt like gold versions of the Apple Watch."

It was among the first of a series of mixed reviews of Ive's design.
The largely male cohort of tech reviewers praised the design as beautiful, not quite as stylish as those of established players such as Rolex and
Omega but elegant, innovative, and in line with Apple's legacy of making products that transformed categories. In contrast, the women-led
fashion press discounted it as a computer for the wrist and found the
array of strap options overwhelming. It was too big for some women's
wrists. Both groups agreed that it was a nice product to have but not as
indispensable as their iPhones.

Their views were best captured by Bloomberg's headline "Apple
Watch Review: You'll Want One, but You Don't Need One."

The reviews echoed the debates that had raged inside Infinite
Loop in the months before the watch's launch. Few questioned the
beauty of Ive's design, but an array of marketers and engineers continually wrestled with its purpose.

DESPITE THE LUKEWARM REVIEWS, Cook had pushed for ambitious
sales goals. His encouragement had led Apple's forecasting team to
estimate that the company would need to make 40 million watches
in its first year in order to fulfill unprecedented demand for Apple's
new product category. It was an ambitious figure and far exceeded
the number of iPads that Apple had sold in the year after the tablet's
2010 debut. With Apple's customer base growing, the company was
confident it could hit its aggressive sales target.

The early sales results said otherwise. Upon rising each morning in Palo Alto, Cook would have analyzed the latest data on the watch's performance with some dismay. The aluminum model, stainless-steel version, and luxurious gold edition posted numbers that fell short of his expectations. Their weak performance stirred unease inside Apple that the watch wouldn't be a breakout hit.

As purchases lagged, Apple's operations team slashed production. It cut the number of watches it planned to make by 70 percent shortly after launch and cut it another 30 percent a few weeks later. The customer apathy led UBS's Milunovich to lower his expectations for the watch by 25 percent, to 31 million. He found posts online from some early buyers who were critical of the product, complaining that it provided insufficient battery life and delayed notifications. The complaints were the very shortcomings that made Ive worry that the watch wasn't ready. Milunovich questioned whether the watch could generate the type of word-of-mouth enthusiasm that had fueled iPhone sales. "Its main strength is as a watch, which is not super compelling," he wrote.

Rancor boiled over at Infinite Loop. The sputtering launch supported Apple's sales leaders' skepticism about Deneve's strategy. The sales team pushed to expand distribution into major chains such as Best Buy. They urged Cook to revert to a more traditional approach and abandon the efforts to keep the watch exclusive. They warned him that if the company waited too long, the product could become a zombie device.

WITH CONCERNS MOUNTING, Williams grilled Deneve about the watch's marketing and sales strategy. The pressure was reminiscent of how Williams had begun pressuring Forstall in meetings to fix Maps after it floundered. Though Williams and Cook had approved the watch's fashion-focused rollout, they had made that decision more out of trust than comfort. The fashion world was foreign to both Williams and Cook, practical men who drove inexpensive cars and wore basic clothes. Cook had deferred to Ive on that strategy and sought to balance it with the input of the marketing team led by Schiller. As Williams dug in, he pressed to expand sales across traditional, well-known retailers.

Deneve resisted the push to broaden distribution. Like Ive, he subscribed to the philosophy that new product categories take time to develop. Sales of the iPod and iPhone had started slowly before exploding. He encouraged patience.

AS THE PROBLEMS with the watch simmered, Ive stood his ground. He insisted that with time the watch would prove doubters wrong, just as skyrocketing sales silenced early critics of the iPhone and iPad.

Privately, he sometimes told a different story. He complained to friends, colleagues, and board members about his general unhappiness with how the project unfolded. He allowed, in some conversations, that the watch may have been launched prematurely. The company had thrust it into the market to ease its overreliance on the iPhone and rebut critics who had questioned Apple's innovation chops. The product's shortcomings reflected those business pressures. The notification issues and complaints about batteries had surprised no one, including Ive, who had been worried about them on the day the watch was unveiled.

During the product's development, Ive had played the role of Jobs and himself, overseeing industrial and software design, as well as directing marketing. The work had pulled him out of the design studio and into more and more meetings. The influence he had once sought over all aspects of a product's development had burdened him with endless obligations and stress that took a physical toll. He fretted that he'd missed time with his eleven-year-old boys. He grew sick and caught pneumonia.

Compounding his frustration about it all was a feeling that he had shouldered many of those responsibilities alone. Jobs had visited the studio almost daily and supported the designers' work, giving them direction and urging them onward. Cook, on the other hand, seldom came by, and when he did, it was only briefly.

In a few years, Ive had gone from being Jobs's favored disciple to being one of many leaders in Cook's egalitarian world. He decided that he wanted out.

THAT SPRING, Ive conveyed how he felt to Cook. He let the CEO know that he was tired and wanted to step back from the business. His creative energy had waned, and he wasn't functioning at the level he wanted to in a job that demanded more than ever. He raised his frustration about his increased management responsibilities over the growing design studio and a software design team that had swelled into the hundreds. He had found the fights with marketing over the direction of the watch draining. He had flourished when he had focused on leading his elite team of twenty designers while Jobs had bulldozed all semblance of bureaucracy out of their path and made key decisions by decree. He felt outnumbered by colleagues who challenged his ideas, exhausted after his decades of work, and weary from the grief that had followed his friend's death. He wanted to recalibrate and gather his energy.

The complaints exposed the folly of Cook's decision three years earlier to sharpen the company's functional structure after firing Forstall. Ive had wanted a voice in software design, not necessarily oversight of a new division. A good corporate soldier, he had assumed that duty willingly, only to later regret it. Rather than protect Ive and provide him with space to be creative as Jobs once had, Cook, who lived to work, had asked Ive to do the same. He had squeezed more out of the artist than the artist had to give.

Ive wanted to leave.

The news unsettled Cook, who didn't want to be remembered as the CEO who had lost the world's leading industrial designer. He feared that Ive's defection would also pose a financial risk. Investors might dump shares if Ive departed, concerned about the company's future without the man Jobs had called the most important person at Apple. Some observers speculated that the sell-off could sink Apple's share price by as much as 10 percent, or more than $50 billion in market value, more than all of that of FedEx. Cook didn't understand what Ive was going through and turned to colleagues for advice.

Working with Apple's top executives, he came up with a plan to allow Ive to shift into a part-time role. They reached an agreement for Ive to stay at Apple but step back from day-to-day management and spend more time working on future projects, as well as the new campus Apple was building. He would also work on remodels of Apple Stores in cities around the world, a project that Jobs would have managed. Fulfilling the late CEO's legacy would be his main priority.

In Ive's absence, two of his lieutenants—industrial designer Richard Howarth and software designer Alan Dye—would be elevated to vice president and given day-to-day responsibility over the design teams. Both executives would report to Cook. Ive would continue to weigh in on designs but would not need to be at Apple every day. To make his plan to step back look positive for shareholders and the public, Apple's leaders proposed promoting Ive to chief design officer, a new title at the company. The promotion could be announced in the press and enable Cook to put a positive spin on the truth behind the exchange to staff and investors.

Only a few people knew the truth: Ive was burned out.

IN ADVANCE OF THE CHANGE, Ive arranged to grant an exclusive interview for an article in the *Telegraph* by his close friend Stephen Fry.

The British actor and writer, a self-professed Apple fanboy, penned a glowing profile that described his friend Ive as a "wonder boy" whom Cook had empowered to free Apple's software from its skeuomorphic past in favor of a "brighter, clearer set of exquisitely designed images."

"Cook quite clearly adores Jony," Fry wrote, "not just as the goose who continues to lay his golden eggs (solid gold in the case of the Apple Designer Watch), but as a colleague and a person. Everyone does. It is impossible not to get delightedly caught up in the earnest halting way he expresses his highly focused passion."

In the interview, Ive explained his promotion and the significance of Dye's and Howarth's new roles: "Having Alan and Richard in place frees me up from some of the administrative and management work which isn't . . . which isn't . . ."

"Which isn't what you were on this planet to do?" Fry asked.

"Exactly," Ive said.

WITH THE RESTRUCTURING COMPLETE, Cook returned his focus to the watch's struggles. He called a meeting that summer with some of Apple's top executives and marketers, including Jimmy Iovine, who flew up from Los Angeles. They gathered in the company's boardroom, where the watch had become a ticking time bomb inside an otherwise buzzing business.

After everyone took a seat, Cook acknowledged that the reception of the watch had been disappointing. He wanted the group's best answers to a basic question: How do we turn sales around?

As they talked, the conversation continually returned to the idea of changing the way they marketed the watch from fashion to fitness. FitBit's sales were booming behind its emphasis on tracking people's exercise. It showed that there was an appetite among users for devices that supported workouts. Apple needed to move its emphasis on the watch from runways to running.

The new strategy gradually took shape. The company's product marketing team would work with Nike to develop a co-branded watch that would provide a fitness halo for the entire product, and Iovine would work to get the watch onto the wrists of athletes such as tennis star Serena Williams.

"All you gotta do is get Serena to wear it," Iovine said.

The group agreed that fitness was in and fashion was out.

Ive wasn't there to object.

Chapter 14

Fuse

Morale was high when Apple's top executives gathered for the company's annual retreat. Though the Apple Watch had disappointed, the iPhone business was rocketing upward in late 2014 and the China Mobile deal, along with the unprecedented demand in the world's largest smartphone market, put it on track to post a 52 percent increase in sales in 2015. For the first time since Steve Jobs's death, the conference room at the Carmel Valley Ranch overflowed with confidence.

As those present settled into their seats, Tim Cook opened the gathering with a series of slides that looked toward the company's future. Apple was in the process of developing wireless headphones, broadening the availability of Apple Pay, and introducing new features to the Apple Watch. But the comment that got everyone's attention addressed recent news reports that Apple was working on a car.

"Yes," Cook said, as he clicked to a slide addressing the press coverage. "We're doing one."

He said that the team at work on the car didn't know what size or shape it would be, but they were hiring aggressively and forging ahead with plans to enter one of the world's largest and most competitive markets around 2019. Code-named Project Titan, it was the type of ambitious, inspired bet that energized the company's workforce, infusing everyone with swagger and confidence that they could do what others deemed impossible.

But as dreams of disrupting the future spread across the room, Cook concentrated everyone's short-term focus on a project he believed could transform the company's present-day fortunes. He was excited about an effort to expand Apple's footprint in music. It would broaden its business from building devices to one that would develop new, world-class services. It would all start with an initiative code-named "Fuse."

APPLE GAVE EVERY PROJECT a code name, a special designation that filled its employees' work with mystery and secrecy. The tradition dated back to the late 1970s, when the company had launched a project to develop an inexpensive computer under the code name of an engineer's favorite apple, the McIntosh. Some names were more practical than imaginative. The music effort took its name from the idea that the company would fuse the recently acquired Beats business with its existing Apple music business.

Shortly after the deal with Beats Music closed, Apple's leadership gathered more than two hundred members of the Beats Music team at Infinite Loop to talk to them about the combination. The first thing Apple's engineers wanted their new colleagues from Beats to know was that Apple was close to finished with a streaming music service of its own. Their service had never launched partly because Apple still favored the sale and ownership of music over renting every song ever made for $9.99 a month. Its existence humbled the Beats Music team, which thought they'd been acquired to build that service.

Instead of defining a new business, they realized, they would be combining two separate concepts.

Jeff Robbin, Apple's vice president of consumer applications and the lead engineer on iTunes, assumed responsibility for developing the streaming service, while Nine Inch Nails leader Trent Reznor, who had joined Apple through the Beats deal, played a lead role in representing Beats, helping staff from both companies think through a service that offered critic-curated playlists and a radio station with artist interviews.

As designers and engineers advanced ideas for the service, Robbin often pushed back because he considered their concepts too ambitious. He favored a stripped-down design for the streaming music app, similar to the one Apple had developed for its never-released service, with limited album art, muted colors, and a spreadsheet-style list of songs. The newcomers from Beats and some designers at Apple disliked it and surreptitiously developed a more visually dynamic app with album covers, trendy fonts, and vibrant colors. They kept their option secret until a few days before a scheduled review of their work with senior leaders. When they revealed their alternative design to Robbin, he was furious. "We can't build any of this," he scoffed.

The designers pushed back, arguing that their artistic concept was more in line with what customers would expect than the more practical engineering style that he favored. To resolve the disagreement, they took both designs to Eddy Cue, Apple's senior vice president of services. They laid out poster boards showing the muted app style Robbin favored and the colorful app style his team championed. Cue scanned the images and considered his options. "It's obvious," he said, pointing to the one developed by Beats' designers. "We're doing this one."

Cue's selection validated Cook's rationale for buying Beats.

Without Jobs playing conductor, the competition between the teams from Apple and the teams from Beats helped push the streaming service forward with a design that had more imagination.

COOK NEVER GOT INVOLVED at a granular level in the music product's development, but he took an interest in its business plan.

As the project gained momentum, he joined Cue and Jimmy Iovine for a presentation from the former Beats marketing team about their subscription goals. It was Apple's first subscription service other than iCloud, so there was nothing to benchmark it against. Ian Rogers, the former CEO of Beats Music, and Bozoma Saint John, its head of marketing, thought the new Apple version of the service could acquire about 10 million subscribers. The figure was a hundred-fold increase from the 100,000 subscribers Beats Music had amassed in the months before Apple had bought it. As Saint John floated the goal, Cook listened expressionlessly.

"That's nice," he said flatly. "Can you do more?"

Cook had more faith in Apple's distribution power than the newcomers did. Apple shipped about 200 million iPhones annually. He knew it would be able to preload the new music app onto those devices, putting it directly in front of a huge network of potential customers. The projections of the Beats team had failed to take that into account. He thought they should be much more ambitious. After Cook's challenge, the group floated the idea of doubling the goal to 20 million.

The new target unnerved Saint John, whose discomfort showed, but Cue, her boss, reassured her that it was feasible. With a single question, Cook had squeezed more commercial ambition out of the Apple Music team than they had volunteered.

Cook wanted the new app to be at the forefront of an emerging strategy to grow the company's revenue by squeezing more sales out of its massive iPhone business. For years, he had watched as Apple facilitated the distribution of software through the iPhone's App Store, which vetted and approved every app available for download to iPhones. The company's role as gatekeeper was lucrative: Apple collected 30 percent of the sale price of every app it sold and a similar amount each month from subscription apps. The sales had made the store a fast-growing contributor to Apple's bottom line, providing the majority of its $18 billion in services sales. But Cook saw an opportunity in the growing app economy to make more money by transitioning from distributing to making apps.

The music service would become his test case for building a new empire on the back of his predecessor's revolutionary invention.

IN THE EARLY DAYS of summer 2015, the engineers and designers stressed about the music service. Their deadline to deliver a working product was early June, and as the days ticked by, many facets of the app weren't working.

Reznor endorsed a feature called Connect that would enable artists to share songs, photos, and videos directly with their fans. He and colleagues thought the internal social media network would differentiate their service, Apple Music, from streaming's biggest player, Spotify. But it stoked unease inside Apple, where a 2010 social network, iTunes Ping, had been canceled because it had become plagued by fake accounts and spam. Chastened by that past experience, Apple's leadership didn't want to enable community conversation on Connect, a decision that made some of the Beats team fear that there wouldn't be enough content to make the feature worthwhile. The diverging visions fostered concern that the feature would flop.

Meanwhile, the Beats engineers were adjusting to a proprietary coding language that only Apple used. They had developed their previous app with a widely used coding language common to thousands of app developers, but Apple wanted to use an exclusive code similar to what it had used to develop iTunes. The Beats engineers thought it loaded features more slowly than their native app, and they were scrambling to find a way around it.

Similar pressure was mounting for Iovine. The longtime label executive was responsible for ensuring that Apple finalized the licensing agreements it needed with record labels to provide a complete catalog of music at launch. The negotiations were challenged by Apple's plan to offer a free three-month trial to the service, and it wanted the labels to join with it in forgoing any licensing fees. The sales pitch was that if they would forgo their fees for three months, Apple would bring in millions of new subscribers who would listen to more songs and allow Apple to funnel more money back to the labels and artists. Though large labels like Sony and Universal agreed to deals, independent labels balked, leaving Apple without music from popular artists such as Adele and Radiohead. Their absence threatened to overshadow the launch by fueling stories about what Apple Music didn't have instead of what it did.

Two days before the launch event, the app wasn't finished and the deals weren't negotiated. Everyone was on edge, unsettled by one question: Is this going to work?

ON A WARM, SUNNY MORNING in early June, more than five thousand software developers lumbered toward San Francisco's Moscone Center, their backpacks and shoulder bags heavy with laptops, lanyards with ID cards swinging from their necks. A two-story white image of the Apple logo on the exterior of the convention center acted as a beacon calling them inside for a three-hour software showcase.

Tim Cook led the faithful through a presentation of promises that drew applause and whistles, hoots and howls. With the faithful entranced by the Apple spell, he swaggered across the stage and promised one more thing. It was the second time in a year he'd used Jobs's magical turn of phrase. This time, though, he summoned it for a baby of his own making, something he had bought as much as Apple had built.

"You know, we love music," he said. "And music is such an important part of our lives and our culture."

His words concealed what Apple had lost. A decade earlier, Steve Jobs had stalked the same stage and introduced the iTunes Music Store, where people could buy songs digitally. The innovative service had revolutionized music and put Apple at the vanguard of culture. Billions of dollars in sales and millions of customers had followed. The success had made the company complacent as Spotify had led a new wave of disruption, siphoning away customers by giving them access to an almost infinite catalog of songs for a monthly fee. Cook and the audience knew that Apple's position as a music industry leader had faded. It needed a second act.

Whereas Jobs had led with innovation, Cook imitated. He said that Apple was introducing its own Spotify-like subscription service called Apple Music. "It will change the way that you experience music forever," he said. He called on Jimmy Iovine, a man with a longer musical pedigree than himself, to explain.

Iovine took the stage wearing a silk-screened T-shirt with an image of the Statue of Liberty, an homage to his girlfriend, the English model Liberty Ross. The diminutive music mogul's charisma as a speaker usually came from his spontaneity and restless energy, which poured forth through a nasally Brooklyn accent that transported listeners to 1970s New York. Street-smart, he bragged that he usually didn't prepare for public speaking engagements. But the image-obsessed Apple had put Iovine onstage with a script and converted his conversational style into a book-on-tape monologue. Recalling Jobs's iTunes introduction a decade earlier, he flatly told the audience that his first impression of the service had been: "Wow, the ads are real. These guys really do think different."

For those accustomed to the animated Iovine, the scripted speech seemed inauthentic. He seemed uncertain and uneasy as he stared out over the thousands of people before him.

"Technology and art can work together," he said, reading from the teleprompter. "At least at Apple," he added.

Apple Music would be different because songs in playlists would be curated by people rather than algorithms, he said. "Picture this, you're in a special moment, you're exercising, or in some other special moment." He paused and looked toward his longtime business partner Dr. Dre, seated near the stage.

"Right, Dre," he said with a wink. He looked back at the audience. "He exercises a lot," he said, pausing just long enough for the audience to grasp that he was referring to Dre's active sex life. Iovine grinned as the crowd began to laugh. "Your heart's pumping," he said, smiling now that everyone was in on the joke. "And you're about to turn up the reps, and the next song comes on."

"Eh'nt!" he shouted. "Buzzkill!"

Iovine explained that bad transitions between songs on playlists occur because most are programmed by algorithms. Much like Beats Music, he said, Apple Music would have human curators for its playlists, so that each song built on its predecessor. The crowd was underwhelmed.

THE POP STAR TAYLOR SWIFT was on tour in Europe that June as Apple barreled ahead with its new music service. She received a text from a music industry friend with an image of an Apple Music contract saying that there would be 0 percent compensation to artists.

Upset, she penned a letter in the middle of the night that posted to her website early the next morning:

To Apple, Love Taylor

. . . I'm sure you are aware that Apple Music will be offering a free 3 month trial to anyone who signs up for the service. I'm not sure you know that Apple Music will not be paying writers, producers, or artists for those three months. I find it to be shocking, disappointing, and completely unlike this historically progressive and generous company.

. . . I say to Apple with all due respect, it's not too late to change this policy and change the minds of those in the music industry who will be deeply and gravely affected by this. We don't ask you for free iPhones. Please don't ask us to provide you with our music for no compensation.

Taylor

That Father's Day morning, Iovine woke up at home in southern California to a message with a link to Swift's website. He clicked through to find the letter skewering Apple's new music service because it planned to forgo paying artists during its three-month free trial. Swift's independent record label, Big Machine Label Group, led by Scott Borchetta, had actually been in talks with Iovine about making sure that its artists would be paid before licensing songs to Apple Music. The parties hadn't been able to strike a deal before Swift had vented publicly.

Iovine couldn't ignore her acerbic complaint: We don't ask you for free iPhones, please don't ask us for free music.

He immediately called Borchetta. "What is this?" he demanded, his voice rising an octave. "What is this letter?"

"She just sent it to me," Borchetta said. "I didn't put her up to it. But she has a point."

Iovine paused. "Let me call Trent," he said. He hung up, called Trent Reznor, and explained Swift's letter. The Nine Inch Nails front man appreciated Swift's point. Soon Iovine was on the phone with Eddy Cue, who was stunned that Apple's big new product was getting skewered by one of the biggest names in music. He worried the company's emerging strategy around services could have its wings clipped before it took flight.

"This is a drag," Cue griped.

He and Iovine called Cook to discuss how to head off the public relations crisis. Cook decided that Swift's argument was valid. He said that Apple should pay the artists and directed Cue and Iovine to figure out the terms. He and Iovine called Borchetta, who was at a Father's Day pool party in Nashville, and pressed the label executive for a solution. They ignored the sound of people splashing in the pool and listened as Borchetta told them that Apple needed to do the right thing and pay artists. Otherwise, how could it market Apple Music as an artist-friendly service?

"Here's the good news: You haven't launched yet," Borchetta said. "You have time for a fix."

"What's the right rate?" Cue asked, referring to the amount streaming services typically paid for each song played.

Borchetta took a deep breath as he realized that he had been given the power to set the rate for the entire industry. At the time, Spotify was paying artists about $0.006 per stream.

"You know what the Spotify rate is," Borchetta said. "Beat it."

What had seemed like a simple solution threatened to upend Apple Music's finances by creating unaccounted-for costs. The company would need to dig into its coffers—the $200 billion it had on hand—to cover the unplanned payments to artists. Iovine and Cue conferred with Cook and got his approval to cut a deal. If they didn't surrender, they would risk Swift's letter emboldening other artists and triggering a mutiny against the service.

Later in the day, Iovine and Cue arranged a conference call with Borchetta and Swift. "Taylor," Cue said. "I want you to know we've taken your letter seriously. We've decided we're going to pay from stream number one."

Swift thanked Cue for taking time to speak with her directly and for respecting her position. She and Borchetta considered it a huge step forward for the industry.

In the days that followed, Apple signed a deal with a collection of independent labels that included Adele and Radiohead. It later signed with Borchetta's Big Machine Label Group. The terms of the contracts were never disclosed, but in a letter to labels years later, Apple would brag that it paid artists more than Spotify did.

Cue's and Iovine's personal outreach to Swift stayed with the artist. Several months later, she appeared in a commercial for Apple Music, leading some reporters to speculate that the letter had been a publicity stunt orchestrated by Apple. But Borchetta, Iovine, and several Apple Music employees said that that was not the case.

"People think it's too good to be true, but this time it isn't," Borchetta said years later from Nashville. "No one wanted this dirt on their shoes. They were kicking off this new service. It was a PR nightmare."

AS APPLE SMOOTHED the edges of its music business, Cook found himself confronting a series of key decisions about the company's traditional hardware business.

Four years into the job, he was still reluctant to get involved in product development. He continued to believe that he would fail if he tried to mimic Jobs. But with Jony Ive sliding into a part-time role, there was a void in the day-to-day leadership over product. Some of Cook's top lieutenants had begun looking to him for more direction.

During a series of discussions in 2015, hardware engineering chief Dan Riccio presented Cook with a plan for a home speaker that would rely on the Siri voice assistant to field questions and play music. Amazon had popularized the category, known as smart speakers, with the introduction of its Echo devices controlled by Alexa. Apple's engineers had spent years exploring a similar concept, and some had been stunned to see such a sophisticated device come out of a company they considered to be a digital Walmart. Riccio proposed entering the category, and his team developed some early concepts for a smart speaker that would feature the highest sound quality on the market. He brought it to Cook for approval.

During discussions, Cook tested Riccio with questions about what the product would do and how people would use it. He closed by asking for more information. Under the impression that Cook wasn't sold on a speaker, Riccio's team wound down its work. Then, several months later, Cook emailed Riccio a link to an article about Amazon's Echo speaker and asked where Apple stood on its own speaker effort.

Riccio's team lurched back into action, ramping up work on a project that had been abandoned. The push came as Amazon's own speaker began to gain ground with customers, selling about 3 million. Its Echo seemed to have traction and was leaving Apple behind. The episode divided staff. Some saw patience and pensive thought; others saw bureaucratic sloth that had been absent in Apple's nimble years. Whereas Jobs had made decisions on instinct, providing engineers with firm and fast direction, Cook preferred to listen and gather information before proceeding. He suffered from what some called "analysis paralysis."

When the iPhone found itself at a development crossroads that year, he proved more decisive. At the time, Apple's most important product had fallen into a release cadence known as the "ticktock cycle." The company would overhaul the iPhone's design in a "tick" year, triggering a surge in sales, and refine the design in the subsequent "tock" year, when sales would wane. The strategy helped spread the costs of labor and new machinery over two years. But for the first time the company's product road map called for its ticktock tactic to miss a beat.

After overhauling its iPhone design with the iPhone 6 in 2014, the company planned to refine that phone in 2015 and 2016. The resulting ticktock-tock cadence created internal pressure to do something truly dramatic in 2017, a year that would mark the tenth anniversary of the iPhone.

Cook pressed for ideas that would reinvigorate the product. A group of engineers from a recently acquired Israeli chip and technology company called PrimeSense advanced the idea of miniaturizing a technology for gaming consoles. The system, which they had developed, used cameras and sensors to process users' hand gestures. They proposed taking the nine-by-three-inch concept and shrinking it tenfold so that people could unlock their phones with their faces. The facial recognition technology would enable Apple to eliminate the home button and expand the phone's screen from edge to edge, blurring the display into the space around it like an infinity pool.

It was an ambitious concept that would require huge engineering leaps, but Cook endorsed a plan that would minimize the risks. The PrimeSense technology would become part of a premium iPhone that would sell at a higher price. The cost increase would offset the pricier components, but perhaps more important, it would moderate the demand for a product that many feared would be hard to make in the numbers a popular new iPhone would sell, upward of 50 million units in about three months. They would fulfill the excess demand—and hedge against a possible facial recognition failure—by complementing the premium phone with another minor update to the iPhone 6. The plan was a master class in risk management.

EVEN AS THE COMPANY'S MUSIC DIVISION moved past the Taylor Swift drama, new problems arose. Customers and reviewers ripped the new service.

The *Wall Street Journal*'s technology reviewer Joanna Stern was blunt: "I don't love Apple Music." She said it "lacks polish and simplicity" and compared its lists and menus to Russian nesting dolls. The *New York Times* said it was like something Microsoft would make. And the tech site the Verge said it was "messy, slow to load, complicated to setup." Even Walt Mossberg, a longtime Apple supporter, balanced his approval of how the service integrated iTunes with an acknowledgment that the offering fell short of its competitors'.

Customers complained that one of the signature features, Connect, didn't work. It made the app busier than its peers and provided limited value because a lot of artists weren't using it properly. Eventually, tech websites posted guides on how to remove Connect. Inside Apple, engineers discussed killing the feature.

Apple Music fell short in the areas it most prided itself on: simplicity and beauty. The company had risen to prominence by creating intuitive software and hardware that left people saying, "It just works." But three years after its Maps disaster, it had once again introduced a high-profile service that had failed to live up to its standards.

Though Cook likely wasn't pleased with the reviews, he could find a silver lining in Apple Music's subscription numbers. Each day, more and more people signed on for a three-month trial of the product, and when the trial period came to an end, many of them became paying subscribers.

The number of subscriptions to Apple Music showed that the criticisms hardly mattered. Despite the troubles—Iovine's struggles onstage, Taylor Swift's attack, the damning reviews—Apple Music took off. The former Beats marketing team, unnerved by Cook's ambitious subscription targets, watched with wonder as the app proliferated across half a billion iPhones. The three-month free trial brought in millions of new customers, and many of them stuck around. It had 10 million paid subscribers in six months, a milestone that its rival Spotify had taken six years to hit. Within a year, the number would hit 20 million.

Cook could only smile. He knew that Apple had built a distribution machine.

Chapter 15

Accountants

Unbound, Jony Ive took flight. His Gulfstream GV fueled up in San Jose and took off for Hawaii in May, France in June, and the Virgin Islands by year-end. They were part of a rotation of exclusive destinations that Ive reached from a cabin of luxury.

From the tarmac, he would ascend a few steps into a compartment whose curving white walls framed rows of plush leather seats the creamy color of a caffe latte. The tasteful interior had been custom designed by Jobs, one of the few people with a sense of style as refined as his own.

Ive had bought the jet from Steve Jobs's family after his late boss died. The family had been using it since around 2000, when Apple's board had given Jobs the private plane as a thank-you gift for rescuing the company from bankruptcy. Jobs had spent more than a year customizing its interior, insisting on small details such as having the cabin's polished metal buttons be replaced with ones made of brushed metal. For Ive, who had consulted on the jet's interior design, the exacting touches were reminders of a man whose high standards had changed the world.

After the Apple Watch launch, Ive wanted to escape from the

exhaustion of Cupertino. His part-time agreement allowed him to heal through travel while his lieutenants shouldered the management of the hundreds of people working on Apple's industrial and software designs. Ive stayed updated on the work being done but mostly skipped the studio's weekly meetings that had defined two decades of his life. While his industrial designers and software designers debated the curves and colors of future products, he recuperated at his Kauai estate and spent time alongside the electric blue water of the French Riviera.

When he returned to San Francisco, he monitored an ongoing renovation of his Pacific Heights mansion. The team at Foster + Partners, the architects of Apple's new campus, had drawn up plans to turn the $17 million, four-bedroom, seven-bath home into a more customized space for him, Heather, and their two boys. As construction progressed, he sometimes spent time at the Battery, an exclusive social club in downtown San Francisco where Silicon Valley's elite gathered. Rather than meeting in the design studio where he once showed Jobs iPhone prototypes, he would summon designers to the venue forty-seven miles from Cupertino for occasional meetings so he could stay abreast of Apple's continuing projects.

He toggled between being engaged and disengaged, present and absent, in charge but not fully responsible. His time once again became his own.

AS IVE BEGAN TO DETACH, work on the Titan car project accelerated. The company hired hundreds of engineers and academics with expertise in batteries and cameras, machine learning and mathematics. They were lured with promises of developing the company's next great product, one that would speed past Detroit and remake the world.

The new hires found themselves working in nondescript warehouses in Sunnyvale, California, a new outpost shrouded in absolute secrecy. They labored on the most complex project Apple had ever undertaken, an endeavor considered to be as complex as NASA's journey to the moon. To succeed, they would need to develop an operating system that could process information from cameras and sensors that would provide a multidimensional picture of the world outside and determine how a car should travel. The car itself would need sophisticated battery cells that would power it for hundreds of miles. The team also needed to define the customer experience: What would it feel like to sit inside?

The early entrants into autonomous cars had taken a piecemeal approach. The industry leader, Google, had prioritized making an operating system over making a car. It had invested time into gradually improving a system that would enable minivans to robotically navigate the streets of Phoenix. Tesla had concentrated on making electric vehicles that offered limited self-driving capabilities. Apple's leadership wanted to develop an autonomous system and build an electric-powered car simultaneously.

The industrial design team took a leading role. Ive and members of his team began traveling to Los Angeles, where a number of influential auto design studios develop concepts for cars that eventually take the road. They talked about what they liked and disliked about vehicles, sketched ideas, and evaluated ways to enlarge their trademark curves in a more dynamic form.

Though he was no longer working full-time, Ive's role as Apple's tastemaker gave him an outsize voice in the project. He had strong opinions about cars, rooted in years of studying the vehicles around him. In his youth, he had worked alongside his father restoring an Austin-Healey Sprite. As an adult, he had amassed a stable of cars, many of them British makes, including an Aston Martin DB4 and a vintage Bentley Continental S3. His opinions about cars were so strong that he had once bristled at the notion of getting into a

Mercedes S-Class Sedan that was sent to pick him up at his hotel because he disliked the way the automaker bubbled the car's frame around its rear wheels. In his opinion, the line didn't follow the form.

Ive wanted Apple to make a fully self-driving car, replete with a voice assistant and devoid of a driver. His vision differed from the views of hardware chief Dan Riccio and his product designers. They favored building a semiautonomous electric vehicle that would alternate between self-driving and being driven, much as Tesla's cars did. They envisioned Apple doing to Tesla and the auto industry what it had done to Nokia and mobile phones: entering the market late with technology so superior that it would soon become dominant.

As the debate simmered, the industrial design team worked through prototype concepts. They imagined a car interior without a steering wheel. If the car doesn't need a driver, they thought, why include a wheel? They transformed the cabin of the car into a lounge with four seats turned toward one another rather than facing forward. They discussed materials and considered glass that adjustably tinted the sunroof to lessen the heat of California rays. They imagined mechanical doors that would close without a sound. They pressed for transparent windows that would double as augmented reality displays, superimposing on the glass the name of a restaurant or a street outside.

They shared a nostalgia for Toyota's minimalist van wagon, which had been popular in Japan in the 1980s. It featured subtle angles that gave its boxy shape definition and an angular windshield that was unlike anything else on the market. With that design as an inspiration, they mocked up an Apple minivan with a rectangular minimalism softened by their signature Bézier corners. To the engineers, it looked like an egg on wheels, with no angles or edges, just a rolling cabin of curves.

As they had for years, the design team set specifications that were more severe than those of most cars on the market. They wanted a car with almost invisible sensors, a demand that forced Apple's engineers to begin to develop their own technology, known as lidar, because many of the available sensors sat atop the roof like an unattractive prison guard tower.

One day in the fall of 2015, Ive met Tim Cook in Sunnyvale to show him how he envisioned the car working. He imagined that the vehicle would be voice controlled and passengers would climb in and tell Siri where they wanted to go. The two executives entered the prototype of a loungelike cabin interior and sank into seats. Outside, an actor performed as Siri and read from a script that had been written for the fanciful demonstration. As the imaginary car sped forward, Ive pretended to peer out its window.

"Hey, Siri, what was that restaurant we just passed?" he asked.

The actor outside responded. A few other exchanges with the executives followed.

Afterward, Ive exited the car with a look of satisfaction upon his face as if the future was even grander than he'd imagined. He seemed oblivious to the engineers looking on, some of whom were gripped by a worried feeling that the project was as fictional as the demonstration, moving fast but nowhere near its final destination.

EACH MONTH, Apple's future campus pleaded for Ive's attention. He would leave San Francisco and go down to Cupertino, where a hole in the ground was taking shape.

Jobs had dreamed of having an event space at Apple's new headquarters that could be a home stadium for the company's technology spectacles. Before his death, he had floated ideas for connecting it to the main building by an underground tunnel, creating a subterranean walkway from his office to a stage. Gradually, the plans had shifted until it had become a refined theater perched on a hill about a quarter mile from the main campus with a view of an undulating landscape dotted with fruit trees and oaks. It fell to Ive to bring the concept to life as Jobs had intended.

The planned theater would be a twenty-two-foot-tall circle of glass topped by a carbon-fiber roof that looked like a flying saucer. To preserve the illusion that the roof was floating, the glass walls couldn't have posts or columns, so the architects had to imagine ways to conceal the power lines, sprinkler lines, and more inside the joints between the glass panels. The roof would be the equivalent of the hulls of forty-four carbon-fiber yachts bolted together into an eight-thousand-pound circle of silver. All of it would be bead blasted to a MacBook sheen. It would sit over a terrazzo floor in a sunlit room that would feel like an open-air gazebo. The auditorium would be underground, accessible by two curving staircases.

Ive was passionate about ensuring that the interior of the theater would be immaculate, designwise. For the seats, architects from Foster + Partners sourced dozens of leather samples from tanneries around the world. Ive examined each swatch of leather, just as he had with the watchbands. He searched for the right mix of supple, soft, and smooth before picking Poltrona Frau leather, common in Ferrari sports cars. He later reviewed an array of color options, agonizing over the right choice for the two thousand seats inside before picking a caramel color with a reddish hue. Each chair would cost $14,000, but like Jobs, Ive refused to put a dollar sign on good taste.

The seats were supposed to sit above rows of oak flooring. For the wood, Foster + Partners ordered hundreds of samples of oak from around the world in three-inch-wide-by-four-foot-long samples. Ive came in and evaluated each plank, asking how cleaning and maintaining them would affect their appearance over time. He eventually chose an oak from the Czech Republic. He wanted the wood to bow slightly to create an imperceptible curve that would bend toward the stage, a request that could only be satisfied with a custom-designed process. The architects created a ten-by-twenty-foot prototype of the theater with the leather seats for Ive to experience and approve.

Ive found the work on the building invigorating. Most decisions were new and novel, not nearly as dull and repetitive as refining curves and selecting materials for yet another incremental update to the iPhone, iPad, or Mac. He so enjoyed stretching himself creatively that he also weighed in on the renovation and development of Apple Stores in big cities such as Chicago and Paris. He and Angela Ahrendts worked with the architects to turn the stores into what they called "town squares," places where people not only shopped but gathered to attend classes on Apple products, watch movies, and hang out. They adopted many of the concepts of the new Apple campus, with transparent glass. The result was a unified architectural sensibility across all of Apple's real estate, the first major refresh after Jobs.

In the midst of working on the stores, Ive consulted for Ahrendts on a project she had dreamed up to further Apple's business in China. She wanted to introduce a fleet of buses that would become Apple Stores on wheels, traveling around the world's most populous nation selling iPhones in cities and towns where the company didn't have a store. The plan called for the buses to return each night to a depot where they could be cleaned before being dispatched again the next day.

Some members of Ahrendts's team scoffed at the absurdity of enlisting one of the world's great designers to work on upscale Greyhound buses.

WORKING ON THE BUILDING and the car, as well as vetting refreshes to the iPhone, iPad, and Mac, kept Ive engaged, but there were occasional reminders that his influence over the company was diminishing.

Jony Ive (*right*) shows Tim Cook (*left*) the recently unveiled Mac Pro in a ritual of corporate marketing after the 2019 Worldwide Developers Conference in San Jose, California. *Brittany Hosea-Small/AFP via Getty Images*

Tim Cook and his friend Lisa Straka Cooper were known to jokingly sing "The Way We Were," just one of the examples of what classmates called Cook's "corny jokes."

Breahna Crosslin

At Robertsdale High in south Alabama in 1978, Tim Cook was known for his big hair, diligent work ethic, and involvement in the school band.

Breahna Crosslin

While studying industrial design at Newcastle Polytechnic, Jony Ive created his first phone, the Orator, a futuristic device that reimagined the traditional product by shaping it like a question mark. He's shown here discussing it at a *Vanity Fair* event in 2014.

Kimberly White/Getty Images for Vanity Fair

Tim Cook (*left*), shown here with Steve Jobs (*center*) and Apple chief marketer Phil Schiller (*right*), joined Apple in 1998 as Senior Vice President for Worldwide Operations, revolutionizing their inventory management and helping them improve profitability. *David Paul Morris/Getty Images*

Jony Ive (*left*) and Steve Jobs (*right*) dreamed up many of Apple's product designs together, including the 2001 iMac G4, which was inspired by a walk among the flowers in Jobs's backyard in Palo Alto. Jobs told biographer Walter Isaacson that Ive was his "spiritual partner." *Michael O'Neill/Contour RA by Getty Images*

In 2014, Tim Cook landed in Beijing to finalize a landmark deal with China Mobile's Chairman Xi Guohua to bring the iPhone to China's largest wireless provider. The agreement, six years in the making, turbocharged Apple's sales in China. *Reuters/Kim Kyung-Hoon*

After Jobs's death, Jony Ive (*right*) began working more with friend and fellow designer Marc Newson (*left*). In 2016, they designed the lobby of London hotel Claridge's for Christmas, putting the emphasis on a small evergreen tree inside a birch forest. *David M. Benett/Dave Benett/Getty Images for Claridge's*

Jony Ive assembled an elite team of about two dozen designers who developed most of Apple's products, including the Apple Watch, which was unveiled in 2014. *Leander Kahney*

Tim Cook brought music producer Jimmy Iovine to Apple in 2014 with a $4 billion deal for Beats Electronics. Iovine led Apple Music, which became a catalyst for Apple's push to offer more subscription services.

Michael Kovac/WireImage

Faced with the threat of tariffs on iPhone exports from China, Tim Cook cultivated a personal relationship with Donald Trump that helped the company avoid financial penalties. Trump mistakenly called the CEO "Tim Apple" during a meeting broadcast on TV in 2019.

Mike Theiler/CNP/MediaPunch

Jony Ive joined his wife and high school sweetheart, Heather, to walk the red carpet at the Met Gala after Apple signed on to sponsor the Metropolitan Museum of Art's exhibit *Manus × Machina: Fashion in an Age of Technology*.

Neilson Barnard/Getty Images

Tim Cook and Laurene Powell Jobs walk the red carpet at the Met Gala.

Reuters/Eduardo Munoz

Tim Cook brought Hollywood stars to Apple Park in 2019 for the unveiling of Apple's long-awaited TV service, a high-profile subscription offering that was at the forefront of a strategy to shift the company's focus from selling more iPhones to selling more software and services. *David Paul Morris/Bloomberg via Getty Images*

Apple spent an estimated $5 billion on its new campus, Apple Park, which was the last of Steve Jobs's products. The distance across its circular interior is greater than the height of the Empire State Building. *Sam Hall/Bloomberg via Getty Images*

After the completion of Apple Park, London's National Portrait Gallery commissioned a photograph of Jony Ive inside the headquarters he'd spent seven years helping Foster + Partners design. *Andreas Gursky/ARS, New York, Courtesy of The National Portrait Gallery, London*

In 2017, Tim Cook opened the Steve Jobs Theater on Apple's new campus during an event where the company unveiled its tenth-anniversary iPhone.

Xinhua/Alamy Live News

In October, Cook named James Bell to Apple's eight-person board of directors. The appointment addressed the lack of diversity on the all-white board by installing an experienced Black executive who had worked as Boeing's chief financial officer. Reverend Jesse Jackson, a longtime Apple stock owner, had been pressuring the company for years to appoint a Black board member, even attending the company's annual general shareholder meeting to tell Alabamian Tim Cook that there was an "unbroken line from Selma to Silicon Valley—all part of the long journey for equality, human rights and economic fairness." The attacks pushed Cook, a self-proclaimed champion of diversity, to seek a Black candidate, but the choice of Bell irritated his star designer.

Ive supported diversifying the board, but Bell was filling a position vacated by outgoing board member Mickey Drexler, a longtime confidant of Ive and someone Ive believed understood the marketing and tastefulness that Jobs had instilled into the company. Just as Jobs had run Apple on instinct, Drexler had relied on his gut at the Gap and J. Crew to identify and bet on the fashion trends that had turned both companies into retailing giants. His departure meant that the board lost a director with a decade-plus of experience, innate marketing sensibilities, and an ear for creatives such as Ive. In his place, Cook installed a director schooled in operations and finance. It was the second time in as many years that Cook had chosen an operator over a natural-born marketer. After longtime Apple chairman and Jobs confidant Bill Campbell had stepped down in 2014, Cook had replaced him with Susan Wagner, BlackRock's chief operating officer. Wagner and Bell had tilted the balance of the board's expertise toward operations.

The change distressed Ive. He complained to colleagues and friends that Cook should have put Jobs's widow, Laurene Powell Jobs, onto the board, or someone else who had known Jobs well. At the very least, he thought, Cook could have chosen someone with marketing sensibilities. A colleague whom Ive spoke with defended Bell. "He's an underrepresented minority," the colleague said. "He has a great reputation."

"Who cares?" Ive said. "We have to be concerned about the company. He's another one of those accountants."

At that moment, Ive's worry about the company trumped any concerns he had about diversity. He was a self-proclaimed progressive who had spoken out about feminist issues as a British high schooler in Margaret Thatcher's United Kingdom. He was especially keen to see more women on the board. But at Apple, as far as he was concerned, protecting Jobs's legacy and preserving Apple's creative sensibilities took precedence. He told others that Apple's previous CFO, Peter Oppenheimer, shared his belief that the board needed to be balanced between creative-minded and business-minded members.

Below the surface of his frustration was another discomfort: he had failed to influence who would be appointed to the board. When Jobs had led the company, Ive had had the ear of the CEO. He and Jobs had discussed future plans and the state of the business over their regular lunches. His opinion had mattered and had shaped major business decisions about the company's future. With the ascendence of Cook, who seldom visited the design studio, Ive's sway had waned. Cook's consiglieres for most matters were Jeff Williams, his longtime operations lieutenant, and Luca Maestri, the chief finance officer. Ive had gone from influencer to observer, and his choice to step back from day-to-day operations sidelined him further.

FOR IVE, the pain of losing Jobs had never gone away. It afflicted him each time the October anniversary of Jobs's death passed. In the fall of 2015, Sony planned to release a biopic about the late CEO. The movie distressed Laurene Powell Jobs. The film was based on Walter

Isaacson's biography, which she disliked, and focused on Jobs's denial of paternity of his first daughter, Lisa Brennan-Jobs, which could tarnish her late husband's legacy. Before filming began, she tried to hamper the project by discouraging actor Leonardo DiCaprio from appearing in it.

Just days after the anniversary of Jobs's death, Ive was in Beverly Hills for *Vanity Fair*'s New Establishment Summit, where he took the stage to anchor a star-powered panel with *Star Wars: The Force Awakens* director J. J. Abrams and *A Beautiful Mind* coproducer Brian Grazer. Ive sank into the light gray easy chair on the stage and kicked his suede Wallabees out in front of him for what he anticipated would be a relaxed conversation.

He and Abrams, who sat to his left, had become friends and creative muses. Ive told *The New Yorker* that he had advised Abrams over dinner to make a future *Star Wars* lightsaber "more spitty," giving it an unevenness that would make it more primitive and menacing. It was a visual concept that Abrams had felt compelled to try. The result was villain Kylo Ren's ominous laser sword.

Grazer moderated a discussion between the designer and director about the complexity of the creative process. When the time came for audience questions, a man in a suit jacket approached a microphone at the center of the auditorium.

"Jony, I'm curious if you can talk a little bit about sort of the cottage industry of Steve Jobs movies," he said. "Have you seen the Steve Jobs movie? Will you see it? And what, how do you sort of feel about that whole world that's emerging?"

Ive jerked forward, glaring at the questioner. "That is the sweetest description I've heard," he said glibly. He shifted in his seat and raised his eyebrows. "Cottage industry," he scoffed. He peered at the audience with a bit of snark. "You mean the one that involves Sony?

"There's an awful lot I can say," he said. He hadn't seen the film but was troubled by it.

"This is sort of a primal fear of mine, and this touches quite deep, I think, for me, in that how you are defined, and how you are portrayed, can be hijacked by people with agendas that are very different from your close family and from your friends," he said. "And I don't really know what more to say."

He dropped his hands to the arms of his chair in a gesture of defeat.

"There are sons and daughters and widows and very close friends who are completely bemused and completely upset," he said. "We are remembering Steve's life, and at the same time, beautifully choreographed is the release of a movie, and I don't recognize this person at all. And I'm sorry to sound a bit grumpy about it. But it's, I just find it ever so sad. Because, you know, he was a, he had his triumphs and his tragedies like us all, and unlike most of us, he's having his identity described, defined by a whole bunch of other people."

Ive gripped his left knee and sheepishly tucked his right leg beneath his left, as if to curl his emotions back into place. He seldom became expressive onstage, but the question had lifted his emotional lid. With Apple changing around him, he was missing his friend and creative partner in ways he had never imagined.

FAR AWAY FROM CALIFORNIA in New York, a museum curator was mulling over the higher value people placed on what was made by hand than what was made by machine.

It struck Andrew Bolton as archaic for society to automatically designate handmade goods as luxurious and exclusive while it deemed machine-made items mediocre and commoditized. As the leader of the Metropolitan Museum of Art's Costume Institute, he wanted to challenge those preconceived notions with an exhibit that would force people to ask: Can I even tell the difference?

Bolton, who had joined the museum a decade earlier, was fashion's most influential storyteller. His thematic exhibits were credited with having elevated an industry that lived at the intersection of art and commerce. His work had earned him a seat at every major fashion show, and that July, he had found himself inspired by Chanel's show in Paris, where he had watched a pregnant model float down the runway in a white wedding dress made of synthetic scuba suit fabric with a twenty-foot train of pixelated gold. The woman, who looked as though she had walked out of a painting by the Dutch master Jan van Eyck, challenged the notion of made-to-measure fashion by wearing machine-made neoprene and a hand-stitched train. The dress had been the physical embodiment of man and machine.

Bolton wanted to use the concept as the theme for the museum's most important event, the Met Gala. Each May, the museum launched a new exhibit in conjunction with its annual fundraising event, led by *Vogue* editor Anna Wintour. Under her leadership, the Met Gala had become one of the most elite social gatherings of the year. Models, CEOs, artists, actors, and athletes paraded the red carpet as photographers' cameras clicked.

When Bolton told Wintour about his idea and the exhibit title, *Manus × Machina,* she immediately thought of Ive. She called Ive to see if Apple would have any interest in sponsoring the gala. Bolton's idea of challenging the perceptions of machine-made and handmade goods immediately resonated with Ive, who had spent a lifetime trying to make products on a large scale with a handmade-level of detail. He considered the exhibit to be a perfect way to extend the Apple Watch's association with fashion. He approached Cook about a sponsorship, estimated at more than $3 million, and got the CEO's blessing.

That fall, Ive invited Bolton and Wintour to Cupertino to see the studio and discuss the exhibit. They arrived in late October to find a tidy studio space alive with designers focused on sketching and prototyping. Ive ushered his guests toward the studio's waist-high oak tables, where thin black sheets veiled the designers' embryonic concepts. He arrived at one where watches were kept and began to show them the design team's latest creation, a special collaboration with the 175-year-old fashion house Hermès.

The partnership had been born over lunch in Paris, Ive explained. He had arranged to meet the Hermès CEO shortly after the watch's debut at Colette and broached the idea of collaborating on a future watch. The result combined one of Apple's stainless-steel cases with leather straps from Hermès, which were created through a secretive process by tanners who had been passing down their process for decades. It was a union of new technologies and ancient techniques.

After their tour, Bolton and Wintour addressed the designers and discussed the forthcoming exhibit. Bolton shared pictures of dresses that would be in the show, and Wintour fielded questions from the group about the future of fashion. Listening nearby, Bolton was struck by the similarities between tech and fashion designers. Though they operated in separate fields, the practitioners of both disciplines poured untold hours into making something destined for obsolescence. The most breathtaking dress at a fashion show would be phased out a year later, dethroned by a new style. The most mesmerizing iPhone would soon be upended by the arrival of faster chips and better cameras. Both the Ives and Lagerfelds of the world devoted their lives to the pursuit of the next new thing.

WITH THE MET GALA on the horizon, Ive felt assured about the future of the Apple Watch. Tech and fashion were converging, and with the watch Apple had positioned itself to be at the forefront of that cultural collision, just as it had with the iPod and music years earlier. He ignored the worries of colleagues about its sales. He believed that sales would increase as Apple improved the watch's battery life and added health features. He saw no need to fret over how long that would take.

But deep in the executive wing, Cook was less convinced. He was continually distressed by the shortcomings of Ive's fashion-focused marketing strategy. As planning of future fitness promotions evolved, he decided to overhaul the company's marketing and sales teams. He relieved Phil Schiller of his oversight of the group responsible for Apple Watch advertising, despite the marketer and TBWA\Media Arts Lab teaming up to win a top prize at Cannes for its billboard campaign showcasing photographs taken with iPhones. He also endorsed the exit of Paul Deneve, the architect of the watch release strategy, whom the company had lured away from Yves Saint Laurent. Schiller would take a greater role in managing the App Store, while Deneve decided to depart for a possible return to the fashion industry. Both executives had played a central role in the introduction of a product that had fallen short of Cook's expectations.

Ive didn't know it, but the Met Gala would be the last time that Apple would walk a runway. The company he had helped make was changing in ways that even he failed to see.

Chapter 16

Security

Early one December morning in 2016, about eighty staffers from the San Bernardino County Department of Public Health filed into a local municipal building for a day of training and team-building exercises. There were office assistants and clinicians; data analysts and health inspectors; mothers, fathers, brothers, and sisters. They settled into seats in a bland government conference room where a Christmas tree, tucked into a corner, shimmered with holiday cheer.

Just before a midmorning break, one of their colleagues, Rizwan Farook, exited the room. Moments later, the door swung open, and he returned wearing a black mask and wielding an automatic rifle. He stepped inside and opened fire. *Pop. Pop. Pop.*

Some workers sprinted for the exit. Others dropped to the floor and sought cover under tables. A second shooter, Farook's wife, burst inside and joined in spraying the room with a barrage of bullets that pierced the walls, the windows, and a sprinkler pipe. The ceiling gushed water as the assault continued.

A 911 call summoned police, who cautiously entered the building. They made their way through the hiss of sprinklers and found bodies strewn across the floor. They advanced through rooms only to discover that the shooters were gone.

As paramedics cared for the wounded, officers interviewed survivors and learned that the masked shooter had been Farook. Investigators determined that the American-born Pakistani had rented a black SUV, which they traced to a nearby town. Police in the area found the car on a residential street. As they closed in, Farook revved the engine and raced to escape. His wife turned her gun on the trailing cops and fired at police through the rear window. A police cruiser approaching from the opposite direction forced Farook to slam on the SUV's brakes. His wife fired shots at the cruiser as Farook exited the driver's seat and fired shots. More than 150 officers arrived at the scene and unloaded hundreds of rounds of ammo, killing Farook and his wife.

When the shooting stopped, investigators cleared the crime scene. In a Facebook post, his wife had pledged allegiance to the Islamic State before the attack, which killed fourteen people. Agents searching through the SUV found several electronics, including an iPhone, the fingerprints of the digital age. They hoped it might explain that day's chaos and violence.

THE NEXT DAY, Bruce Sewell was at the gym working out as cable news stations continued to provide coverage of a terrorist shooting some four hundred miles south of the Bay Area. When his phone buzzed, the Apple general counsel stepped away from the whir of exercise machines. Tucked in a quiet place, he listened as someone from the company's twenty-four-hour law enforcement desk told him that the FBI wanted to speak with him immediately.

Within minutes, Sewell was on the phone with an agent, who recounted the previous day's events and added a new wrinkle: a search warrant had led the government to an iPhone hidden in one of Farook's cars. The agent wanted Apple's help gaining access to the phone, so law enforcement officials could determine if the shooters were part of a terrorist cell planning more attacks. They needed to move fast.

Sewell called Tim Cook. The CEO could tell by the tone of his general counsel's voice that something was wrong. Sewell was usually calm and measured, courtesy of a steadiness he had acquired while working as a firefighter before law school. He had put out enough literal fires that legal ones seldom flustered him. But his voice shook as he relayed that the FBI had found an iPhone belonging to a presumed shooter.

Apple's law enforcement liaisons were following protocols, he said. They had provided the FBI with options for gaining access and made clear that Apple would provide remote tech support, explaining its software even as it resisted getting directly involved in retrieving information off the device. Per company policy, it would not unlock the phone.

The policy was contentious. As mobile devices became a hub of sensitive data such as health information and communication, its engineers had increased security and encryption to protect users from hackers. Meanwhile, law enforcement wanted more access to phones that might have details in order to solve crimes and save lives. Apple's interest in protecting users and the FBI's interest in protecting communities were increasingly at odds. In 2014, the friction had increased after Apple had introduced features that prevented anyone from accessing an iPhone without a PIN or fingerprint. Police officers and prosecutors thought the feature put privacy over public safety, allowing criminals to communicate with less risk that their correspondence would land in a court of law. Law enforcement labeled the predicament "going dark."

In the wake of the San Bernardino shooting, Sewell was encouraged to hear that San Bernardino County had issued the iPhone 5c the FBI had recovered. The device included software that allowed the health department to control its use. There was a chance that the county might be able to gain access to it without Apple's help. The government could also access Farook's iCloud account, where he might have backed up the phone to Apple's digital storage service. Though Apple wouldn't unlock a phone, it would decrypt backups on iCloud and turn over messages and photos in response to a subpoena, a detail the company didn't advertise to customers but promoted to law enforcement. The security loophole put access to the phone within reach.

But in the days that followed, Apple and FBI optimism about a quick resolution crumbled. The FBI reached the iCloud account only to find that the shooter had last backed up his phone months earlier. It also discovered that the software management system the county was using hadn't been fully implemented, preventing the FBI from using it to access the phone. It issued a search warrant that yielded some emails and messages, but none of the correspondence provided a breakthrough. The answers they sought were on the phone.

IN EARLY JANUARY, nearly a month after the attack, Cook arrived at the U.S. Patent and Trademark Office in San Jose to meet with a delegation from Washington, D.C., including FBI Director James Comey, Attorney General Loretta Lynch, and White House Chief of Staff Denis McDonough. The Obama administration representatives were in Silicon Valley to encourage Facebook, Google, and other social media services to remove Islamic State messages that were radicalizing terrorists. Comey also wanted to discuss encrypted communications services that were complicating criminal investigations. The relationship between the government and the tech giants had soured after Edward Snowden had leaked documents showing that tech companies had aided the National Security Agency's surveillance of

Americans. The public blowback had led the multinationals to become
distant and adversarial. The Obama administration wanted a reset.

Upbeat about the continued strength of the iPhone and rapid
growth of Apple Music, Cook was emboldened and prepared to fight.
He entered a bland room with no windows and took a seat at a conference table alongside his peers, including Facebook COO Sheryl
Sandberg and Twitter chairman Omid Kordestani. The contingent
from Washington sat opposite from them and opened a discussion
that included asking the tech leaders for help hiring social media spe-
cialists to assist government efforts to disrupt terrorist recruitment.
Cook remained largely silent until the conversation shifted to encryption. Then he unloaded.

The Obama administration had shown a lack of leadership on encryption, he said. He called on the government to denounce calls by
the FBI for what he considered to be the equivalent of a technologi-
cal "back door" to the iPhone, saying that special software to provide
government access to devices could fall into the hands of bad actors
who might use it against ordinary people. He suggested that Apple
had a morally superior position grounded in protecting privacy, while
the government wanted to erode those protections.

As Cook spoke, McDonough's ears began to redden. The people
seated nearby could tell that he had not come to Silicon Valley to
be lectured. To the government people in the room that day, Cook
sounded sanctimonious.

Lynch intervened to say that there needed to be a balance between privacy and national security interests. Comey expanded on
the government's position, saying that companies should develop a
system for court-ordered access to devices during investigations.

No one mentioned San Bernardino. The dispute over the iPhone
lurked in the background.

IN SOUTHERN CALIFORNIA, the FBI was playing a game of digital Russian roulette. It could make ten attempts to guess the four-digit passcode of the locked iPhone, but if they all failed, the device could automatically erase itself or be disabled, accessing it would become impossible, and their final lead in the case would be dead.

Faced with that worst-case scenario, Comey strode into the Hart Senate Office Building for an Intelligence Committee briefing about global threats. It was February 9, more than two months after terrorists had killed fourteen people in San Bernardino, and he was frustrated that the FBI still hadn't gained access to the iPhone it had recovered. Senator Richard Burr saw the FBI director's dark mood. "Director Comey, what's the risk to law enforcement and to prosecution if, when presented with a legal court order, a company refuses to provide the communications that the court has ordered them to?" he asked.

"The risk is that we won't be able to make a case and a really bad guy will go free," Comey said, glaring at the senators before him. He explained that the problem, known as "going dark," prevented local law enforcement from resolving murder, drug, and kidnapping cases.

"I think there would be consensus in America that if that's carried out, that if a court certifies that the reason is there, that a company ought to then produce that information. Is that logical?" Senator Burr asked.

"Yes, especially with respect to devices, phones, that default lock," Comey said. He added that those devices had become a massive concern for law enforcement because they often held evidence of child pornography, plans for a kidnapping, or other details that could help solve crimes. He pointed both his hands toward his heart and focused on a personal frustration. "It affects our counterterrorism work," he said. "San Bernardino, a very important investigation to us; we still have one of those killers' phones that we have not been able to open. It's been over two months now."

Newspaper reporters and newscasters jumped on the comments. Comey's condemnation of Apple provided the first public confirmation that encryption was stymieing the San Bernardino investigation, and signaled that Comey was spoiling for a fight.

IN CUPERTINO, Sewell interpreted Comey's comments as the latest outburst from an FBI director who was waging a personal crusade against encryption. Apple's law enforcement liaisons provided Sewell with regular updates about their work with the FBI. He trusted that a resolution was on the horizon.

Hundreds of miles to the south near Riverside, California, Justice Department attorneys plotted to break the impasse. They drafted an application for a court order under the All Writs Act, a 1789 law that could force a company to assist in a criminal case. The rule, just two sentences long, had previously been used to get Apple's assistance in gaining access to the phones of child sex abusers and drug dealers. It was obscure but effective.

On February 16, government attorneys filed a forty-page request under seal in the U.S. District Court for Central California, asking a judge to force Apple to develop software that would disable the iPhone's ten-guess limit. The software would ease the pressure of the Russian roulette and buy the FBI time to pick the phone's lock. The judge granted the request preliminarily and gave Apple just five days to respond.

The ruling aggravated Sewell. He interpreted the Justice Department's move as a legal middle finger. He had limited time to plan a public response to the government's allegation that Apple was aiding terrorists. He fumed that the government's only purpose could be to turn the terrorist tragedy into a battering ram to shatter Apple's brand.

Sewell called Cook and told the CEO about the court order. After his legal team secured a copy of the ruling, he and Cook read
through it together. It began graphically: "In the hopes of gaining
crucial evidence about the December 2, 2015 massacre in San Bernardino, California, the government has sought to search a lawfully-
seized Apple iPhone used by one of the mass murderers. Despite
both a warrant authorizing the search and the phone owner's consent,
the government has been unable to complete the search because it
cannot access the iPhone's encrypted content. Apple has the exclusive
technical means which would assist the government in completing its
search, but has declined that assistance voluntarily."

Cook could see the ledger the government was creating: the good
guys at the FBI were fighting to solve a massacre, while the bad guys
at Apple obstinately stood in their way. The government had framed
the case to create a public relations contest as much as a courtroom
challenge. Apple's position on security hung in the balance.

If Apple created the software the government demanded, Cook
feared, it could be used on any iPhone in the world. As people stored
photos, health information, and financial data on their iPhones, they
needed to be able to trust that the information was protected. Creating a system to unlock the San Bernardino device would shatter
people's confidence in their iPhones. It threatened the magic of the
device that Cook depended on for the majority of Apple's sales. Not
to mention that it would weaken Cook's recent letter to customers
hyping the iPhone as more private and secure than Google's competing Android operating system, which tracked people online. "Our
business model is very straightforward: We sell great products," he
wrote. "We don't build a profile based on your email content or web
browsing habits to sell to advertisers."

Cook assembled company leaders—finance chief Luca Maestri,
marketing chief Phil Schiller, software chief Craig Federighi, and

head of communications Steve Dowling—in the boardroom to figure out how to fend off the government's attack. It was late afternoon, and he wanted a response that would reach all of Apple's customers by sunrise.

Caving in to the government was not an option. Cook had decided long before the San Bernardino shooting that he would fight a court order. As Apple had increased iPhone security, he had discussed hypothetical scenarios with its legal team about what Apple should do if someone was kidnapped and law enforcement said that the only way to rescue the victim was by accessing the kidnapper's iPhone. Cook had scrutinized every angle of the scenario, asking, "Did you think about this?" Ultimately, he had decided that protecting all Apple customers by refusing to create a back door outweighed solving a single crime. He was ready to fight.

Adrenaline coursed through the executives around Cook as they talked through how they would respond. The showdown risked harming Apple's brand.

Sensing his team's unsettled energy, Cook tried to slow everyone down with methodical questions. He suggested, unemotionally, that they start at the beginning. "What do we know about the phone?" he asked.

The precise, simple question forced everyone to focus. Sewell walked Cook through the state of the phone and recapped the FBI's requests for assistance, providing everyone with the history of the predicament.

Cook then turned to what the government wanted Apple to do. "What is this technical fix?" he asked. "How long would it take?"

Federighi dissected the FBI's request for something the company had long ago labeled derogatorily as GovtOS, a custom software system that would allow agents to bypass the iPhone's autolock capabilities. Creating something like that would take six-plus engineers two weeks or more. Once Apple created that software, it could expect an influx of requests from law enforcement seeking entry into criminals' phones. As the custom software spread, the risk would increase that it might fall into the hands of hackers or an authoritarian government, which could use it in ways never intended.

The group agreed that Cook would write a letter explaining Apple's position for the company's website. It would be the quickest way to address employees, customers, media, and lawmakers. Afterward, Sewell and Maestri would host press briefings to answer questions. The strategy simultaneously put Cook in front of the issue and shielded him from a media confrontation.

Cook spent an hour discussing what the letter should say, as Dowling took notes. To counter the Justice Department's effort to make Apple appear protective of the killers' privacy, Cook wanted to make clear that Apple's sympathies were with the victims and their families. He also wanted to reframe Apple's resistance to unlocking the iPhone as a position it was taking to protect all customers, not just the San Bernardino killers. He needed to take an emotionally charged conversation about terrorists and make it into a philosophical one about privacy.

Dowling drafted the first version of the letter for Cook to review. The CEO suggested changes and handed it back to Dowling for revisions. They repeated the process about half a dozen times over six hours as they defined the tone and adjusted individual words.

Meanwhile, Sewell and Noreen Krall, Apple's chief litigation counsel, drafted a legal response with a strident tone. They wanted to play the bad cop in court while Cook played the good cop in public. They made sure that their response addressed gaps in the government's narrative by detailing exactly how Apple already helped the FBI. In a bit of legal one-upmanship, they made sure that it was more pages than the Justice Department's complaint.

"This is not a case about one isolated iPhone," it began. "Rather, this case is about the Department of Justice and the FBI seeking through the courts a dangerous power that Congress and the American people have withheld: the ability to force companies like Apple to undermine the basic security and privacy interests of hundreds of millions of individuals around the globe."

Around 4:30 A.M., Cook's letter was published online. The group had worked through the night without any sleep and were keenly aware that they were taking an unpopular position that could hurt Apple. It was a bet-the-company gamble.

IN THE DAYS THAT FOLLOWED, Apple's showdown with the FBI dominated the news, accounting for about five hundred articles a day and constant discussion on TV. It became a presidential campaign talking point, as Republican candidate Donald Trump blasted Apple and called for a boycott of its products. Public opinion was split, with half the country demanding that Apple cooperate with the FBI and half supporting its resistance. The battle between the world's largest company and the world's most powerful government was enthralling.

On February 25, a week after the letter, a crew from ABC News arrived at Infinite Loop with *World News Tonight* anchor David Muir. The group was greeted by an Apple communications official and ushered across the skylit atrium toward the company's fourth-floor executive wing. Dowling had invited the ABC team to Cupertino with the hope that they would tilt opinion in the company's favor. He had known Muir's producer for years and trusted the dark-haired, forty-two-year-old anchor to relay Apple's perspective to the world. After all, it was "Daddy's network," part of the vast Disney empire overseen by Apple board member Bob Iger.

The group walked through the executive wing to Cook's office, where Muir found a tired and somber CEO. Cook knew that the interview might be his best shot at persuading the country that what Apple was doing was not crazy. The location where they met spoke to the gravity of the moment. The deeply private Cook had agreed to do the interview in his office, as the communications team tried to humanize a man who often appeared robotic.

As he perched on a stool opposite Muir, the camera provided a glimpse of the CEO's everyday workplace. There was a tidy desk topped by manila folders and a silver iMac. The wall behind it featured color photographs from Apple's retail stores not far from a framed copy of an Auburn alumni magazine. Collectively, they told the story of his devotions: Apple and Auburn.

Cook clasped his hands in his lap, his blue eyes puffy from too little sleep, and gazed at Muir. He stiffened atop a stool, wooden and focused.

"I don't think we've ever done an interview in your office before," Muir said.

"I'm not sure I've ever done an interview in the office," Cook said without a smile.

Muir decided to dive right in. "As we sit here, you know, some of the families of the victims in San Bernardino have now come out in support of the judge's order that Apple help the FBI unlock that iPhone, one family reportedly saying 'We're angry and confused as to why Apple is refusing to do this.' What would you say to those families tonight?"

Cook listened intently. "David, they have our deepest sympathy," he said. "What they've been through, no one should have to go through."

He paused and gazed downward. "Apple has cooperated with the FBI fully in this case," he continued. "They came to us and asked us for all the information we had on this phone, and we gave everything that we had. We went further than that and volunteered engineers to help them and gave them numerous suggestions about how they might learn more about this particular case. But this case is not about one phone. This case is about the future. What is at stake here is: Can the government compel Apple to write software that we believe would make hundreds of millions of customers vulnerable around the world?"

The exclusive interview played that night on ABC to an audience of about 9 million viewers. ABC posted the entire thirty-minute conversation on YouTube, as well. Cook's appearance on "Daddy's network" was viewed inside Apple as a win in its public debate with the FBI, a moment when the CEO came off as serious, sympathetic, and fully in command of the issue, advancing without debate a complex philosophical argument that he had summarized with a succinct analogy: the government's request for Apple to write code to make it easier to gain access to the phone was "the software equivalent of cancer."

MEANWHILE, Sewell pushed ahead as Apple's tough guy, aggressively challenging the government inside and outside the court. In the days after the February feud erupted, he had a call with Assistant U.S. Attorney Sally Yates, who was leading the government's case. Yates accused Sewell of being too aggressive.

"You guys are being too aggressive," Sewell said. "We're not going to back off."

The Justice Department escalated its attack on Apple, arguing in one of its filings that the company's refusal to comply with the order "appears to be based on its concern for its business model and public brand marketing strategy." It noted that Apple had previously complied with All Writs Act requests and pointed out that the company, under Cook, had increased its efforts to cast itself as a protector of customer privacy in contrast to its technology peers Google and

Facebook, which collected customer data for their advertising businesses. It pointed to Cook's recent letter and public comments depicting Apple as a white knight battling Silicon Valley's dark forces.

In the eyes of the government and Apple's tech rivals, Cook's privacy position was laced with hypocrisy. Not only had Apple helped the government unlock phones in the past, but it had begun storing some Chinese customers' data on servers in that country, where the government closely monitors its citizens. They reasoned that if privacy was as much of a human right as Cook said it was, he should defy the Chinese government as well. But in that country, where the government could restrict the sale of international brands, Cook had surrendered his moral high ground to protect sales and had explained the compromise away by saying that Apple abided by the rules of the countries where it operated. He had also cut a series of deals with Google worth an estimated $10 billion annually to make it the default search engine of iPhones, allowing Apple to benefit financially from the very data collection practices that he publicly condemned. Plus he allowed the company to encourage people to back up their iPhones to iCloud, for which Apple charged a monthly fee above a data threshold, and didn't overtly tell them that their sensitive information would be vulnerable to government subpoena. In their opinion, Cook's fortress of iPhone privacy was riddled with moneymaking peepholes.

In March, the House Judiciary Committee summoned Comey and Sewell to testify about the ongoing battle. Representative Bob Goodlatte asked Sewell whether Apple was taking a stand against a technology problem or a potential business model problem. Sewell's hands began to shake with frustration. "Every time I hear this, my blood boils," he said. "This is not a marketing issue. That's a way of demeaning the other side of the argument. We don't put up billboards that talk about our security. We don't take ads out that market our encryption. We're doing this because we think that protecting the security and the privacy of hundreds of millions of iPhone users is the right thing to do."

Sewell's strident defense astonished FBI onlookers. They knew that privacy was part of Apple's marketing playbook and were irate at what they viewed as Apple's decision to prioritize it over national security. A few years later, they would feel totally vindicated that Sewell's performance had been more theater than substance after seeing Apple blanket Las Vegas with billboards and blitz TV networks with commercials saying, "What happens on your iPhone, stays on your iPhone."

THE STALEMATE DRAGGED ON for more than a month. In late March, Sewell traveled to San Bernardino County for a trial to determine whether Apple would be forced to comply with the court order. His legal team had spent the past three days rehearsing opening arguments, preparing witnesses, and practicing answers to questions expected from the court. Sewell had spent more than six years battling Samsung, but this was easily the most important trial of his life. He buzzed with adrenaline as the legal team went through its final preparations. Then someone's phone rang. It was the court.

Sewell soon found himself on a conference call with the Justice Department and the judge. He listened as they said that the court would be suspending the hearing for two weeks because the FBI might have found another way to access the phone.

Sewell immediately called Cook, who was in Cupertino, and relayed the news. "You're not going to believe this, but we just got a two-week stay," he said.

Cook immediately began asking questions. He wanted to know everything Sewell could tell him. All Sewell could say was that the FBI had found a third party that would be able to break into the phone, and in doing so, it would undermine the government's argument that only Apple could access the device.

Cook was silent as he processed the latest twist. He listened as Sewell explained that the judge would include in her order that Apple was not in violation of the court during the postponement. For now, Apple was no longer the bad guy.

A FEW DAYS LATER, the Justice Department dropped the case. The government had cracked the terrorist's iPhone without Apple's help by paying professional hackers more than $1 million to break into it. The FBI wasn't happy about the outcome. The agency, which didn't have the capability to break into the phone itself, wanted a more permanent resolution to the dispute over accessing iPhones in criminal investigations. It wanted to be able to force Apple to unlock phones by court order when need be, but it had lost that option by finding an alternative way of gaining entry.

Cook and Sewell had been prepared to go all the way to the U.S. Supreme Court, a process that would have prolonged the debate and lengthened the amount of time the FBI accused it of putting profits ahead of public safety. It would have been disastrous for the Apple brand. Though the core issue of the case was left unresolved, the damage it inflicted on Apple was limited.

The resolution freed Cook to focus on a more pressing concern: the state of Apple's business.

During the monthlong saga, the sales of Apple's iPhone 6s, launched in September 2015, had sputtered, especially in China. The watch, which Cook had hoped would benefit from new fitness-focused marketing, wasn't contributing enough revenue to offset the falling iPhone sales. For the first time in more than a decade, the company reported a decline in quarterly sales.

On April 26, Cook faced questions about the results by worried

Wall Street analysts. He sounded weary as he broke down the details of a challenging three-month period when Apple had sold 10 million fewer iPhones than a year earlier. As he spoke, the company's share price plunged 8 percent, and its market value sank by $46 billion.

Its troubles, like tidewaters, rolled out to reveal a long-standing corporate vulnerability: Apple's future rested on a product from the past. The public expected the company to continue inventing products that were transformative and new, lest it risk entering a period of stagnation or, worse, irrelevancy. For Cook, the pressure was unrelenting.

Chapter 17

Hawaii Days

Jony Ive arrived in Sunnyvale for a scheduled review of a car project stuck in neutral. It was early 2016, and he quickly grew agitated at the lack of progress.

Software development for the fully driverless car he had imagined was lagging behind because of a lack of data and the complexity of building an autonomous system from the ground up. The hardware effort was making progress but trailing the company's ambitious timeline. There was no way a fully autonomous car would be ready by the self-imposed deadline of 2019.

Ive erupted. It was clear to everyone involved that the project was suffering under the weight of its ambitions. Ive's vision for a fully autonomous vehicle had contributed to the build-out of a massive team of programmers and sensor experts, while hardware chief Dan Riccio's focus on creating an electric vehicle had led to the development of a massive team of battery and automotive experts. The project had three leaders who appeared more focused on building out their corporate fiefdom than on moving a unified project forward. The challenges were reminiscent of the corporate infighting that plagued the Apple Watch.

Lavish spending compounded the woes. The project costs had ballooned to a staggering $1 billion a year. Project Titan leaders had hired autonomous-vehicle researchers at $10 million each and invested in the development of lasers to aim into passengers' eyes in a bid to reduce the motion sickness caused by a car's abrupt movements. Apple's R-and-D expenses mushroomed, nearly doubling to $8.1 billion by the end of 2015. It was loose change for a company with $200 billion in cash, but the engineers viewed the warehouse as the latest example of a Silicon Valley giant spending big with nothing to show for it.

In a fit of frustration, Ive pulled the entire design team off the project, so they could return their focus to other work. He no longer considered the car worth their time. Apple's ambitious project would have to keep moving without its most vaunted division.

Members of the Project Titan team shared his displeasure. The special project group had swelled to a thousand-person organization with a mutant culture that blended the grinding determination of Apple executives hell-bent on doing the impossible with the experienced skepticism of outsiders savvy about the challenges of self-driving vehicles. Old-timers and newcomers alike knew that Jobs had built the iPhone using a different approach: he had relied on a lean team of mostly existing employees in various divisions whom he had guided. But in Tim Cook's new collaborative kingdom, Apple's CEO no longer led product development, and that void continued to challenge the company's efforts to innovate.

The Ive-Riccio blowup was the culmination of growing frustration at a company unaccustomed to failure. In its wake, a February car demonstration to the board of directors was canceled. The future of the undertaking was in doubt. The race to Apple's next big thing slowed as its leaders dropped a caution flag.

WITHOUT THE THRILL of a new endeavor, the distance between the artist and his longtime workplace widened. He turned his focus away from the familiar. His interest couldn't be fully contained by continual improvements in the curvature of future iPhones or reductions in the thickness of future laptops. His nourishment came from exploring new ideas and fulfilling unexpected curiosities. Without Jobs bringing order to Apple's creative process and clarity of thought to its ambitious projects, Ive found himself adrift.

In 2016, as he looked elsewhere for fulfillment, it came as no surprise that he turned his attention to Apple's involvement in the Metropolitan Museum of Art's coming exhibit, *Manus × Machina*. Its focus on the intersection of technology and fashion satisfied his desire for discovery. His years working to imbue the Apple Watch with timeless style had roused an interest in learning more about the art of the couturier. On the first day of May, Ive exited the Carlyle Hotel on New York City's Upper East Side and headed to Fifth Avenue to trek north to the city's largest art museum for a private tour of Andrew Bolton's new exhibit. The show marked a moment of arrival for Ive and Apple. Nearly two years after showing Anna Wintour the watch at Carlyle for the first time, he was returning to the city not as an outsider from the land of technology but as an accepted contributor to the world of fashion.

Bolton greeted him inside the limestone-walled museum and led him into its Robert Lehman Wing, a triangle-shaped extension that jutted into Central Park. The skylit gallery typically displayed three hundred paintings, including works by Italian Renaissance masters. But in recent months, it had been transformed by the construction of a building within the building, with white scrims erected to convert the interior into a colorless Gothic-style cathedral.

Ive followed Bolton through the white corridors, where a series of churchlike alcoves displayed a collection of ensembles that pushed the boundaries of fashion. Bolton had found 170 examples of dresses and designs that challenged the notion that handmade clothing was superior to that made by machine. The exhibition ranged from a traditional-looking cream suit with navy trim designed by Gabrielle "Coco" Chanel to a cascading evening dress made of pink bird-of-paradise feathers designed by Yves Saint Laurent. Floating above it all was the heavenly sound of layered keyboards and soaring synthe-sizers from British musician Brian Eno's song "An Ending (Ascent)."

The tour culminated at the centerpiece of the exhibition: a circular room with a domed ceiling where Karl Lagerfeld's Chanel wedding dress was on display. Its scuba-suit body flowed into a twenty-foot-long train topped by a gold foliate pattern and hand-stitched gems. Bolton explained that the train had required 450 hours of workmanship.

"It's haute couture without the couture," he said.

Ive chuckled at the play on words. Indeed, the dress was custom-made, as couture should be, but its use of a synthetic scuba suit broke with the doctrine that haute couture must be handmade. For nearly ten minutes, Ive examined the dress, marveling at its blend of familiar and unfamiliar, formality and informality. Lagerfeld's creative vision left him inspired.

The following day, Ive returned to the museum for a press preview of the exhibit ahead of that evening's Met Gala. He strode across its marble floors and made his way to a breakfast with about a hundred reporters in the Carroll and Milton Petrie European Sculpture Court. He sought out the architect of that day's affairs, Anna Wintour, who wore her black sunglasses indoors. They chatted briefly until the press began to take seats arranged before a lectern. Soon Ive walked to a microphone and placed several printed pages beneath it. He gazed at the group gathered before him and began to explain why a designer of iPhones was opening an exhibit on fashion.

"When Anna and Andrew first talked to me about the exhibition, I was particularly intrigued that it would stimulate a conversation exploring the relationship between what is made by hand and what is made by machine—that it would challenge the preconception held by some that the former is inherently more valuable than the latter," he said.

He surveyed the room. Few of those present appreciated how unusual it was for Apple to sponsor an event. Even with riches that exceeded the wealth of many countries, it shunned the corporate tradition of lending its brand to entities that it couldn't completely control. Part of the reason Apple had broken with protocol was that the event was important to Ive. It would have been difficult for Cook to tell him no. But Ive dispensed with that factor of the sponsorship and focused the reporters on how he felt drawn to an exhibit that mirrored his own creative beliefs.

"In the design team at Apple . . . many of us believe in the poetic possibilities of the machine," he said. He added, "Our goal has always been to try to create objects that are as beautiful as they are functional; as elegant as they are useful."

He said that some contemporary designers had lost their curiosity about how things are made. "With a father who is a fabulous craftsman, I was raised with the fundamental belief that it is only when you personally work with a material with your hands, that you come to understand its true nature, its characteristics, its attributes, and I think—very importantly—its potential."

He let the words about his father linger.

"Deep care is critical to determine authentic, successful design," he said before noting the beauty of the fashion on display. "Regardless of whether it has been made by hand or by machine, it is creation led by great consideration, rather than driven by a preoccupation with schedule or price."

The audience applauded as he picked up his notes and returned to his seat. The speech put into words his philosophy of design. In his view, just as it had been for Jobs, art should lead commerce, not the other way around.

After the press preview, Bolton gave a tour to Laurene Powell Jobs and Tim Cook. As they walked from one ensemble to the next, Jobs asked questions about the displays while Cook shuffled along silently with his eyes scanning the white walls and alcoves of the pop-up cathedral. During Apple's work on its new headquarters, Cook had developed an engineer's interest in architecture and asked Bolton how he had constructed the building within a building around them.

That evening, Ive and Cook donned tuxedos and white ties and prepared to attend the Met Gala. They expected a spectacle unlike any other. Models clad in futuristic chrome masks would usher some of the world's most influential artists, actors, executives, and politicians through an archway made of three hundred thousand fuchsia roses to candlelit tables topped with crystal glassware. Eventually, Ive and Cook would arrive at a table near the front of the room where they would be seated close enough to the stage to almost shake the hand of the Weeknd as he performed his R&B ballad "Tell Your Friends."

Before all of that, though, they had to navigate the pandemonium of paparazzi.

They arrived at the Met to find a scene chaotic with photographers two rows deep cramped behind green barricades. The cameramen bellowed at passing stars such as Beyoncé and Nicole Kidman. A cacophony of clicking cameras rose above their shouts as the divinely dressed celebrities posed.

Cook walked through the commotion alongside Laurene Powell Jobs. Midway up the red carpet, he stopped to speak with Uber CEO Travis Kalanick. Both Silicon Valley executives looked confused as photographers called for them to face the cameras and smile. They looked in different directions before grasping what was expected of them during their unfamiliar turn as socialites.

Ive made a quiet entrance and posed for photographs alone. He tucked his hands into his pockets and raised his stubbled chin toward the photographers. He didn't smile, but his eyes were wide with enthusiasm as he confidently stood on the red carpet. The son of the Chingford craftsman had arrived on society's biggest stage.

AT INFINITE LOOP, Ive became a ghost. He resisted returning to a place where he could get roped into meetings and looped into reviews of updates to a line of products that had preoccupied him for years.

The hardware and software he worked on had settled into a rinse-and-repeat phase. The iPhone continued to drive the majority of Apple's sales. The company attempted to excite customers by adding new colors, faster chips, and better cameras, yet its shape and features remained roughly the same. The iPad, Apple Watch, and MacBook stuck with their basic form factors as well. Future leaps in design would require advances in engineering. For some, the predicament prompted boredom.

Rather than wallow in the monotony of Cupertino, Ive often met members of the design team in San Francisco for briefings on their work. He would reserve the Musto Bar at his social club, the Battery, and hold design reviews inside the library-themed speakeasy-style space.

The intermittent gatherings changed the group's cadence. For years, they had met three times a week at the same kitchen table in the studio. They had reviewed the latest update to every ongoing product and discussed how the prototypes could be improved. The regularity of those sessions had allowed them to make incremental adjustments in a span of days to refine their work, like sculptors chiseling away at raw marble. When he had been alive, Jobs had guided those changes. Later, Ive had approved each adjustment and refinement.

Initially, the design team had done well in Ive's absence. They had swiftly defined the direction of the tenth-anniversary iPhone, due in 2017. With the planned facial recognition system set to replace the home button, the group quickly coalesced around creating a nearly full-screen display that included an indention at the top where they could inlay the camera system that would recognize a user's face and unlock the phone. Form followed function and won Ive's approval.

But his absence from the day-to-day routine could create challenges, especially because he wanted to maintain control over product direction and insisted on having final approval for products. He got his wish to step back, but he struggled to actually let go. The design team and engineers would work all month and make decisions, then wait for him to show up for a few days every few months to approve them. The dysfunction created discord for the harmonious design team.

After Cook had approved making a smart speaker, the design team had worked hard to define how it should look. The speaker, known as the HomePod, was a cylinder the size of a coffeepot topped by a dark cap with touch screen volume controls. Late in its development, product designers recall Ive reviewing the design and insisting that the edges of the double-layered, diamond-embossed fabric fit seamlessly under the speaker's display cap. A member of the textile team poured hours into reworking the design to Ive's specification. The difficulties reminded some members of the design team of how Apple had struggled to adjust its operations after Jobs's death. The difference was that everyone knew that Jobs would never return, whereas Ive could show up at any moment.

Sometimes word would spread through the studio that he was unexpectedly coming into the office. Junior staff compared it to old footage of the 1920s stock market crash with papers being tossed into the air and people scurrying about the usually calm studio in a rush to get materials and prototypes ready before Ive's arrival.

Some of them took to calling that period "the Hawaii days." With Ive seldom around, it was more romantic to assume that he was spending most of his time at his Kauai estate, seated beside an outdoor pool and surrounded by palm trees, than imagine him an hour up the interstate in nearby San Francisco.

Similar inefficiencies tormented the software designers. Staff considered the approval of software concepts by Alan Dye, whom Ive had tapped to lead the division, as temporary authorizations. Ultimately, they wanted Ive's assessment.

The dynamic led everyone to look forward to a monthly "design week," when Ive promised to spend an entire workweek in the studio, reviewing and discussing their work. The problem: he rarely, if ever, showed up.

Ahead of one design week in late 2016, Johnnie Manzari, who was in charge of Apple's Photos app, stood before more than a dozen eleven-by-seventeen-inch images of changes he planned to pitch. He was reviewing his work when word trickled through the studio that Ive wasn't going to come.

"What am I going to do now?" Manzari said to a colleague with disappointment.

It wasn't that Manzari or anyone else on the team needed Ive to make every decision, but most of them craved time with him. He had some of the world's most refined eyes, and he always challenged them to do better.

IVE TURNED OUTSIDE APPLE for creative nourishment. That November, his Gulfstream jet arrived in England, where he was scheduled to work with Marc Newson on the design of an immersive

Christmas tree installation at his favorite London hotel, Claridge's in Mayfair.

The old-fashioned institution had an almost mythical reputation. Founded in 1812, it was known as an extension of Buckingham Palace because it regularly hosted members of the royal family. It had a high-ceilinged foyer and walls lined with art deco mirrors. Each year, the hotel tapped a leading creative person to transform the lobby with holiday cheer. The previous year, Christopher Bailey of Burberry had designed a tree of golden umbrellas.

Ive and Newson were among the first industrial designers to take on the tradition. As they discussed how to approach the design, Ive imagined a room that would testify to his lifelong pursuit of authenticity and simplicity. He and Newson wanted to bring a forest of tranquility to the lobby, complete with tufts of snow and groves of birch trees. They imagined a lighting system that would mirror the hours of the day, warming the wintry scene with bright light at noon before declining and fading into twinkling starlight at night.

Just before the holidays, their vision came to life as slender silver birch trees were placed in the hotel's entryway behind towering green pine trees. The trees stood before a wallpaper backdrop of a forest that appeared to stretch on infinitely. Tucked in front of it all was an evergreen sapling.

Ive told people that the slender tree symbolized the future, standing beneath a spotlight, alone.

Chapter 18

Smoke

Samsung was perfecting the art of aggravation.

By 2016, the South Korean company had barreled past Apple's lawsuit and allegations that it had copied the iPhone. It had appealed court verdicts in Apple's favor and continued to dominate the smartphone market. Its Galaxy lineup, a series of premium phones with vibrant displays, had earned it praise from tech reviewers with new features such as bigger screens and better cameras.

Samsung also sought an edge by preempting Apple's product announcements. In early August, a month before the iPhone 7's release, Samsung booked New York City's Hammerstein Ballroom and invited press from around the world to gather for the unveiling of its newest Galaxy. The event aped the presentations popularized by Apple with D. J. Koh, the head of Samsung's mobile communications business, taking the stage to show off a new device with bright lights, slick videos, and a flourish of instrumental music. Koh looked like a buttoned-up Steve Jobs in a blue sport coat and cream trousers. He joked that he felt like George Clooney. The star of the show was the Galaxy Note 7, the world's first phone capable of verifying a user's identity by scanning his or her eye.

The so-called iris scanner dampened Apple's aura of originality, and the phone itself posed a serious threat to iPhone sales, at least for a few days. Soon after it went on sale, a customer named Joni Barwick bought one. A marketer based in Marion, Illinois, she thought the phone's giant display would make it easier for her to evaluate advertising materials and access an array of Google products she used for work. The phone was a multitasking powerhouse. She used it throughout the day and plugged it in beside her bed to recharge each night.

One morning around three o'clock, she awoke to the crackle of a firework sparkler. She turned in bed to see orange and red flames shooting across her bedside table from her Galaxy phone. Acrid smoke filled the air. Her husband, John, grabbed the phone by its leather case and rushed it downstairs to the kitchen. He dropped it onto the counter, slipped on oven mitts, and carried the phone toward the door. Melting plastic dropped onto the floor as he dashed toward the backyard, afraid that the house was going to catch on fire.

After the blaze burned out, John called Samsung to report the incident. He estimated that the destruction to the nightstand, hardwood floor, and carpets totaled $9,000. The company said it would call him back within twenty-four hours. It never did.

Meanwhile, the fire spread. Around the world, Galaxy phones began combusting. U.S. consumer protection authorities received ninety-two reports of burning Note 7s in the weeks after its release. Though the cause of the fires was unclear, experts assumed the battery was the culprit. The rechargeable lithium-ion batteries in smartphones have a razor-thin separation between their positive and negative components. When it is breached, a battery can explode. The fires posed enough of a safety risk that Samsung said it planned to delay shipments of its new smartphone. It knew the problem was radioactive.

TIM COOK'S SUPPLY CHAIN MASTERY made it impossible for him to take joy from Samsung's woes. Worry, not celebration, hung over the first Monday executive meeting after Samsung's fires. Cook wanted to know what had caused the combustion of the phones and whether iPhones were susceptible to the same problem. The company was days away from launching its new iPhone 7. He had time to head off a similarly embarrassing problem if he was armed with the right information.

With his typical precision, he questioned a team of battery specialist and supply-chain experts to determine if the iPhone was vulnerable. They explained that Samsung relied on a Chinese supplier named ATL for about 30 percent of its smartphone batteries, with the other 70 percent of the supply coming from a Samsung subsidiary, SDI. Apple relied on ATL to provide batteries for some of its iPhones but didn't use SDI. Fortunately for Apple, investigations determined that the problem for Samsung was most likely coming from SDI. Its iPhones were safe.

For Samsung, the fires raged on. The Note 7 turned out to be especially susceptible to combustion on airplanes. The low cabin pressure caused phones to catch fire on several flights. As Samsung moved to recall the phone, the Federal Aviation Administration moved to block passengers from turning on the phones during flights or stowing them in checked baggage. Flight attendants began singling Samsung out before takeoff. "The use of a Samsung Note 7 is prohibited," they would say. "The device should be switched off completely."

The hazard warnings brought smirks of delight to the faces of traveling Apple executives. Confident that their devices were safe, they could sit back and enjoy knowing that they were reaching millions of people a day with free advertising. Each announcement doubled as a reminder to buy an iPhone and avoid being branded as dangerous.

REASSURED, Cook turned his attention to Apple's upcoming event in San Francisco. He posted a photo to Twitter at 7:37 A.M. from outside the Bill Graham Civic Auditorium. The rising sun cast shadows on the center's granite facade. A giant Apple logo, towering fifteen feet above the sidewalk, emitted a white glow from the center of an arched window.

"Big day!" Cook wrote.

When the CEO took the stage, he walked with the measured calm of a conqueror. Cook would never match Steve Jobs's ferocious showmanship, but by now, after five years at the head of the company, he was developing a cool, collected stage presence of his own, with an earnest, less polished confidence that almost qualified as bravado. Gazing out over the throngs of the faithful, he reeled off a litany of proofs that Apple ruled the world. He boasted about iPhone sales, crowed about the nearly 20 million customers who had subscribed to Apple Music in its first year, and bragged about how the App Store was generating double the revenue of its nearest competitor. Then he turned the audience's attention to the company's newest triumph, the Apple Watch.

What Cook didn't mention was that in the year and a half since its release, sales of the Apple Watch still lagged behind the original projections. The company had sold an estimated 12 million smartwatches in the first year, more than the original iPhone after its launch but several million fewer units than the iPad after its debut. Given the way Apple's customer base had swelled, many on Wall Street viewed the Apple Watch as a disappointment, especially because the company needed new businesses to offset its declining iPhone sales. The estimated $6 billion in revenue the watch generated was a poor substitute for the nearly $20 billion decline in iPhone revenue over the past year.

Onstage, Cook ignored all those facts. Instead, he went on with Apple's preferred version of reality, focusing on the fact that the company had finished the year second in the world in watch industry revenue behind only Rolex. He declined to disclose numbers and instead emphasized that the smartwatch was number one in customer satisfaction. He then turned the stage over to operations chief Jeff Williams, who unveiled the Apple Watch Series 2, a device that looked exactly like its predecessor but had a series of new features, including water resistance, so that it could be worn for swimming or surfing, and built-in GPS, so that runners and walkers could accurately track their mileage and pace. Williams then announced that Apple had partnered with Nike to make a co-branded watch that would include a special perforated band and a Nike running app. Nearly a year after Apple's top executives had huddled to discuss how to salvage the watch, they were making their planned shift from fashion to fitness.

Though Cook had worried in previous months about the prospects for the forthcoming iPhones, he betrayed none of those concerns when he strode back onstage to introduce them. The iPhone franchise might be facing challenges, he said, but he was confident about its future. "There's a reason why you see so many iPhones everywhere you look," he said. "We've now sold over a billion of them. This makes iPhone the best-selling product of its kind in the history of the world."

He then introduced the newest models, the iPhone 7 and 7 Plus. The phones looked just like their predecessors, the iPhone 6 and 6s, with a few minor changes, including the addition of a second camera on the rear of the Plus and the removal of a headphone jack on both phones. The second camera, combined with a new chip and software, powered a new photography feature called portrait mode. It instantly combined two separate images to create a focused portrait of a person against a blurred background. The subtle but significant leap in camera photography drew applause. The elimination of the headphone jack elicited silence. It was a key component of every smartphone, one that people used all the time to listen to music or place calls. Eliminating it raised the obvious question: Why?

"The reason to move on," Phil Schiller said after taking the stage, "it really comes down to one word: courage."

It was an audacious claim that might have stirred cheers coming from Jobs but stirred snickers coming from one of his successors.

Apple's next new thing was wireless earbuds called AirPods, Schiller said. Ive's design team had first come up with the idea during brainstorm sessions in 2013. The watch could free its wearers from their phones only if they had a way to connect to it wirelessly to make calls or listen to music. Out of that necessity had come the pursuit of a new product.

Creating the wireless earbuds had been an odyssey. In meeting after meeting, engineers and battery experts had worked with the designers to make something that would be small and look acceptable in people's ears. A combination of battery power constraints and the limits of Bluetooth technology, which enabled the wireless connection, led them to explore a design in which the buds connected through a cable to a battery that hung behind the head. Dissatisfied by the clunky design, the design head, Rico Zorkendorfer, and the engineers pushed to create smaller and smaller batteries. The quest led to the acquisition of a small startup called Passif Semiconductor, led by two music-obsessed hardware engineers named Ben Cook and Axel Berny, who had dreamed for years of truly wireless headphones. They had developed a chip that would consume less power and separately receive a signal in two independent headphones, as if they were a single unit. After acquiring the company, Zorkendorfer and the engineering team overhauled their ugly prototype with designs inspired by Passif. They cut the cords connecting the two earbuds and created a case that recharged the independent headset. Knowing that the case would need to fit into people's pockets, Zorkendorfer sketched a slim case the size of a Zippo lighter. He then collaborated with the engineers to create a magnetic closing mechanism that flipped open and snapped shut with a satisfying click.

Into the case they slipped two cordless earbuds as white as Q-tips. A new chip, the W1, helped justify the earbuds' premium price tag, $159, by enabling them to play a distinct audio feed to each ear. The earbuds connected to an iPhone the moment they were removed from their case with what Schiller called it-just-works magic.

"That," he said, "is the breakthrough."

THE SALES PITCH FELL FLAT. Within days, Apple's AirPods were being mocked online.

Schiller's comment about courage was ridiculed. He had oversimplified something Jobs had once said, eliminating a nuance that the CEO would have used to characterize a tough product decision such as eliminating the headphone jack. A year before his death, Jobs had highlighted the company's history of jettisoning some popular technologies such as floppy discs in favor of emerging ones such as CD-ROM drives. He believed that customers wanted Apple to make those choices for them and would reward the company if it was right by buying its products. "People call us crazy," he recalled. "We have at least the courage of our convictions to say 'We don't think this is what makes a great product. We're going to leave it out.'"

The comedy site CollegeHumor created an iPhone 7 parody video that featured an actor with a British accent faking a Jony Ive voiceover explaining Apple's product changes. "We've done something that at first seems counterintuitive and then is," he said. "We've made it worse."

"We removed the headphone jack," said an actor playing Tim Cook, complete with glasses and gray hair. "That's all. That's the newness right there. It's just the lack of a thing that was there and now it's not. It's gone. It's not there anymore. Ta-dah!"

The comedian Conan O'Brien poked fun at AirPods. In a reference to the old iPod ads, he showed silhouettes of people with white AirPods in their ears dancing to high-energy music against a bright yellow background. As they shook their heads, the $159 wireless earbuds exploded out of their ears and fell down a street drain, forcing them to buy new ones. The ad ended with the tagline "Apple AirPods. Wireless. Expensive. Lost."

Days after the launch, Cook sat down with Robin Roberts of ABC, "Daddy's network," to dispel fears about the pricey new gadget. He said that he had worn them while on treadmills, walking, using the phone, and listening to music and had never had a problem. "I have never personally had them fall out since I've been using them," he said, defensively.

Cook stopped short of telling viewers the truth: the much ballyhooed AirPods weren't yet ready. Back in Cupertino, Apple's engineers were still trying to get the antennas that connected to the phone to work. Its software and hardware teams feuded as they tried to determine the cause of the issue. Each used different testing processes to try and improve the antennas' performance. The dispute was emblematic of the problems corroding Apple's previously well-oiled product development process. In accordance with its commitment to secrecy, the separate divisions withheld information from each other. Jobs had encouraged that and brought together each group's contribution into a single product. But Cook declined to get involved and expected division leaders to fill Jobs's role as integrators.

Reality said otherwise. When it came time to ship AirPods before the Christmas shopping season, engineering and manufacturing problems persisted, and Apple missed out on millions of dollars in sales. The loss led its human resources staff to do an autopsy of the project and push a new concept that built on Apple's legacy "Think Different" campaign. They encouraged employees to prioritize the collective over the individual and came up with a new slogan: "Different. Together."

THE PUBLIC BLOWBACK against the iPhone 7 and AirPods was overshadowed by the ongoing turmoil engulfing Apple's biggest competitor. Samsung recalled 2.5 million phones with faulty batteries and replaced them with phones with batteries from a different supplier. Then the replacement phones began overheating, and Samsung was forced to issue a second recall, an embarrassment that led some inside the company to darkly joke that the Note 7 was a topic too hot to touch. The mistakes cost Samsung at least $5 billion plus tremendous damage to its reputation.

For months, flights across the United States began with announcements that lumped Samsung's phone together with other airplane hazards. Eventually, Samsung pulled the Note 7 off the market. In a span of two months, the latest Galaxy went from anticipated to unavailable.

Cook had caught the luckiest break from his biggest competitor. Twice.

In a year when Apple's own smartphone lineup was weak, Apple had forged ahead and anticipated a downturn in sales. It had capped production of the iPhone 7 because it expected soft demand for a phone that looked like its two predecessors. Abiding by Cook's maxim that inventory is evil, its executives tried to limit the risk that they would produce more than they could sell. Upon its release, though, the iPhone 7 Plus flew off the shelf, with customers clamoring for its dual rear camera and portrait mode. It took months to catch up with demand.

Cook had moderated customers' expectations for the iPhone 7, but Samsung's unforced error helped Apple's latest release become the world's best-selling smartphone. The Galaxy Note 7 wasn't in the top five. Samsung's smartphone business would struggle to recover. The threat it had posed to Apple's business faded as the bad karma of copying products and events caught up with the company.

WITH IPHONE 7 SALES exceeding expectations, Apple's share price began an upward climb. The company's stock price had been depressed for years as investors worried that its core iPhone business was running out of juice. The stock's disappointing performance attracted the attention of a leading value investor.

Ted Weschler, an investment manager at Warren Buffett's investment company, Berkshire Hathaway, had been following Apple for several years. He considered the iPhone to be more effective than Coca-Cola at creating a loyal customer base. Once people bought an iPhone, they seldom switched because they didn't want to have to learn a new operating system. Apple's lock on those customers meant that it could charge them to store photos on iCloud and listen to songs on Apple Music, and collect fees on the apps they purchased. He recognized that Cook was taking the ecosystem that Jobs had created and wringing more revenue out of it, turning the iPhone into a subscription-based business that would generate cash for years to come. He also liked that Cook continued to repurchase shares, something he couldn't fathom Jobs doing. He quietly accumulated a $1 billion stake in Apple when the price was about $27 a share.

On a visit to New York, Weschler talked with Berkshire Hathaway board member David "Sandy" Gottesman about his interest in

Apple. The ninety-year-old Gottesman had become a billionaire after founding the investment advisory firm First Manhattan Co. and
befriending Buffett. He had been an investor in Apple for years and
loved its products. He told Weschler that he took his iPhone with
him everywhere and had been devastated when it slipped out of his
pocket in the back of a taxi.

"I felt like I lost a piece of my soul," he said.

When Weschler relayed the story to Buffett, his boss perked up.
Buffett was struck that a friend his age felt that way about a piece
of technology and decided to dig into Apple's business. The Oracle of Omaha, as he is known, had a strong aversion to investing in
technology companies. He made investments in businesses that he
understood, and he considered many tech business models as foreign.
He also had a poor track record with the industry, most notably with
a 2011 investment in IBM that had performed poorly. But after hearing Gottesman's story, he began to pay more attention to the iPhones
being used around him.

Buffett wondered what it would take for an iPhone owner to
switch from Apple to Samsung, the battery-plagued brand. During
his Sunday trips to Dairy Queen with his grandchildren, he noticed
that they were always absorbed in their phones. He realized that
Weschler was right: the iPhone wasn't tech, it was a modern-day
Kraft Macaroni & Cheese. The product had a grip on users and
popular culture that should endure for years. At his direction, Berkshire added to its position in Apple, bringing its total investment
to $7 billion and eventually making Apple one of the firm's largest
holdings.

When news of the share purchases hit the press, the reaction at
Infinite Loop was mixed. Product engineers worried that Berkshire's
stake in Apple would make the company cautious. There would be no
more risks, no more crazy ones, no more round pegs in square holes.

Too much wealth could be lost. They feared that Cupertino would become the Boca Raton of engineering.

But Cook was giddy. He viewed Berkshire's stake as the ultimate validation of his leadership. He described having Buffett as a shareholder as an honor and privilege. It suggested that the world's savviest investor saw Apple as Cook did, as a technology company that rivaled Coca-Cola in its consumer appeal.

Wall Street sided with Cook. Buffett had turned $174,000 into $80 billion over four decades with an investment strategy that had focused on long-term value. Main Street investors followed and copied his every move in hopes of enjoying similar success. Many of them snatched up positions in Apple, rejuvenating its stock.

Cook marveled as Apple's share price surged. He told CNBC that Buffett's investment had been the highest compliment. "And I don't mean that in a lighthearted kind of way," he said. "I mean, wow, it's Warren Buffett is investing in the company."

WITH APPLE'S SHARE PRICE on the rise, Cook was free to focus his attention elsewhere.

In a bid to restart the company's car effort, he encouraged the return of its former top hardware executive, Bob Mansfield. Mansfield, who had retired, spent much of the summer reviewing the work under way before scheduling an off-site meeting.

Early that fall, hundreds of people working on the project piled onto charter buses that ferried them to a Silicon Valley hotel. They filed into an expansive meeting room where Mansfield was waiting. The heavyset engineer with close-cropped hair had a background in semiconductors and had risen to the top of Apple by delivering breakthroughs on products such as the MacBook Air. In Apple's hierarchical structure, he commanded respect.

Mansfield addressed the group bluntly, making clear what most people in the room already knew: the project was a mess. Though he acknowledged that he didn't fully appreciate the technical challenges of autonomous vehicles, he planned to use a Thor's hammer approach to get the work back on track. He announced that there would be layoffs—some two hundred of the staff would be let go—as he streamlined the operation and shifted its focus. It was clear to him that Apple had no business plowing ahead with the construction of a car until it determined how to structure the underlying software that would enable it to navigate roads without a driver.

You're doing too much, he said. From that point forward, he said, they would be shifting their emphasis to developing an operating system that could make a self-driving car reality. Everyone in the room knew that it would be a years-long undertaking.

The goal of a 2019 launch was abandoned. Across the company, the initiative in Sunnyvale shifted from being viewed as Apple's hottest project to a never-ending research experiment.

COOK'S FIVE-YEAR EXPANSION of the Apple empire also hit a wall. Despite Buffett's vote of confidence and the improved sales of the iPhone 7, the company continued to struggle in China. The iPhone business, which had tripled in size after Cook's China Mobile deal, gave up some of its gains. Sales plunged 17 percent from their peak as a customer base obsessed with status symbols put off buying the 6s and 7 phones because they looked nearly identical to the iPhone 6 launched two years earlier. The very market that had fueled Apple's expansion was contributing to its contraction.

The problem wasn't with Chinese consumers alone. The country's government abruptly shut down sales of iTunes movies and books, closing off critical aspects of Apple's growing services business. Staff based in China had tried to warn Cook and Apple's leadership that the country's autocratic leader, Xi Jinping, was starting to take a harder line with Western companies. He was cracking down on Western ideology and favoring local tech giants he could control, such as Huawei and Tencent. Shortly after the iTunes shutdown, Xi met with some of the local tech leaders and told them that China must ensure that online content created a healthy, positive culture.

The partial iTunes suspension unsettled Cook. He had spent a decade building a business in China and was now being told by his people on the ground that his vision for the future there may be unrealistic. Advisers warned him that the government was preparing to attack Apple. The Chinese Communist Party was known to punish foreign companies it deemed too large or powerful. Often it did so with an invisible hand by unleashing China's *shuijun,* or "water army," a group of government-backed influencers who shaped public opinion about brands on Chinese social media. Equally important, China contained a workforce of more than 3 million people who toiled to crank out Macs, iPads, and iPhones. At Cook's direction, Apple had concentrated its manufacturing in the country and needed the government's support to make and export products.

To avert a larger move against Apple's business, Cook worked with the company's policy team to improve its relations with the Chinese government. They outlined a strategy that shifted the way the company talked about its business there. Instead of emphasizing the number of phones it sold, its local messaging began to highlight the number of developers it supported and people it indirectly employed. He began to meet with some of those developers when he visited, shining a personal spotlight on the way Apple was knitted into China's economy.

In October, Cook flew to Shenzhen to announce plans to create a $45 million research and development center there. The public affairs team had suggested offering it as an olive branch to the Chinese leaders. They told Cook that it would convey that Apple was committed to China and would assist in its efforts to become a more technologically advanced nation. He met with Chinese premier Li Keqiang and other leaders during the trip.

The investment and visit bought Apple goodwill in its second largest market, which Cook desperately needed. Though he didn't know it, China's place in the global economy was on the brink of unimaginable disruption.

BACK IN THE UNITED STATES, the Republican nominee for president, Donald Trump, was whipping up populist sentiment across the country, railing against companies for outsourcing production to China, and promising to bring jobs back to the United States.

Cook was target number one.

"We're going to get Apple to start building their damn computers and things in this country instead of in other countries," Trump told a packed arena during a campaign appearance at Liberty University in Lynchburg, Virginia. "We're going to make America great again!"

For Trump, the country's road to greatness ran through Cupertino. He promised his supporters that if he was elected, he would slap a 45 percent tariff on imports from China, a levy that would cripple the iPhone business.

Cook was a political shape-shifter. Like many other things in his life, his allegiance was hard to pin down. He had registered to vote as a Republican in the 1990s but donated to both Democrats and Republicans. In the face of Trump's campaign attacks, he threw his weight behind Hillary Clinton, cohosting a fundraiser for her in Los Altos, California. He contributed $268,500 to her campaign. She was the barrier between Apple and a four-year assault on the efficient manufacturing machine he'd built in China.

On Election Day, as the first results rolled in, he was at the office. Polls all week had heavily favored Clinton, with the *New York Times* giving her an 85 percent chance of winning. Trump campaign officials told CNN that it would take a miracle for him to come out on top. But as darkness fell in California, the forecasts began to shift, with more votes than expected coming in for Trump in North Carolina and Florida. The Trump campaign also started to tell networks that it had the edge over Clinton in Pennsylvania, a state that hadn't gone Republican since 1988. Trump was declared the victor in Ohio as well. By 9:00 P.M. Pacific time, the race appeared to be over, and Cook, like much of the rest of America, was stunned.

A force for predictability, steadiness, and calm, Apple's CEO now faced an uncertain future with an unpredictable president. The ultimate antagonist, a king of chaos, was entering the White House.

Chapter 19
The Jony 50

The summons went out in January. With anticipation mounting on Wall Street for a tenth-anniversary iPhone, Jony Ive invited the company's top software designers to gather at the Battery for a product review.

Around 11:00 A.M., a group of twenty designers and some Apple security officials carried black Pelican cases containing unreleased iPhones to the exclusive San Francisco social club's fifth-floor penthouse. They exited into a 6,200-square-foot room with exposed steel beams and floor-to-ceiling windows that offered views of the hulking Bay Bridge, arching four and a half miles from San Francisco to Oakland. Opposite the windows, a gas fireplace sat inside a slate-colored wall. The room was minimalist and modern with tasteful touches that would have appealed to Ive. It featured glass tables hemmed in by translucent orange chairs reminiscent of Apple's candy-colored iMacs.

The group began spreading out eleven-by-seventeen-inch printouts of design ideas they had been waiting for weeks to share with Ive. They were in the process of redefining how people would navigate an iPhone that eliminated the home button in favor of a full-screen display. Without a button to press, Ive wanted to place a slender white bar at the bottom of the screen that people would swipe upward to reach the iPhone's dashboard. Its introduction left a bevy of other decisions to be made, including about the appearance of the lock and home screens, as well as how to display video on a device with a chinlike notch. The wait for Ive's direction had led some members of the team to nostalgically recall their weekly meetings with Steve Jobs and Scott Forstall, who had regularly provided guidance during an iterative design process. In its stead, they had adjusted to infrequent meetings with Ive, whose reviews could stop a design cold or point it in an entirely new direction.

After arranging their work, the designers took seats on nearby couches and waited. As the clock ticked toward 1:00 P.M. without Ive appearing, some of the designers grew hungry and grabbed plates of sushi provided for lunch. Deputy software design chief Alan Dye assured the team that Ive was still coming and would be there soon. Some stewed as they pecked on their laptops and thumbed through their phones, waiting. Many wondered, *How did it come to this?*

Just before 2:00 P.M., nearly three hours after the meeting had been scheduled to begin, Ive exited the elevator and saw the team spread across couches in the penthouse's lounge. He didn't apologize or comment on his tardiness. Instead, he walked toward the tables holding the printouts and began to examine the work. Dye led him through each project like a master of ceremonies.

As Ive slowly considered each design, he offered feedback, but he didn't make any final decisions. He wanted more time to think. The three hours everyone had spent waiting for him to start the meeting were only the precursor to a longer wait to come.

IN THE MONTHS THAT FOLLOWED, Ive joined Apple's top leaders for the company's annual corporate retreat. He had been to dozens of the multiday meetings over the years and understood as well as anyone else how the mood of the gathering was colored by future products and current sales. That year, amid the car project setback and the iPhone sales downturn, unease hovered over the conference.

As the presentations began, Ive stood outside to catch some fresh air. He was loitering near the entrance of the five-star hotel when a recent hire named Peter Stern stepped before the group and began to offer an update on the company's iCloud service. Stern had joined the company a few months earlier from Time Warner Cable, where his success at growing the subscriber base had led Apple to recruit him for a role developing the company's subscription offerings. In the wake of the Apple Music release, Cook wanted to find more ways to squeeze sales out of the iPhone from services the company created. Stern explained why.

Standing before everyone, he clicked to an image of an X-shaped chart that showed Apple's hardware profit margins declining and its services profit margins rising. His message to the attendees was that Apple's legacy business, the one most associated with Ive, had become a drag on its performance as the costs of adding more cameras and components to the iPhone had risen while the phone's sales price remained flat. Meanwhile, services such as iCloud subscriptions were lifting the company's bottom line because they had relatively fixed costs and more and more people were signing up to pay monthly fees for them. Stern's job was to find a way to generate more of that easy money.

The presentation alarmed some people in the audience. It depicted a future in which Ive—and the company's business as a product maker—would matter less and Cook's increasing emphasis on services—such as Apple Music and iCloud—would matter more.

AS THE COMPANY around them changed, Ive had begun contending with growing unrest within his team. The elevation of Richard

Howarth to vice president of design had created tension as Howarth had gone from ordinary member to leader of the close-knit group of about twenty. Ive had spent more than a decade working under Jobs to become one of the most powerful people at the company. His word was final. But Howarth didn't have that standing. Ive's absence created a vacuum and other leaders at the company tried to fill it. For all Howarth's gifts as a designer, he could become defensive and passionate in this turbulent time when engineers challenged him. Such outbursts increased as operationally minded executives and engineers with seniority sought to increase their influence over designs.

The team he led had spent a year on a complete redesign of the iPad. Designer Danny Coster had led the effort. The Kiwi had been instrumental to the creation of the translucent iMac and had helped birth the name "Bondi blue" after the beach in Australia. He had developed a refreshed iPad with more refined curves and a lighter body that felt natural in people's hands. Some of the product designers working on it considered it so elegant that they said it would be the first model they would gladly purchase at retail prices. However, Apple's operations team determined that making the iPad would require building several new features from scratch. The first-time costs of new machinery, a new logic board, and other components would amount to billions of dollars, an investment that would take years to recoup. Those so-called nonrecurring engineering costs led Apple's business division to suspend the iPad.

Such cost-conscious decisions frustrated some members of the product team. In the wake of it, Coster decided to leave Apple and join the action camera company GoPro as the head of design. It was the first high-profile exit of one of Apple's core design team members. He had been at Apple since 1994. His would not be the design team's last defection.

As work on the HomePod concluded, the lead designer on the

project, Chris Stringer, decided that he was ready to move on from Apple. He had joined the company in 1995 and had reached the point where he was no longer as energized by the work as he had been over his last two decades of service. He approached Ive in February to advise him of his plans to leave. In addition to his fading interest, Stringer found the HomePod dissatisfying because Apple treated it as a hobby, depriving it of the cross-division focus it lavished on core products such as the iPhone and iPad. Its development limped along, partly because Apple's digital assistant, Siri, couldn't order products, food, or an Uber as Amazon's rival Echo could. In the back of his mind, he imagined the possibilities of a more sophisticated speaker. It was a project he knew Apple would never pursue. Speakers would never clear Tim Cook's threshold and become a $10 billion business, so Stringer eventually set up his own audio company.

Like many of his peers, Stringer could afford to leave or retire. The company had begun granting the fifty-two-year-old shares in Apple when its stock price was about $1, a figure that had appreciated over the years, particularly under Cook, to more than $133. The company's success had made him a multimillionaire, with homes in the Bay Area, on Lake Tahoe, and in southern California. He could "vest in peace," as they said at Apple, a joking reference to the growing number of early retirees who colleagues called VIPs.

AMID THE DEVELOPMENT OF the tenth-anniversary iPhone, a similar restlessness pervaded the software design team. Imran Chaudhri, one of the top software designers, started plotting his own exit. The British American, who shaved his head and dressed in black T-shirts and jeans, had joined Apple as an intern in 1995 and solidified his role at the company as part of the team that had developed the iPhone's multitouch technology. He had spent years working under Scott Forstall before being tapped by Ive to join a small group that developed the Apple Watch interface. He also presented at one of the company's recent developer keynotes. Over time, he began to wrestle with how the company seemed to make fewer innovative leaps.

Feeling somewhat creatively unfulfilled, he decided the time had come to leave Apple. Following common practice at the company, he told Ive and Alan Dye that he planned to depart in a few months after he collected equity shares that he was due to earn as part of his compensation. Such an arrangement had become more common at Apple under Tim Cook. It was a contrast to Steve Jobs, who had punished deserters, refusing to rehire them and treating their departure like a scorned lover would.

A month before he was set to leave, Chaudhri wrote an email to colleagues announcing his planned departure. He told them that he would not be in the design studio but available by email until his last day. He reminded them of what they had done together at Apple to make products to empower people and told them that it was an honor to work alongside many of them. He was fond of a line from the Persian poet Rumi, who said, "When you do things from your soul, you feel a river moving in you, a joy." Playing off that line, Imran wrote, "Sadly, rivers dry out, and when they do, you look for a new one."

The email alarmed Ive and Dye. They feared that the message Chaudhri sent could be interpreted to mean that Apple's best days had passed. Its river had run dry. It was one thing for outsiders to say that the company was no longer innovative, but another thing altogether for that critique to come from someone who had helped birth multitouch technology for the iPhone. They worried it would poison morale and moved to contain the damage.

Shortly after the email, Dye fired Chaudhri.

The move had crushing financial ramifications. Chaudhri would no longer receive his shares. Stung, he complained to friends about the dismissal, telling them that Ive and Dye misunderstood his comment about the river. He explained to those people that the email was a personal reflection on his own lack of joy, not a comment on Apple.

Yet, inside a company wrestling with its own insecurities, it was interpreted as a personal attack.

THE CREATIVE BRAIN DRAIN at Apple was disappointing but unsurprising to Ive. He shared many of the same frustrations as those voiced by his outgoing colleagues. He knew that their decisions hadn't been easy. Two decades of devotion to a company makes it a part of a person's identity. It takes fortitude to walk away from that, fortitude that Ive had been unable to muster.

In March 2017, he organized an intimate gathering at one of his favorite San Francisco restaurants, Quince, to celebrate his fiftieth birthday. He dined regularly at the heralded contemporary Italian restaurant, which boasted three Michelin stars. He, his wife, Heather, and their sons arrived at the historic brick establishment tucked into the cobblestone streets of the city's Jackson Square neighborhood. Inside, they greeted Laurene Powell Jobs and Steve Jobs's son, Reed, as well as Ive's best friends, the designer Marc Newson and the Tokyo-based music producer Nick Wood. Staff ushered them through a dining area with white tablecloths and a Murano glass chandelier. A champagne cart visited every table with complimentary glasses of Ive's favorite drink. As the group posed for a photo afterward, Ive clasped his arm around the neck of one of his sons and stared wide-eyed at the camera.

Milestone birthdays often spur people to reflect on life, decisions made, and opportunities missed. A quarter of a century had passed since Ive had left England for the United States. He'd spent nearly half his life working at Apple and had two boys who considered California home. He had accumulated more wealth than he could have imagined. It was not the life he had envisioned when he had spent weeks agonizing over Apple's job offer in 1992. Since then he had lost his creative partner, boss, and friend and shepherded the creation of Apple's only new product in the years that followed. He, too, wanted to move on but had promised to see one more project to fruition.

SOON THE QUIET, REFLECTIVE EVENING gave way to a more raucous affair.

That spring, invitations reached Ive's friends and family. They outlined a multiday extravaganza that would begin in London, include a two-day celebration among the limestone estates of the Cotswolds, and culminate with a flight to Venice for lunch along the canals, where some would stay for a night of fun at the luxurious twenty-four-room Aman hotel.

It was an extravagant event. The host in the Cotswolds, Matthew Freud, wanted his friend's fiftieth-birthday bash to be exceptional. The founder of a global marketing and communications firm, grandson of Sigmund Freud, and ex-husband of Elisabeth Murdoch, he loaned his twenty-two-room, $7 million mansion known as Burford Priory for the event. Ive's guest list included friends and colleagues who supported his work at Apple, including Robert Brunner, Jimmy Iovine, Paul Deneve, and the twenty-person industrial design team. Few of Apple's business leaders were invited; he largely limited the gathering to the company's creative core.

After Jobs's death and Ive's being knighted, his network of friends had expanded to include British comedians, directors, and musicians. They arrived alongside the Apple crowd at Burford Priory, which had been converted into a countryside carnival. There were bumper cars and red-and-white carnival booths, including one where the artist Damien Hirst, Ive's friend, judged guests' spin-art paintings. Eventually, everyone drifted into a tent for a catered dinner and a birthday roast.

Ive sat beside Heather and his boys in a rounded leather chair perched above the stage. He wore a vintage yellow Apple T-shirt and light blue sport coat on which someone had affixed his name tag for the evening: Chubby.

Freud stepped onstage and approached the microphone. "Welcome to Burford Priory on this most auspicious occasion of Jony's big party," he said. "It's huge and it's generous and it's brilliant and it's over the top, and with the best of intentions, it's all got a bit out of control."

Ive's actor friend Stephen Fry took the stage to play the master of ceremonies. He began his roast at the beginning, in Chingford, where he said Ive's father had held up his firstborn just after birth and said, "It's a be-be-beautiful, beautiful, fragile object. We made him just small enough to hold . . . with just one belly button. It took ever so much, ever so much, ever so much care to refine his little bezel."

The crowd cackled at his parody of Ive's fanciful description of Apple products.

"Jony—or as he preferred to spell it, Joan-y," Fry said, poking fun at the stylistic way Ive wrote his name rather than the commonplace "Johnny." "Why have none of us confronted him about that!?"

Few laughed harder at Fry's joke than Ive. His seriousness at Apple events and at work concealed his tremendous sense of humor among friends. He howled as Fry recounted his early work designing white toilets at Tangerine. Fry said that that work had set off a career of making bathroom-inspired products at Apple. "Everything that came from Apple from now on was essentially a white slab with thick, smooth, rounded corners, just like those wonderful baths, basins . . . and bidets with their sweeping, shiny white curves that had troubled and tortured his fevered imagination for so long," Fry said. "And the rest, of course, is history."

Ive joined everyone else in applause as Fry finished. The actor Sacha Baron Cohen followed, taking the stage with a microphone and slideshow clicker and hosting a pretend Apple presentation about the Jony 50 model, 20 percent wider with 85 percent less hair.

A smattering of video toasts played on a big screen behind the stage, including one from former president Barack Obama and another from the actor Ben Stiller. In his video, Stiller stood before a white wall and explained that he and Ive's families had become fast friends after meeting a few years earlier. In the process, he said, he had come to admire Ive's humility and heart.

Ive grew emotional as Stiller spoke. Heather reached over and grabbed his left arm.

"I'd like to raise a glass to you," Stiller said. He stepped away from the wall and turned a corner into what the Ives realized with a shock was their kitchen in Kauai. The audience rocked with laughter in their seats as Stiller approached Ive's refrigerator in search of tequila.

"Hey Richard, Richard!" Stiller shouted.

"Yes, Mr. Stiller," the housekeeper said.

"Where's the rest of the tequila?" Stiller asked.

"I believe you drank it all last night," the housekeeper said.

Stiller asked him to go out and get some more. Then he slipped out the kitchen door, pulled off his shirt, and dropped his pants. Totally nude, he dived into Ive's pool.

EVENTUALLY, the parody and laughs gave way to more serious remarks. Laurene Powell Jobs took the stage in an elegant black dress. "Jony and Steve shared a remarkable, deep bond," she said. "It was through their trusted partnership that Steve saw the greatest work of his life."

She nodded as she spoke. "I witnessed one of their creative epiphanies," she said. "They were working at our house, walking around the garden in all of its summer splendor, and they strolled through the paths that were bordered by an abundance of flowers and fruit trees, and within a year, the latest iMac design had a long neck and swivel screen inspired by the flowers."

She smiled and fixed her eyes on Ive. "Without question, Jony, you have democratized great design," she said. "What makes him such a brilliant friend also makes him such a sublime artist. Jony has an unparalleled ability to transform abstract ideas and technical concepts into deep, human, emotional experiences because of his own emotional depth. When you work on products that he's designed, you feel better about yourself. He thinks of others when he creates things. That's why they're so special."

AFTER DINNER, the lights were dimmed and U2 took the stage.

The Edge started to strum his guitar as Bono gripped the microphone stand and slung it down toward his shoulder. They launched into "Desire," off their eighth album, *Rattle and Hum*, which Iovine had produced. Ive, who had been college age when it had been released, leaned down to whisper into his son's ear. He then began to bounce up and down and sing along, as the friends around him joined in, testifying to the fullness of his life outside Apple.

Late in a concert full of hits, Bono stopped to introduce a song that he said the band had been gifted when it was at a musical crossroads. He recalled a moment in the early 1990s when he and the Edge had wanted to make songs tinged with electronic dance music, while bassist Adam Clayton and drummer Larry Mullen, Jr., favored sticking to rock. The tension nearly broke up the band. Then, in the studio one day, the Edge had started to play some melancholic chords that Bono put to lyrics.

"This song brought us together," Bono said. "It's about how difficult relationships are."

As he finished, the Edge started to strum his guitar and Mullen began to tap the cymbals. The organ kicked in with a bluesy backbeat as Bono began to sing "One":

> *Did I ask too much? More than a lot.*
> *You gave me nothing, now it's all I got.*

Ive's friends and creative colleagues from Apple swayed in front of the stage. They reveled in a song that had kept the band before them together, even as Jobs's heirs were breaking apart.

Though Apple's top business leaders had worked with Ive for decades, they were not there. Not Eddy Cue. Not Phil Schiller. Not even Tim Cook.

Chapter 20

Power Moves

As Donald Trump took the presidential oath on the Capitol steps, Tim Cook was at work in Cupertino, watching closely for clues about the future. Cook's office was a tidy testament to the values he carried. Robert F. Kennedy looked over his shoulder from a bronze bust atop the file cabinet behind his desk, and Martin Luther King, Jr., gazed at him from a photograph hung on the wall near the door. Cook considered the two men to be representatives of the best of America, both crusaders for justice in the 1960s. He had read their biographies, dissected their speeches, and included their quotes in emails to staff. He admired their idealism and resilience as they had directed a flawed America toward a brighter future. Nearly fifty years later, though, he found himself tracking the words of a reality show star who portrayed the country in a darker light.

The new president huffed and spoke haltingly as a cold wind whipped across an overcast sky. The America he described was a land of rusted-out factories scattered like tombstones, a place where greedy businessmen shipped working-class jobs overseas, a land of gangs and violence, drugs and poverty. It was a place as bleak and terrifying as a horror movie. "This American carnage stops right here and stops right now," Trump said with a scowl, as the crowd cheered his ominous words. He vowed to bring back jobs, revive economic opportunity, and always—always—put America first.

"We must protect our borders from the ravages of other countries making our products, stealing our companies and destroying our jobs," he said.

To the CEO of the United States' most valuable company, the speech must have sounded apocalyptic and personal. The power of Apple's business flowed from the immense profits it reaped from China's endless supply of cheap labor. Cook had built that outsourcing machine. In Trump's speech, he fit the profile of a greedy American businessman, exporting what could have been American manufacturing work overseas. He justified those business practices by telling politicians that China was the only country in the world where factories could hire hundreds of thousands of seasonal workers to crank out the iPhones that Apple sold. He persuaded Barack Obama and others that there weren't enough people or manufacturing engineers to do that work in the United States. But the new president was unmoved by such practical considerations. His speech harkened back to a campaign promise that he would demand that Apple build factories in the United States, a threat that would break Cook's machine and send the company's share price into a tailspin.

Cook would need to develop a stronger relationship with Trump to head off the risk of a calamitous presidential order. Protecting his corporate nation-state would require Cook, the businessman, to become as wily and charming as the politician he would battle.

SIX YEARS INTO HIS LEADERSHIP OF APPLE, the operator was outgrowing the grids of his spreadsheets. His showdown with the FBI had forced him to become a crisis mitigator. The iTunes disruptions in China had demanded he play diplomat. The Met Gala in New York had forced him to be a patron of the arts. Through it all, he had shown versatility transitioning from legal dispute to geopolitical conflict to red carpet debut. Yet he continued to be dogged by investors and critics worried about Apple's dependency on the iPhone and still fretting over the question: What is next?

Already, 2017 was shaping up to be a challenge beyond anything Cook had ever faced, a challenge that required the soft-spoken CEO to become more of a power player than ever. It wasn't just the existential threats being tweeted by the White House's disruptor-in-chief; it was balancing those threats against the competing demands of the Communist Party leaders in China, who could shut down Apple's supply chain in an instant if they so desired. Somehow Cook would have to learn to placate both the American president and China's politburo, summoning a mastery of global diplomacy that even Kissinger would have admired.

At the same time, Cook had to pull off another seemingly impossible task: figuring out a way to defuse the constant pressure for Apple to invent another device that would transform the world yet again. The demand was both unrealistic and unrelenting, and the clamor was only growing, especially in the wake of the latest numbers. Early that year, sales from the company's iPhone business were rebounding but still shy of their peak two years earlier. Apple was on track to report iPhone sales that were 4 percent lower than it had posted in the first half of fiscal 2015. Rising Apple Watch sales were not enough to make up the difference. Cook needed to show Wall Street that Apple could continue to increase its total revenue, a daunting challenge given that the company was recording about $220 billion in annual sales. As he looked across Apple's business for a solution, he identified what he thought would be the perfect answer. It was a radical idea that reinvented Apple's entire business strategy. Instead of defining the company entirely through its dazzling products, Cook wanted to focus more attention on the promise and potential of the services offered through those products.

The App Store had already become a major revenue pipeline. Apple took a 30 percent cut of the price of every app it sold and a similar cut of apps that charged a subscription fee. The company kept the costs of distributing those apps low by maintaining a lean team of reviewers to vet them. Meanwhile, developers were popularizing mobile games such as Fortnite, whose customers spent money to buy weapons and superhero powers. Apple took a cut on every purchase. Some 80 percent of that was estimated to be pure profit. The number of apps being downloaded to iPhones showed no sign of abating. At its current growth rate, Cook could see the App Store doubling the size of the services business. He wanted Wall Street to recognize the value he could see so clearly.

AT THAT POINT, investors viewed Apple as "the iPhone company." They considered the business to be mature and expected product costs to rise as sales contracted. The combination meant that Apple had a stubbornly low price-to-earnings ratio, the calculation of a company's share price relative to its projections of future profits. Hardware companies such as Apple receive a lower price multiple than software companies because they are hit-driven businesses: an especially popular product such as the iPhone 6 could cause a sales surge, while a dud such as the iPhone 6s could pummel profit. The fear that Apple was one poor product away from irrelevance meant it had a fifteen times earnings ratio, less than half that of Google or Facebook. That bothered Cook because it depressed the company's market value. He wanted Apple, then valued at $650 billion, to be a $1 trillion company.

That January, Cook looked to remove the hardware noose constraining the company's valuation by focusing investors' attention on its growing software sales. He knew a higher multiple would follow. After detailing the strong performance of the iPhone 7 during a call with analysts, he said the company's services business—formerly known as iTunes, software, and services—delivered $7.2 billion in sales. The App Store accounted for a third of it. The rest had come from iTunes, Apple Pay, Apple Music, and more. By the end of the year, Cook said that services would be a Fortune 100 business, posting nearly $27 billion in sales, about the same as Facebook, one of the apps that Apple distributed. "Our goal is to double the size of our services business in the next four years," he explained.

The promise ricocheted across Infinite Loop, where no one could recall Cook publicly setting a financial target. Whereas Jobs had once wowed Main Street by unveiling revolutionary new products, Cook was putting his marketing know-how into wowing Wall Street by spotlighting an existing and growing business. Inside the hierarchical company, few knew how Cook planned to fulfill his promise, but no one doubted that he would deliver.

IN THE WAKE of the new financial targets, Cook met with Peter Stern, a former Time Warner Cable executive, and Eddy Cue, head of services, to map out how to proceed. The company's biggest service at the time was iCloud, which charged as little as 99 cents a month to back up people's photos. Stern, who led the iCloud business, proposed increasing the number of subscribers and the amount they paid by bundling iCloud with other subscription apps. The strategy mirrored Amazon, which created a Prime delivery service and attracted subscribers by including access to its Prime video app with TV shows and movies. Stern championed the idea of an Apple bundle.

Cook recognized the services possibilities were endless. Apple could create a fitness app with yoga classes, design a news service with magazines, or build its own Netflix. Those apps would cost relatively little for Apple to develop but potentially bring the company millions of subscribers, creating what financial gurus called recurring revenue, a steady stream of monthly payments that would fill Apple's piggy bank as reliably as a child's allowance. Subscribers would pay Apple more over the lifetime of their memberships than the $1,000 cost of an iPhone. It could be a business breakthrough.

In the meeting, Cook embraced Stern's strategy but challenged the group with a question. "Will there be anything good in this bundle?" he asked.

Accustomed to Cook's Socratic style, the group understood the CEO's message: Don't be lazy about this, he was saying. Be sure to create services with real value.

Cook conveyed a different message in a series of meetings with Apple's finance team. He began meeting with them monthly to review the App Store, iCloud, and Apple Music performance. The sessions were as intense as the "Date Night with Tim" marathons he had held on Fridays to review iPhone, iPad, and Mac sales. Finance staff prepared to answer an array of questions, including how iTunes sales declines compared to Apple Music subscription gains; how iCloud costs compared to subscriptions revenue; and what best-selling apps were driving App Store sales. To help, they created a list of upcoming app releases, so that he could keep tabs on new software with revenue potential.

The exercise marked a departure from the way Jobs viewed the App Store. The seventy-thirty split for distribution introduced by Jobs was designed to cover the costs of storing and delivering apps. The late CEO had never expected the store to be a profit center; he had expected it to help Apple sell more iPhones.

But under Cook, the focus had shifted. He was bending Apple toward a future where it focused less on the sale of devices and more on wringing additional money out of the iPhone by selling an array of software across it.

COOK ARRIVED IN WASHINGTON, D.C., late in January to find the nation's capital convulsed by controversy. In the days after moving into the White House, President Trump had sparred with the media over the size of his inaugural crowd, claimed falsely that 3 million illegal votes had been cast for Hillary Clinton, and blasted General Motors for outsourcing jobs to Mexico.

Eager to get on the right side of the combative president, Cook scheduled dinner at Ristorante Tosca with the president's son-in-law and daughter, Jared Kushner and Ivanka Trump. He was joined by Lisa Jackson, Apple's vice president of government affairs and a former head of the Environmental Protection Agency under Barack Obama. She was a politically savvy operator with deep expertise in the nation's bureaucracy. People close to Trump advised Cook that he would be able to work with the president's son-in-law and daughter, who were known Apple admirers.

The trip to Washington contrasted with the approach of his predecessor. Steve Jobs had been antipolitical. He had believed that if Apple made great products, it would have more political and cultural influence. He had kept the company's Washington staff small—for many years it had been just two people—and discouraged it from hiring outside lobbyists. When Laurene Powell Jobs had tried to arrange a meeting for him with President Obama in 2010, Jobs hadn't wanted to do it. He had told his biographer, Walter Isaacson, that he had no interest in giving the president the satisfaction of broadcasting a token meeting with a CEO. But his wife had persisted. Eventually, Jobs had gone to the White House and lectured President Obama about being unfriendly to business, contrasting the ease of building a factory in China with the endless red tape of building anything in the United States.

But Cook enjoyed his trips to Washington. The tax hearing years earlier had taught him the value of political influence. Since then, he'd prioritized making regular trips to walk the halls of Congress and personally meet with senators and congressmen. He traveled often with Jackson and had made her part of his inner circle. He supported her effort to build out Apple's D.C. office, steadily increasing the size of the team to more than a hundred people. The growth reflected Cook's realism. As it had become bigger, Apple had become a target for federal officials frustrated with the company's positions on everything from iPhone security to taxes.

At Tosca, Cook and Jackson followed a host through the high-end Italian restaurant to a table where they were joined by Kushner and Ivanka. Over a menu that included lobster bisque and rustic lamb ragù, they chatted about how the couple was adjusting to Washington before turning to policy matters and what the administration might be prioritizing. The conversation was pleasant, leaving Cook confident that he could work with the couple and the administration.

But the following afternoon, Trump signed an executive order barring immigration from seven predominantly Muslim countries. The order sparked protests across the country, including in Silicon Valley, where more than two thousand Google employees rallied outside their offices alongside company cofounder Sergey Brin.

Cook was blindsided. Over dinner, Kushner and Ivanka hadn't mentioned immigration. If the ban had been in effect decades earlier, Jobs, the son of an immigrant from Iran, might not have been born and Apple might not exist. Cook's inbox overflowed with emails from alarmed staff. He hurried to write a note to all employees, reassuring them that he heard their concerns. "Apple is open," he wrote. "Open to everyone, no matter where they come from, which language they speak, who they love or how they worship."

After he returned to Cupertino, Cook's staff briefed him on

the order's fallout. Apple had hundreds of employees on H1B visas, many of them deployed around the world, and the company's human resources, legal, and security teams were scrambling to track them down. Everyone they contacted was upset and frightened. A few from the countries on the banned list were outside the country or had family traveling whom they worried wouldn't be able to get back into the United States. They were distraught and wanted Apple to take a more forceful stand. Human resources wanted Cook to meet with a small group of affected employees to get a better sense of how they felt. After listening to the dozen employees talk about how devastating the order was to them, he resolved to take action.

Cook assured Apple's staff that he contacted the White House and conveyed a clear message, saying, "The order should just be dropped."

IN THE WAKE OF THE IMMIGRATION ORDER, the Trump administration's plan to attack businesses that outsourced to China began to take shape.

The administration installed two protectionists to lead trade policy. U.S. Trade Representative Robert Lighthizer took office eager to use tariffs to stop Chinese theft of U.S. technology, while Peter Navarro, a trade adviser, called China an economic parasite. They wanted to return portions of the global supply chain to the United States, a perilous prospect for Apple.

Cook needed to find a way to show Trump that making an iPhone required more than just assembling it on a Chinese factory floor. In fact, U.S. companies supplied many critical iPhone parts. An idea bubbled up through Apple's communications team. Every year, Apple spent billions of dollars on new machines and assembly processes for U.S. manufacturers that later delivered custom components for future products. What if some of the planned spending was promoted as coming from a special fund to support U.S. manufacturing?

Visions of headlines followed: "Apple Pledges to Spend Billions on U.S. Factory Jobs." It was the type of supply-chain window dressing that people inside the company thought would win Trump over.

In late May, the communications team invited CNBC anchor Jim Cramer to 1 Infinite Loop for some special news. The *Mad Money* host met Cook for an interview in Apple's courtyard. Cook rocked on a stool opposite Cramer and fielded what seemed like a planted question: What was Apple going to do to create jobs in the United States?

The CEO boasted that Apple was responsible for 2 million American jobs, including 1.5 million app developers and about a half million suppliers' employees. Cramer didn't point out how much smaller that was than China, where Apple supported about 4.5 million workers, including 3 million in factories and 1.5 million developers. Instead, he focused on the United States, asking Cook if he would put money behind creating jobs there.

"We would and we will," Cook said. He said that Apple was creating an "advanced manufacturing fund" with $1 billion that it would invest in U.S. suppliers. "We can be the ripple in the pond," he said.

Cook's lofty language belied what some inside the company considered to be the truth: the advanced manufacturing fund was a public relations stunt. The company had already planned to spend $1 billion on U.S. suppliers. In fact, it had been spending more than that domestically for years; it just hadn't felt the need to promote the payouts. But with a White House obsessed more with appearance than detail, the company gussied up the bore of everyday business with a flourish of corporate stagecraft.

APPLE'S BREAKNECK GROWTH had enlarged its ranks to more than 120,000 employees. The iPhone empire was twice the size of the one Cook inherited. It had offices scattered around the world and the United States. Keeping up with it all was burdensome. Yet Cook remained the same cost-conscious CEO he had always been, prone to flying commercial, so the board of directors intervened.

In 2017, they began requiring that Cook travel on private jets rather than commercial airliners. His time was too valuable and Apple's footprint was too sprawling for him to waste minutes clearing airport security. Plus, the challenges posed by the new president meant he would need to travel to Washington more than ever.

Cook was summoned to the White House in mid-June for a tech summit designed for Trump to show the country his command over its most powerful industry. News of Apple's advanced manufacturing fund had reached the administration and improved the president's view of the nation's most high-profile outsourcer. When tech leaders assembled in the state dining room a month later, Cook found himself seated at the right hand of Trump.

Lest anyone think that Cook was becoming too close with the president, though, Apple confirmed a report that morning by the news outlet Axios that Cook planned to confront the president about the immigration order. Anxious White House aides leaned forward every time Cook spoke, afraid he would trigger a confrontation with their voluble boss. But to their relief, Cook didn't say a word.

As the discussion came to a close, Cook approached Trump privately. "I hope you will put more heart into your immigration policy," he said. Then he exited. He spoke so quickly that the comment barely registered with Trump. But the brief interaction was almost immediately leaked to Axios, which cast the conversation as a confrontation under the headline "Tim Cook to Trump: Put 'More Heart' in Immigration Debate."

Members of the White House marveled at Cook's maneuvering. He had given the president what he wanted by showing up and sitting at his right side. But he saved face with his staff in Cupertino by strategically leaking that he had challenged Trump on immigration, even though he had done it delicately and in private.

But it was Trump who had the last word. During a later interview with the *Wall Street Journal* in the Oval Office, Trump brought up Cook and said the Apple CEO had promised to return some manufacturing to the United States.

"He's promised me three big plants—big, big, big," Trump said.

"Really?" asked a reporter. "Where?"

"We'll have to see," Trump said. "You can call him. But I said, Tim, unless you start building your plants in this country, I won't consider my administration an economic success, OK? And he's called me and he says, you know, they're going forward, three big, beautiful plants. You'll have to call him. I mean, maybe he won't tell you what he tells me, but I believe he will do that."

The claim roiled Apple's leadership. Cook had never said anything about "big plants" to Trump, but when reporters called seeking comment, Apple's spokespeople declined to contradict the president. Cook and his advisers feared that calling Trump a liar would ignite a "tweet war," spark threats of tariffs on Apple products, or worse, inspire a call for a boycott of Apple, so the company remained silent.

In the quiet, unease permeated Cupertino. If Trump was willing to lie about that, Cook and his colleagues could only wonder what he might do next.

AS COOK LOOKED for more ways to improve his standing with Trump, Apple's robust lobbying office threw itself behind a Trump administration push to rewrite the tax laws. Its goal was to reduce the tax rate on overseas profits that led companies to keep cash offshore. When it passed in late 2017, it signaled an end to the years-old controversy over Apple's tax practices. It also presented another opportunity to flatter the president.

Working with Apple's finance and communications team, Cook looked for a commitment that Apple could make to the U.S. economy that would get Trump's attention. The new tax law required that Apple pay a onetime tax of 15.5 percent on its overseas profits, approximately $38 billion. The company planned to build a new customer support campus and data centers, as well as start other construction that would cost about $30 billion. Plus it was spending about $55 billion annually with U.S. suppliers and hiring about five thousand new employees annually in the United States. The company could thus claim that in the wake of tax reform it would be making a $350 billion direct contribution to the U.S. economy over the next five years, as well as adding twenty thousand new jobs. Those were the types of big, simple numbers that Trump loved.

When Cook called Trump to announce the company's sizable commitment, the president was unimpressed. He thought Cook had said that Apple was committing $350 *million. That's a nice-size plant,* the president thought, *not the greatest.* Then Cook reiterated that the contribution was $350 *billion.*

"That is something," Trump said.

On January 17, 2018, Apple issued a press release headlined "Apple Accelerates U.S. Investment and Job Creation: $350 Billion Contribution to U.S. Economy over Next Five Years." The release didn't state that about 80 percent of that commitment would have been made as part of Apple's ongoing business, with or without tax reform. But Trump, never keen on details, wasn't likely to do the math.

During his State of the Union address, he pointed to Apple and its $350 billion commitment as evidence that his America First policies were working.

THE WHIPLASH from Washington kept Cook on his toes. Each time he took a step forward, the administration delivered a setback. In the spring of 2018, the challenge veered beyond his control.

During trade talks in D.C., U.S. and Chinese negotiators clashed after the United States demanded that China cut the trade surplus between their economies by $100 billion, stop stealing intellectual property, and end government subsidies to state-owned companies. It was a direct attack on China's economy, accompanied by a two-hundred-page report of grievances. It threatened to create a full-blown trade war.

At a news conference, Trump threatened to impose a 25 percent tariff on $60 billion worth of Chinese imports. The stock market shuddered, and the Apple share price tumbled 6 percent as investors worried that the production of iPhones would suffer collateral damage.

The empire Cook had built hinged on the company's maintaining good relations with the Chinese government. The country acted as the factory floor for almost all of Apple's products. Its 1.4 billion people had become Apple's largest pool of customers, especially after Cook struck the China Mobile deal. The trade tussle between Washington and Beijing endangered Apple's business model. If the Trump administration slapped tariffs on goods imported from China, the cost of iPhones could rise. If the Chinese retaliated under Xi Jinping, they could block or slow exports of iPhones from factories, as they had already blocked imports of Ford vehicles. On the customer front, they could unleash the *shuijun* "water army" to turn public opinion against Apple on social media. Apple's senior leadership in China issued a warning to Cook: Things could get bad. They implored their aloof and robotic CEO to strike a balance between America's volatile reality-TV star and China's unpredictable autocrat.

Amid the controversy, Cook landed in Beijing for the 2018

China Development Forum. The annual event was designed to be the Chinese Communist Party's answer to World Economic Forum meetings in Davos. Geopolitical tensions were high as a car carried Cook through Beijing's traffic-clogged streets to the Diaoyutai State Guesthouse, a diplomatic venue that had hosted President Richard Nixon and Premier Chou En-lai in 1972. Cook was there on a diplomatic mission of his own: he had signed on to cochair the three day-event and was slated to speak three times. The taciturn master of supply-chain efficiency had become the most important business leader in the world; no one on earth had more to gain from open trade or more to lose from a trade war. He knew that his comments would be tracked on both sides of the Pacific to determine whether his loyalty in the trade war lay with China or the United States.

On Sunday, he stepped to a lectern in front of a red backdrop to open the event. He gazed out at the ballroom full of Party leaders and company chiefs including Google's Sundar Pichai and called on them to stay united in supporting free trade.

Few other U.S. business leaders touched on the topic, but Cook addressed it every day of the summit. He encouraged officials in the United States and China to let "calm heads prevail." During a panel discussion, he was asked what message he would send to Donald Trump. "Countries that embrace openness, that embrace trade, that embrace diversity are the countries that do exceptionally," he said. "And the countries that don't, don't."

On the final day of the event, Chinese premier Li Keqiang called on the business leaders there to protect free trade and oppose protectionism. "There is no winner in a trade war," he said.

The premier's and Cook's comments seemed born of the same briefing notes. To anyone listening, it was clear that Cook had held the Party line.

THE BREWING TRADE TROUBLES increased the urgency of Cook's plan to shift Apple's business strategy. Developing more services would diversify the company's revenue and blunt the fallout from tariffs on its hardware operations. At the company's office near Los Angeles, his vision for the future took on new dimensions.

Jimmy Iovine, the most restless member of Apple's leadership team, wanted to find a way to make Apple Music unique. The two-year-old service had about half as many subscribers as Spotify and was struggling to differentiate itself from its rival. An early ploy to attract subscribers with exclusive albums from Drake and others had flamed out after Kanye West had blamed it for "fucking up the music game." Record labels and artists decided they owed it to their fans to make their music as widely available as possible. The demise of exclusives meant Apple Music and Spotify were just different-colored apps with the same catalog of songs. Leapfrogging rivals would be impossible unless Apple added something unique. Iovine decided the solution was adding original TV shows.

The music mogul tapped into his extensive network in Los Angeles to connect with Hollywood agents. He also began lobbying Cook and Cue to start making TV shows. "I'm pushing these guys," he would say. "They gotta be in the content space. They gotta be in the content space."

Eager to show how that would work, Iovine developed a six-episode, semiautobiographical show about Dr. Dre that featured a single character, with each episode focusing on a different emotion such as anger and how Dre's character deals with it. He and Dre enlisted the support of well-known actors such as Sam Rockwell and started filming. Iovine urged Cook to watch an episode.

When Cook screened it, he was alarmed. There were characters doing lines of cocaine, guns, and an extended orgy scene with simulated sex in a Hollywood mansion. It was a departure from the material in the reserved executive's favorite shows, the unhip political drama *Madam Secretary* and family-friendly *Friday Night Lights*. There was no way Apple could release it. The company had a pristine image that was central to its sales of iPhones and Macs. A show with sex and violence would torch the brand.

Cook told Iovine: We're not making this. It's too violent.

Disappointed, Iovine didn't give up. Instead, he offered up more family-friendly fare with a show called *Planet of the Apps*. Modeled after *Shark Tank*, it followed aspiring entrepreneurs as they created a mobile app and sought funding for it from a panel of celebrity judges. Cook, who approved of the project, liked how it celebrated the app developers who had become big contributors to Apple's business. The early episodes featured colorful language and tense moments where developers under pressure cursed in frustration. Cook, Cue, and others sent back notes asking that the foul language be cut. They wanted the show to be inspirational and positive, not sullied by the offensive dialogue.

When the show debuted, TV critics panned it for its lack of realism. *Variety*, one of Hollywood's leading trade publications, called it a "bland, tepid, barely competent knock-off of 'Shark Tank.'" The *Guardian* described it as "grating" and called on Apple to rise to the higher standards set by Netflix with its pioneering originals.

The harsh reviews were unheard-of for a company accustomed to being heralded for making perfect products. Cook recognized that a move into TV risked tarnishing the company's reputation for excellence. Hollywood programming had the potential to add some star power to Apple's services push, but the company would need to do more than experiment. At various points over the years, the company's leadership team had discussed the possibility of buying Disney, Netflix, or Time Warner, which owned HBO. But the rocky integration of Beats showed how difficult it could be to import companies into Apple's rigid culture. Cook favored proceeding alone. His preference led to what became known inside Apple as Project North Star, a $1 billion bet that Apple could make its own Netflix.

Cook sought to learn as much about Hollywood as possible. He wanted to understand the industry, the players, the process, what worked and what failed. He and Cue summoned an array of experts to Cupertino, including a group of agents from Creative Artists Agency. They met the team from CAA in Apple's boardroom and explained that they were educating themselves on the entertainment industry. Cook put the group at ease by tucking one leg underneath the other in a casual, meditative pose. Then he and Cue began asking questions: What does it cost to make a TV show? How is a show made? How are actors paid?

Everyone in the room knew that television was in a period change. People were dropping cable TV in favor of Netflix and Hulu. The unspoken question was: Can Apple join the fray?

The agents explained how the industry worked and what Netflix had done to succeed. The former DVD-rental service had launched a subscription streaming business that had taken off after it had released two critically acclaimed shows in 2013: the political drama *House of Cards* and the dark prison comedy *Orange Is the New Black*. The edgy programming had filled a void in the TV landscape and subscriptions had soared, proving that millions of people with broadband internet would pay for TV shows from an app just as they had paid cable operators for HBO. Four years after its original shows debuted, Netflix's market value had surged fourfold to $83 billion. The formula for its success was simple: "You just need two hit shows," one of the agents said.

Cook recognized that success in entertainment would require an experienced Hollywood hand. At the exact moment Apple needed help, two of the top names in the business at Sony Pictures were entering a contract year.

Iovine didn't know Zack Van Amburg and Jamie Erlicht, two executives at Sony, but they came recommended by friends. He loved that they had produced one of his favorite shows, *Breaking Bad,* and invited them to his home in Holmby Hills, an opulent neighborhood of $10 million mansions with lush green lawns and towering privacy gates.

Van Amburg and Erlicht were uncertain about what to expect. They had never met Iovine and were generally unimpressed with what Apple had done so far with *Planet of the Apps.* Iovine welcomed them into his living room and talked to them about his vision of creating an MTV inside Apple Music. He didn't know if they were the right fit, but he needed someone who knew the entertainment business better than he did. Van Amburg and Erlicht spoke about their approach at Sony Pictures, where they had played a critical role in developing shows such as NBC's *The Blacklist* and FX's *Rescue Me.* They were understated about their success, which struck Iovine as the right attitude to have at Apple, a place with a low tolerance for outsiders with big egos.

Afterward, Iovine called Cue and encouraged him to meet them. He, too, was impressed. When the duo exited their deal with Sony, Cue signed them on and entrusted them with leading North Star.

Within a few months, they struck a deal for a series starring Reese Witherspoon and Jennifer Aniston called *The Morning Show.* The drama was set at a television news morning show, where Steve Carell would play a morning anchor ensnared in a sexual harassment scandal.

The star-laden show signaled that Apple was serious. The company agreed to pay Aniston and Witherspoon more than $1 million an episode each, bringing the show's total cost to $100 million.

Believing that talent would help it woo more talent, Cook also rallied the team to push to bring Oprah Winfrey back to TV on Apple's new service. In mid-2018, they arranged for her to come to

Cupertino for a private tour of Apple Park. The Auburn football fan Cook did his best impersonation of the coach showing a top recruit the facilities. He eventually brought her to the Steve Jobs Theater, where he and other leaders showed her the space and then asked her to watch a video. Inspiring music filled the darkened theater as words played across the screen. The video said that the turbulent world missed her inspiring voice. Apple's emotional call for her to return to TV made her cry. Not long after, she agreed to join the team.

With its deep pockets, Cook showed that Apple was willing to buy the star power it needed to supercharge its services.

THE HEADACHES of the U.S.-China trade dispute demanded that Cook spend time traveling between the capitals of the world's largest economies. In April 2018, a month after his public appearance in Beijing, he arranged to visit the White House for a private audience with the president.

More than a year after Trump's arrival, an unspoken distrust divided the two men. Cook, a lord of discipline, struggled to keep pace with the president's shifting statements and fluctuating priorities. Trump, a master of volatility, lumped Cook together with the liberal leaders of Silicon Valley who conspired to thwart his agenda. Looking to close the divide, Cook reached out to the newly appointed director of the National Economic Council, Larry Kudlow, and humbly said, "Help me."

A longtime host on CNBC, Kudlow had admired Cook's business acumen and was sympathetic to his precarious position. He appreciated Apple's exposure in China and helped schedule time for Cook with the president.

Trade tensions were escalating as Cook entered the White House that spring. In the days leading up to his visit, Trump had threatened to add another $100 billion in tariffs on a range of imported goods, including cars, smartwatches, and smartphones. The Chinese retaliated with a tariff list of $50 billion of goods imported from the United States. It was the kind of tit for tat that Cook feared would subsume Apple's business.

Cook strode through the cramped, bustling hallways of the White House and made his way to Kudlow's office on the second floor of the West Wing. He traveled without an entourage, favoring instead a low-maintenance style and personal touch that gave White House officials confidence that they were dealing with Apple's ultimate authority.

Cook breezed into Kudlow's wood-paneled office and eased himself into a seat across from the economic director with a quiet confidence that said "I've been here before." He was relaxed and informal, with a smoothness that impressed Kudlow, who was accustomed to stiff CEOs who were anxious about meeting the president. Cook jumped into the matters he wanted to discuss, including India, where foreign-direct-investment rules in that country prevented Apple from opening stores there. They also talked about Apple's ongoing tax case in Ireland, where the company was mired in a years-long battle with regulators over its tax status there. Finally, Cook raised the issue of intellectual property theft. Kudlow recalled that the CEO shared the administration's concerns and agreed with its policy positions. But Cook downplayed China's role, saying the country didn't cause intellectual property issues for Apple. He was more concerned in that regard, he said, about India.

The CEO was playing a deft political game. He had signaled to the administration that he approved of Trump's efforts, while subtly nudging the policy makers away from an escalating tariff war with China, which could have roiled Apple's business model.

Cook's new warmness toward the administration was a consequence of emerging issues in China. The country had recently passed a cybersecurity law requiring that all Chinese phone customers' data be stored on the mainland. It had forced Apple to open negotiations with a state-owned company in Guizhou that would build and operate an Apple data center. The plan had frustrated some members of Apple's privacy and security team. They couldn't reconcile Apple's public refusal to help the FBI in the San Bernardino case with its quiet compliance in China. Instead of his high-minded promise to protect customers' privacy, Cook had capitulated to the demands of a government known for surveilling its citizens, only to later ask for help from the very U.S. government he had once defied. The practical Cook seemed to lose his moral compass when faced with pressures in the market he had built.

After their meeting, Kudlow led Cook into the Oval Office, where they found the president behind his desk. They walked to a pair of chairs facing him. Cook smiled and delivered his opening line. "Thank you for your tax reform, Mr. President," he said enthusiastically.

The tension in the room evaporated. Cook listened intently as the president talked glowingly about the tax changes and the benefits he expected them to bring the economy.

The meeting provided Cook with a template for dealing with Trump in the future. Afterward, he called Trump directly by phone. He also went out of his way not to criticize or correct the president, including when Trump introduced him at a White House meeting as "Tim Apple." When a reporter later asked why Trump had seemed to connect with Cook but not with other executives, Trump said, "He calls me and others don't."

DESPITE THE INTENSIFYING TRADE BATTLE, Apple's business was humming. Cook reported another record quarter in late July and assured investors that Apple expected strong sales in the months ahead.

Revenue from its services was up 40 percent, and its bread-and-butter iPhone had raked in nearly as much cash in the period as it had at its peak in 2015. Profits had hit an all-time high.

Perhaps more important, Cook told Wall Street analysts that the new trade tariffs hadn't affected Apple and he didn't anticipate that they would. Fresh from his White House visit, he was optimistic that the trade tensions between the United States and China would be settled.

Encouraged, investors sent the share price of Apple up 9 percent over the following two days. Cook watched as Apple's market value inched toward $1 trillion. On August 2, 2018, it became the first U.S. company to reach that milestone. Cook had tripled its value in seven years, transforming a company that had once been on the brink of bankruptcy into one that was worth as much as Exxon Mobil, Procter & Gamble, and AT&T combined. Cook was so excited that he wrote a note to staff marking the occasion.

When the Trump administration issued an updated tariff list in September, investors were alarmed to see the Apple Watch and Air-Pods on the list. The company's share price tumbled. Its legal team scrambled to file a protest with the U.S. Trade Representative. Wall Street analysts predicted sales would plummet.

Throughout it all, Cook remained calm. He contacted the White House and spoke with the president. Within days, the Trump administration updated the tariff list. There were no Apple products on it.

Chapter 21

Not Working

The video was vetted at the highest levels of Apple.

One early September evening in 2017, Jony Ive and Tim Cook gathered in the bowels of the company's new performance hall to make a critical decision about whether a film featuring Steve Jobs's voice should open the new theater. They settled into the darkened space and watched as the screen filled with the words *Welcome to the Steve Jobs Theater*.

Then came the voice.

"There's lots of ways to be as a person," Steve Jobs said. "One of the ways that I believe people express their appreciation to the rest of humanity is to make something wonderful and put it out. And you never—you never meet the people, you never shake their hands. You never hear their story or tell yours but somehow in the act of making something with a great deal of care and love, something is transmitted there. And it's a way of expressing to the rest of our species our deep appreciation. So we need to be true to who we are and remember what's really important to us. That's what's going to keep Apple, Apple, is if we keep us, us."

Ive and Cook had an important decision to make: Should the

video play at the beginning of an event to open Jobs's namesake theater?

In previous days, the video had been edited to the millisecond. It was sped up and slowed down until the production team thought the audio struck the perfect emotional note. Yet Ive and Cook worried about using it.

Their indecision spoke to the challenge of honoring the legacy of the man they both loved. In 1997, Jobs had rejected the idea of a version of the company's famed "Think Different" campaign that would have featured his voice because he thought that using his own voice would make it about him instead of it being about Apple. Two decades later, his successors recalled that decision and worried that playing a different video with Jobs's voice would make the theater's opening event about the company's late founder rather than the campus that he had imagined.

Eventually, Ive and Cook reached an agreement: Jobs would open his own theater and remind the world what he and Apple believed.

IVE WOKE UP ENERGIZED on the morning of the event. The chief designer and his team had spent nearly a decade designing everything at the new theater from its bead-blasted ceiling to the recessed handrails along its staircases. They had been a major proponent of a rotating glass elevator inside that twisted as it descended, spiraling downward to open its doors in the opposite direction from which people entered. Usually anxious about releasing a product to the world, Ive was excited to show off the building.

Before the crowds arrived, he gave a private tour of the venue to Virgil Abloh, the influential artistic director of Louis Vuitton's menswear. They had become friends as Ive had burrowed into the world of fashion. He led Abloh across the venue's terrazzo floor and passed beneath its circular space gray ceiling that weighed eighty tons. They entered the underground auditorium, with its floors of curved oak and the Poltrona Frau leather seats. Ive knew every detail of the architectural and interior design.

Guests invited to the show began to amass outside a nearby visitor's entrance. A winding walk guided them up a man-made hill flanked by oak trees and newly spread black mulch. At the summit, the arriving crowd got their first glimpse of the campus's gem: a perfectly clear building with a carbon-fiber roof that rested on a twenty-two-foot-tall glass cylinder. It looked like a supersize MacBook Air.

The building bewildered Apple cofounder Steve Wozniak, who was among the throng. As he stopped and gazed up at it, he thought, *This is not normal.* He scanned the exterior of the building, where the glass walls had been stitched together to conceal the electrical wires, data cables, and sprinkler systems, ensuring that the building looked like an unbroken ring of glass. To Wozniak, it perfectly reflected Jobs's austere design sensibilities.

"It's what you don't see that makes this building incredible," he said, standing in the shade of its metallic roof. "The beauty and openness of the windows is like a German design style. It's so clean. It's minimalization."

After admiring the building, Wozniak spoke to a scrum of reporters about the tenth-anniversary iPhone that everyone anticipated Apple would release later that morning. The company's flagship product needed to be rejuvenated. Its sales had slumped by 9 percent from their 2015 peak. Wall Street analysts expected the new iPhone to arrest that fall, but Wozniak was skeptical that it could. He told the reporters that the smartphone had peaked. New iPhones looked like their predecessors and offered fewer compelling new features. For the first time, he said, he might not buy the latest model.

IVE SETTLED INTO his usual place near the stage beside Laurene Powell Jobs and listened as the video he had approved began to play. The voice of his friend and creative partner filled the room, reminding everyone of Jobs's total devotion to making products. After the video ended, Ive watched as Cook stepped onto the stage beneath an auditorium-size image of Jobs. The CEO walked to the center of the stage and stood before everyone in his predecessor's shadow.

"It was only fitting that Steve should open his theater," he said with a smile. He brushed away a tear and continued. "It's taken some time, but we can now reflect on him with joy instead of sadness."

Cook put his own spin on Jobs's legacy: "His greatest gift, his greatest expression of his appreciation, would not be a single product. But rather, it would be Apple itself."

Over the next two hours, Ive watched as Cook tried to conjure up their late boss's magic. The company was once again facing questions about whether it could innovate. Still fatigued after the watch project and foiled by the complexity of making a car, Ive couldn't rebut the skeptics and reinvigorate the faithful by delivering a new category. Instead, he let Cook provide a new spin on an aging device.

"No other device in our lifetimes has had the impact on the world that the iPhone has," Cook said. He recounted the features it had introduced, from touch display to the App Store. "Now, ten years later, it is fitting that we are here, in this place, on this day." Cook paused. Then, raising his voice to a near shout, "To reveal a product that will set the path for technology for the next decade!"

His raised voice aimed to energize a crowd that found itself staring at yet another glass rectangle. The iPhone that appeared on the screen behind him featured a full-screen display and an indentation for a facial recognition system made up of two cameras, a laser, and a miniature projector. The unit sprayed thirty thousand invisible dots across a user's face and took an instant photo that the phone compared with an image of the owner's face. If the images matched, the phone unlocked.

In a video, Ive said that the device fulfilled one of the design team's long-standing goals: "to create an iPhone that is all display, a physical object that disappears into experience."

The iPhone X eliminated the home button and replaced it with a system of swipes developed by his software team. But unlike with previous models, he said little about its external design or materials. Instead, he spoke about its camera system and the A11 Bionic chip that powered the phone. It harkened back to the days at Apple before Jobs had returned when Ive would despair that the company was more focused on how powerful its chips were than how beautiful its products looked. Now, some two decades later, the artist himself was emphasizing engineering.

A WAVE OF STICKER SHOCK followed as chief marketer Phil Schiller took the stage and revealed that the iPhone X would cost $999.

The price tag was a 50 percent increase from the starting price of the previous year's iPhone 7. It defied the laws of technology, which tended to see prices decline as products matured. At a time when Apple was selling fewer iPhones, Cook intended to try to wring more revenue out of Apple's most important product by charging $350 more for each phone it sold. The increase would more than offset the higher costs of the phone's pricier display and expensive facial recognition system. It was a shrewd strategy, the type that would simultaneously dishearten Main Street and cheer Wall Street.

In a departure, the new phone would be made available in November rather than late September. Schiller didn't explain why, but executives in the audience knew manufacturing problems were to blame. Its engineers had unearthed a performance issue with the facial recognition system that had forced them to reevaluate the technology and delay the phone's release. The company had also run into an imbalance in the supply of two facial recognition parts, code-named Romeo and Juliet. Romeo, the projection unit, took more time to assemble, forcing Juliet, the camera unit, to wait. The issues cost the company six weeks of sales.

Apple lessened the damage by releasing another new model, the iPhone 8, in late September. It featured a home button and a design similar to that of the previous year's model. Planned as a contingency phone since 2015, it provided a much-needed safety net.

AFTER THE EVENT, Ive approached Cook outside the theater for a manufactured moment of collaboration. Photographers captured images of Ive as he peered over Cook's shoulder to look at the new iPhone. The emptiness of the tableau made some colleagues grimace.

Ive's part-time work arrangements—and Cook's approval of them—annoyed Ive's peers. It was out of step with the culture of a company that prioritized its workers being in the office more than its Silicon Valley peers did. His part-time schedule had coincided with several product delays. The setbacks had increased scrutiny of Apple's processes. Among the biggest impediments to its once-efficient operations was the absence of its chief tastemaker.

The release of the new iPhones helped Apple post record revenues in November. During a call with Wall Street analysts, Cook said that early orders of the iPhone X had been strong enough to put it on track to deliver Apple's most profitable year.

But the phone's success didn't temper the growing unrest about Cook's arrangement with Ive. Along with the frustration about Ive's meetings at the Battery and the related delays, irritation grew across Apple's top ranks that the company was paying him more to do seemingly less than everyone else. His salary had long been a source of resentment. Under Jobs and later Cook, Apple had paid the members of its ten-person executive team equally, about $25 million in total annual compensation apiece. Their pay was reported publicly in compliance with the SEC's Section 16 law, which required companies to report compensation of officers who oversaw specific business units. But Ive had received a pay package that exceeded his peers' and concealed it by being one of the only executives not listed as a Section 16 officer even as he worked part-time.

Other examples of Ive's abusing his position surfaced. Following a recent remodel of his Gulfstream V, he had found a flaw in the performance of the custom aluminum soap dispensers that had been installed. Apple's computer engineers were asked to find a solution. Instead of working on future Macs, a member of the team had spent several weeks fixing Ive's soap dispensers. Shareholders have no idea, his colleagues joked.

The design studio's expenses were mounting as well. After the photographer Andrew Zuckerman had completed work on the company's 2016 book, *Designed by Apple in California*, Ive had tapped him to capture video and photographs of Apple Park's development for a planned documentary. Zuckerman had been working at Apple since at least 2010, when Jobs had approved using him to shoot a commercial about FaceTime. The photographer's work had resonated with Ive and Jobs because he shared their obsessions with perfection and minimalism. He created arresting images for galleries and museums of people, animals, and flowers against white backgrounds that accentuated the subject's color and texture. He also directed movies, including a short film starring his friends Maggie Gyllenhaal and Peter Sarsgaard.

For the Apple Park documentary work, Apple agreed to pay him $3.5 million a year. It was the type of sum that team members said Jobs would have ignored. His question would have been "Are the photographs and video any good?" Artistic skill would have trumped commercial considerations. But Jobs no longer controlled the purse strings.

The billings eventually caught the attention of Apple's finance team, which had been emboldened under Cook and CFO Luca Maestri to scrutinize spending on outside contractors. The work Zuckerman was doing made him a target.

Over the years, Zuckerman had become friends with the designers and Ive. In a message exchange that included Apple designers, Zuckerman made a remark that some at Apple deemed offensive. The company monitors its employees' phone records and text messages, and Zuckerman's comment caught the company's attention. Apple's finance team used the message partly as justification to audit Zuckerman's work, a process its contract allowed. The process, which included reviewing years of billings, was invasive and exhausting. A similar financial review of another outside consultancy had been so stressful that the firm's CEO had had a heart attack in the middle of the process, even as the audit had found no improprieties.

At the conclusion of the review, Apple's finance staff determined that it had overpaid Zuckerman for his services. They demanded that he repay as much as $20 million that he had billed over the years. It was an enormous sum that represented much of what Zuckerman had earned through his work on the book and other projects. Desperate to avoid financial disaster, Zuckerman begged Ive for help.

"Sorry," Ive said. He explained that Cook had been behind the audit. "There's nothing I can do."

It was not the only time he had to apologize for the behavior of Apple's finance department. Despite having more than $200 billion in cash, the company had rejected legitimate billings filed by its architecture firm, Foster + Partners, which had worked on Apple Park and Apple Stores. When one of the firm's partners had told Ive, the designer had been furious and fought back. He couldn't fathom why the company would stiff its vendors. But his energy for conflict had waned.

AS IVE CAME TO TERMS with the limits of his power, Cook grew concerned about his absence from day-to-day management. He conveyed that their part-time arrangement was not working.

For Cook, it had become clear that if Ive were gone, the people he left behind would feel empowered to make their own decisions. If Ive were engaged, the people would feel confident knowing he was in charge. But the half-out-half-in situation that had taken root over the past two years had left the product side of the company in leadership purgatory. It was obvious that Ive needed to resume his day-to-day management of design.

Around the same time, his team staged an intervention. A group of designers tried to persuade him to return, threatening to leave if there wasn't some sort of improvement in leadership. Ive agreed. The recent defections of designers Danny Coster, Chris Stringer, and Imran Chaudhri had exposed how much the team was struggling in his absence. He wanted to come back and restore the sense of order that had been lost.

In late 2017, he flew much of his team of twenty designers to Washington, D.C., for a Smithsonian event where he would speak about the future of design. Though he was the one who had been asked to speak, he wanted the group there because everything they had made and achieved had been done together. He pointed to the group throughout the conversation, saying that he would look back more fondly on the way they had worked than on what they had created.

"We have so much trust as a team that we don't censor our ideas because we are nervous and scared that they would sound absurd," he said. "When you have trust, it's not a competition. What we're interested in as a team is genuinely trying to figure out how we can make the best product possible."

The public appearance signaled Ive's return to the fold. Once back in California, he operated on his own terms. He set up a regular schedule to visit the design studio, and he began to meet regularly with the designers at places around San Francisco rather than in Cupertino.

Apple's executive team welcomed his return. They were optimistic that his day-to-day oversight would speed decisions and improve product development. They told friends outside the company that there had immediately been signs of improvement.

Soon after reimmersing himself in the business, Ive pushed for a redesign of the forthcoming iPhone 11. The plans called for an ultrawide camera to be added to the rear of the phone, and the early design for the phone showed the cameras stacked three high inside a slender I-shaped bump. Ive pressed to rearrange the lenses inside a small square with two cameras atop each other and another one centered to their right, creating an equilateral triangle. The resulting design had a balance that minimized the thickness of the hardware.

It was the kind of tasteful touch that had been the hallmark of Ive's career.

IN 2018, Ive's newest product debuted for staff. Employees began to relocate from the dated Infinite Loop campus to the futuristic Apple Park headquarters. The four-story ring had taken seven years to develop and cost an estimated $5 billion, making it one of the most expensive and most discussed corporate campuses ever constructed. The building was essentially a sixty-floor skyscraper bent into a continuous circle, four stories tall and a mile around. Its seamless curved glass exterior ran from floor to ceiling on each floor and bathed the walkways in sunlight. Beyond the windows were rolling hills topped by apricot, apple, and cherry trees. A ripple pool in the center of the ring rolled gentle waves of water over potato-size stones with a meditative hum.

Ive and the designers had poured years of their time into defining everything inside the building from the curvature of its elevator buttons to the badge readers outside its office doors. Echoes of Apple products could be found everywhere. Whereas most buildings are made of ninety-degree angles, Apple Park was never-ending curves. The eight hundred glass panels that ran around the building were perfectly bowed to create a three-quarter-mile circle. The interior of the elevator had rounded corners rather than square edges. The stairwells, made of custom white concrete that looked like marble, featured steps that ended with a slight arc. Each bend harkened back to the carefully considered curve on every iPhone, looping on endlessly unlike any workplace in the world.

The circle was broken into eight identical segments separated by glass doors that were so clear that they made the loop look as though it continued forever. For staff, the arrangement could be as deceptive as a hall of mirrors. Shortly after moving in, an engineer on Apple's Siri team walked into a glass door and broke his nose. Blood ran down his face. He wouldn't be the building's last victim.

Over the next few weeks, Apple Park security called 911 to report a host of similar incidents. One employee cut his eyebrow. Another, possibly concussed, bled from the head. A third needed paramedics. The calls became so routine that security knew how to patch the dispatcher straight through to the injured employee.

"Tell me exactly what happened," a dispatcher said after one injury.

"Um, I walked into a glass door on the first floor of Apple Park when I was trying to go outside, which was very silly," the employee said.

"You walked through a glass door?" the dispatcher asked.

"I didn't walk *through* a glass door," the employee said. "I walked *into* a glass door."

"Okay, one second. Did you injure your head?"

"I hit my head."

To avoid looking as though they had just stepped out of a boxing ring, staff began walking around the building with their arms held out like zombies, hoping their fingers would hit the glass before their faces did. Apple rushed to address the issue by ordering miles of black stickers to apply around the building. Senior executives and members of Apple's business strategy team, among the first to move into the new building, assisted maintenance staff in what employees called "an emergency sticker job."

The black dots stood out as the only visible imperfections in a space where indoors and outdoors blurred together, courtesy of the never-ending glass. Staff began calling the stickers "Jony's tears."

THE AMBITIOUS HEADQUARTERS divided the staff. Some loved it. They would crowd around the ripple pool and type software code as the water rolled before them. Others would linger in the four-thousand-seat cafeteria to marvel at the backlit silhouettes of colleagues who looked as though they were walking on air across bridges that spanned the café. People would meander along the trails that had been built to encourage walks like the ones that Jobs had taken with colleagues around the hills of Palo Alto. Some employees even found joy in small, thoughtful touches such as the privacy afforded by the floor-to-ceiling toilet stalls. They couldn't imagine working anywhere else.

But the love wasn't universal. A faction of employees considered the campus to be the physical embodiment of Ive's late-career tendency to favor form over function. In its uncompromising beauty, it created unnecessary hardships they were forced to endure. Ive and Cook had said that Apple Park had been built to bring everyone together in one space, so that people from different divisions might serendipitously bump into each other and find ways to collaborate. But although its larger cafeteria and parkscape encouraged interaction, the interior discouraged it. The segments inside the building were carved into self-contained wedges of office space accessible only by badge. Staff complained that going to a meeting on the same floor in an adjacent wedge required walking down two flights of stairs and ascending a different stairwell to a room that was practically next door. It made the building feel like a city of one-way streets. They called the locked-down maze "Space Prison."

Noise was a greater nuisance. The building's interior walkway ran alongside curved glass panels that ferried sounds across great distances like a science museum's whispering wall. People's chatter filtered into offices through the seams between the glass panels, leading some employees to fill the gaps with colored pieces of Styrofoam. Eventually, Apple installed white-noise machines to dampen the hallway racket.

Then there were the same kind of early headaches as those that plague other newly completed construction projects. Rats took up residence in the debris of Hewlett-Packard and became the first residents of campus, scurrying across the grounds. Inside the building, steam pipes pushed water out of the freshly poured concrete in beads, leaving rust stains on the walls. Some staff taped up hotel towels to capture the water, leading engineers to joke that the building needed industrial-size Depends.

Gallows humor became a way of getting by. One engineer created a system for sharing "Today at Apple Park" memes about the absurdities. Colleagues poked fun at their commutes, which took fifteen minutes longer than before as they trekked across the 175-acre campus to their desks. Others joked that Cook had found a new way to make a buck by banning employees from picking fruit off the trees on the property. Someone at Apple would collect the fruit and the cafeteria would use it in pies that the company sold to employees. A few quipped that the only imperfection in the building's sloping roof was above the design studio, where pipes jutted into the air to release paint fumes, a building code requirement that Ive and the architects had battled but couldn't overturn. The engineer who coordinated the staff exchanges voiced disappointment that colleagues with seeing-eye dogs had no dedicated place to take them to the bathroom. He enjoyed watching the Labradors being escorted out onto the sculpted hillsides to take a crap.

"It was poetic justice," he said.

FOR SOME, the new headquarters felt haunted. Everything from the hallway noise to the glass-door injuries reminded people of a company that had spun off its axis.

Under Jobs, Apple had balanced art with engineering to create its beautiful and innovative products. His judgment had driven Apple to omit floppy disk drives from the original iMac in favor of CD drives, a choice that had simplified the design process for a product that had become a sensation. His instinct that the original iPad needed a curve at the base had made it easier to lift the company's first tablet off a table. His harshness—"You suck!"—toward people who were responsible for imperfect prototypes and unsuccessful commercials had driven the brilliant work that had made the company great. It was impossible to walk around the headquarters he had dreamed up and not see features that he might have pilloried or improvements he might have made. For all its beauty, it was not quite perfect.

As the employees prepared to move, Cook decided to squeeze more of them into the building, boosting the total number of staff working there from the original plan of twelve thousand to fourteen thousand people. The decision to cram more staff into the same amount of space was a stroke of operational efficiency. In the three years since construction had begun, Apple had increased its workforce by a third, from 92,600 to 123,000 people. Putting a third more people into the same space meant that the open-floor plan contained more desks and engineers enjoyed less space. The cramped floor plan was a daily reminder of the high-tech cider press Apple had become.

OFF CAMPUS, Apple Park became a curiosity. Academics and historians considered it the most lavish in a series of architectural projects being undertaken by tech giants. Google, Facebook, and Amazon were also in the process of trading the bland, low-slung office buildings predominant in Silicon Valley for corporate palaces that evoked their booming market values.

Jobs had sparked the architectural bonanza in 2010 when he had Foster + Partners begin work on the closed loop that seemed like a physical manifestation of Apple's culture of secrecy and control. Facebook had followed, with the architect Frank Gehry designing a campus of community coffee shops and offices lined with plywood, an informal homage to its hoodie-wearing CEO, Mark Zuckerberg. Not to be outdone, Google's parent company, Alphabet, had tapped the Danish architect Bjarke Ingels to imagine a soaring glass canopy over a public walkway, a nod to the accessibility of information its search engine made possible.

The glitzy headquarters followed a long line of monuments to wealth and power that dated all the way back to Egypt's pharaohs. Their arrival seemed fitting for businesses that had become the dominant forces of modern capitalism, with platforms as indispensable for bankers on Wall Street as for villagers in Bangladesh. There were no limits to their growth. Their tentacles could stretch from smartphones, search engines, or social media into seemingly unrelated industries such as finance and health. It was natural for them to indulge their growing self-importance with distinctive buildings, even if they knew that every shrine could also be a tombstone.

Previous temples of capitalism had presaged reversals of corporate fortune. Companies flush with cash had often built with bravado during a boom only to discover later that they had planted the steel at their pinnacle. In 1970, American Can Company had moved to a 155-acre campus in Greenwich, Connecticut, before beginning a series of layoffs and divestitures. Enron had been in the process of building a fifty-story corporate headquarters when it had filed for bankruptcy.

Silicon Valley, a business landscape of disruptive triumphs and speedy retreats, defined the trend. The 175 acres Apple had bought had been abandoned by Hewlett-Packard after the PC market stagnated. Facebook had taken over the fossilized remains of Sun Microsystems' headquarters, which had been completed in 2000, just as the dot-com implosion had devastated Sun's business. After taking over the campus in 2011, Mark Zuckerberg had left Sun's sign visible to remind employees about the risks of becoming comfortable with success.

Apple's high-priced headquarters raised fears that the same thing might happen to it. The park the iPhone built was completed ten years after the company's best-selling product had debuted. The device still accounted for two-thirds of the company's sales, and the Apple Watch, AirPods, and other new products had yet to achieve comparable unit sales. Academics wondered: *Will Apple's palace prove unwise?*

IN THE MIDST of the design team's preparations to move to the new campus, the *New York Times* published an obituary for tech's tie-up with fashion. The convergence of the two industries had been driven largely by Ive and the Apple Watch, but the *Times* fashion critic Vanessa Friedman said that the love affair had cooled. She called the Apple Watch a "yawn."

That condemnation added to the pressure that Ive and the team faced as they settled into their new workspace. They were the last to move to Apple Park. Staff joked that it was because they wanted the rats to be under control before they relocated. In truth, the company needed time to install the heavy machines it used to make prototypes. To no one's surprise, the designers had one of the best views on campus, a fourth-floor perch with a view across the park's interior.

Ive hoped that the new workspace would encourage cross-division collaboration. For the first time, the software and industrial designers would be on the same floor and share a common area. He envisioned them bumping into one another and feeding one another ideas that would improve the way devices looked and how people would interact with them. He knew that achieving that goal would take time. After years of working apart in a secretive company, the groups would need encouragement to see themselves as members of the same team. But he was optimistic about the possibilities.

On one of his first evenings in the design studio, Ive saw a group of designers gather at one of the wall-size windows looking onto the park. He walked over to see what had caught their attention and found that they were watching the sunset. He stood alongside them as the sky changed colors. In their decades together, it was the first time he could recall having stopped work to look at the sky.

IN LATE JUNE, Ive traveled to London for an event at the Royal College of Art, which had recently appointed him chancellor. The honorary role required him to attend an annual dinner for students and supporters of the United Kingdom's leading art and design school. Dressed in a powder blue suit and suede Clark Wallabees, he took a seat at a tech-meets-fashion table with Laurene Powell Jobs, Marc Newson, Naomi Campbell, and Tony Chambers, the former editor in chief of the influential design and architecture magazine *Wallpaper*.

As the group settled into their chairs, Newson grabbed a waiter and asked that their champagne flutes be replaced by open wineglasses.

Newson, Ive, and Chambers were all friends of the chef de cave at Dom Pérignon, who had taught them that flutes deaden the richness of champagne. They told the table that a traditional wineglass allowed champagne to breathe and opened up tasting notes and aromas. It was something that they considered a detail of design.

After the wineglasses arrived and the champagne was poured, Ive turned his attention to Chambers. The two had known each other for years, with Chambers regularly interviewing Ive after new product releases. Ive had even served as guest editor of *Wallpaper* in 2017, designing a blank white cover with only the magazine's banner in rainbow hues.

Recently, Chambers had left the publication, and Ive asked what he had been doing.

"Consulting," Chambers said.

Ive listened as Chambers explained that he'd opened his own firm and begun to build out a small business. The former editor said that opening a graphic design firm was something he'd dreamed of doing when he had graduated from art school. Instead, he had found himself at *Wallpaper* and had risen to the top of its masthead. Then, one day, he said, the dreams of his youth had started to niggle at the back of his mind and he had begun to wonder: *Should I leave* Wallpaper *before it's too late?*

It had been an unsettling question, he confessed. He'd had job security and could have run the magazine until he retired. But he'd felt the creative urge to do something new, so he'd decided to step back from the business and spend a year working part-time. Then, one day, he had announced he was leaving.

Ive looked at him with a mixture of surprise and understanding. "Oh, yeah," he said. "I'm thinking of doing something cheeky like that."

Chapter 22

A Billion Pockets

In a remote corner of Utah, a contemporary hotel blended into a wind-carved mesa. The sharp angles of its stone walls complemented the desert landscape around it.

In late November, Tim Cook arrived there alone.

The Amangiri hotel and resort was an aspirational destination for affluent adventurers. Its thirty-four suites cost about $2,200 per night. Each opened onto a private patio with a fireplace and an infinite view of clear night skies.

Nature inspired and motivated Cook. He considered hiking to be the ultimate form of meditation. Visiting national parks ranked among his few hobbies outside of work. He had supported naming the company's new conference rooms after the outdoor meccas, including the Grand Canyon Room just outside his office. From Amangiri, he could reach Zion National Park, one of his favorites, where blue sky and brilliant green cottonwood trees framed towering red, pink, and salmon sandstone canyons. A little over a year after Ive and his friends had held a fiftieth-birthday celebration at the Aman in Venice, Cook had come to its sister hotel in Utah to unwind by himself.

On Thanksgiving, he took a seat at a table in the dining room, whose floor-to-ceiling windows looked out across empty plains. The restaurant hewed to the culinary traditions of the American Southwest with a standing menu that included pasture-raised chicken confit and salmon with saffron cream. As Cook quietly worked through that evening's meal, a young girl seated nearby noticed that he was alone.

"Should we ask him to join us?" she asked her mother.

Touched by her daughter's thoughtfulness, the woman looked over at the man and prepared to extend him an invitation to join her family. But just before she spoke, a realization came to her: *I know that man. In fact, everyone knows that man. That's Tim Cook.*

The CEO of the world's largest company finished his meal in solitude. The humble workaholic would spend his remaining days in that rural outpost recharging on peaceful walks and visits to the spa. Hard times lay ahead. Thanksgiving fell right before the busiest period on Apple's calendar, when it booked about a third of its revenue on Christmas purchases of iPhones, iPads, and Macs. Hiking around Amangiri would provide peace ahead of the hustle, and, as he told a fellow guest, "They have the best masseuses in the world here."

THE MANDATE FROM CUPERTINO was urgent: Cut production. Now.

As 2018 drew to a close, Apple slashed its orders for the components that went into its three latest iPhone models. Though the iPhone X had breathed life back into the iPhone franchise with its higher price tag lifting sales to a new record, its successors, the XS, XS Max, and XR, looked identical and cost just as much. The XS models started at $1,000, while the XR, with its lower-quality display, cost $749, a $100 increase from the previous year's entry-level model. Cook and company had had high expectations for the XR, particularly in China, where Cook had marketed the phone to his million followers on Weibo, the country's equivalent of Twitter, saying,

"Wonderful to see so many people in China enjoying the new iPhone XR." The message reflected the company's hope that the device would become popular in the world's largest smartphone market. Instead, it was a dud.

China's competitive landscape had changed. Huawei, the country's largest smartphone maker, had hit the market with a series of phones that had better features and lower prices than Apple's, including the P20, which cost a third as much as the XR but had more storage, a better camera, and a bigger battery. Cost-conscious Chinese customers were buying Huawei phones in droves. Phone users in China lived inside a superapp called WeChat that made Apple's slick software less compelling because it could be used for everything from messaging and payments to social media and car hailing. The rising trade tensions between the United States and China added another challenge. Negative headlines in the news about tariffs had hurt Apple's brand. Once the market leader in China, over the next several years the company would drop to become the fifth biggest phone seller there.

As boxes of iPhone XRs went unsold, Cook and company rushed to overhaul their manufacturing plans for the phone. Their directive to cut production rippled through the supply chain. Foxconn warned investors about the capsizing smartphone demand and scrambled to cut $2.9 billion in expenses to salvage its bottom line. Chipmakers, display providers, and a laser supplier all cut their quarterly profit estimates. It would have been rare for one company to do so; it was unprecedented for so many businesses to do so simultaneously. The iPhone empire appeared to be collapsing.

Cook turned to his sales and marketing teams for solutions. Their investigation found that the iPhone woes went beyond China to Europe and the United States, where customers who had previously bought a new phone every two years were delaying upgrading. The change in their purchase habits testified to the facts that the iPhones they owned generally performed the functions they needed and carriers were no longer offering subsidies that reduced the price of a new model. Apple tried to combat the sticker shock of the full-priced iPhones by providing a subsidy of its own: it developed a trade-in program that reduced the price of a new iPhone by as much as several hundred dollars when people turned in an old model. The company sold the older phones to middlemen, who unloaded the used phones overseas.

In early December, Apple began advertising the iPhone XR for $449 with an asterisk, noting that the discount price was available only to people who traded in an older iPhone. The marketing team sweetened the deal by boosting the trade-in value of older iPhones by $25. In essence, Apple was paying people to buy new phones. The push turned the usually relaxed Apple Store environment into something like a car lot where staff were encouraged to tout the trade-in discounts alongside computer monitors that showcased pictures of the XR beneath its "limited time" discount price.

On Wall Street, the fallout was swift and punishing. Apple lost more than $300 billion in market value in late 2018, slightly more than what Walmart had been worth at the turn of the year. Its nearly decade-long run as the world's most valuable company ended when it was supplanted by its onetime nemesis, Microsoft.

COOK HAD MISREAD THE MARKET. Apple had failed to move the number of iPhones he had expected when the company had provided Wall Street with sales projections for the Christmas period. As he scrutinized the daily sales figures, he could see the gap widen between what had been promised and what would be delivered. Securities laws required that he disclose the shortfall.

Shortly after the markets closed on January 2, 2019, Apple released a letter from Cook to investors, reducing the company's quarterly sales forecast for the first time in sixteen years. He said that Apple now expected sales of $84 billion instead of the $89 billion it had forecast, a surprise cut that swung the company from projected revenue growth to a 4.5 percent decline. Cook blamed weak iPhone demand and China's slowing economy for the shortfall, adding that Apple hadn't expected the magnitude of the slowdown.

"We believe the economic environment in China has been further impacted by rising trade tensions with the United States," he wrote. "As the climate of mounting uncertainty weighed on financial markets, the effects appeared to reach consumers as well, with traffic to our retail stores and our channel partners in China declining as the quarter progressed."

For Cook, it was easier to blame the economy and trade sanctions than acknowledge the reality that customers were tiring of incremental improvements to the iPhone. He omitted that China accounted for less than half of the decline in iPhone revenue. Sales declined in Europe and the U.S., as well. But investors didn't learn that until later when they pored over the numbers the company released. The $9 billion decrease in iPhone revenue had been the largest drop in a single quarter since 2007. The company's most important product was running out of steam.

That afternoon, Cook sat down with CNBC reporter Josh Lipton at Apple Headquarters to explain what had gone wrong. The CEO, who often smiled, joked, and spoke energetically during TV interviews, slouched forward on a stool and tightened his lips before speaking slowly and gravely about the problems Apple was confronting.

"We saw, as the quarter went on, things like traffic in our retail stores, traffic in our channel partner stores, the reports of the smartphone industry contracting," he said. He tried to reassure investors that Apple had a plan to shore up sales: "We're not going to sit around waiting for the macro to change. I hope that it does, and I'm actually optimistic, but we're going to focus really deeply on the things we can control."

THE FOLLOWING DAY, the Apple share price fell 10 percent and the company lost $75 billion in value. The single-day decline was Apple's biggest in six years and sank its valuation to a level it had not seen since February 2017. It shook the U.S. economy. The company had become one of the most widely held institutional stocks, included in mutual funds, index funds, and 401(k)s. Thanks in part to Warren Buffett and Berkshire Hathaway, everyone from grandmothers in Florida to autoworkers in the Midwest had an interest in Apple's business. They all suffered.

Cook sought to turn the tide by reviving foot traffic at Apple Stores. He held a series of meetings with his executive leadership team, including head of retail Angela Ahrendts, about stores' failure to draw customers. The former Burberry chief executive had become a polarizing figure during her five years at Apple. Initially heralded for bringing style to Apple's all-male executive team, the fashionably dressed executive had fallen out of favor with some of her peers. Rumors spread among her colleagues that her compensation package included large sums for a hairstylist and a private driver to chauffeur her in a luxury car. (A spokeswoman denied that this was part of her compensation.) Such perks were common across corporate America but had been frowned upon at Apple while Jobs had been alive. She had struggled to win her colleagues over with ideas as well. They had snickered at her yearlong effort to create a fleet of buses that would wheel across China as traveling Apple Stores. When she had raised the idea with Cook, he had cut her off and said there was no need. Finding it impractical, he had killed it.

Throughout her time at the company, colleagues said, she had struggled with Cook's expectation that executives be able to both provide long-term vision and attend to minuscule details. Balancing those two duties as Cook peppered staff with question after question could be exhausting. After one meeting, she had gone into a women's bathroom, closed the door behind her, and taken a deep breath. She had looked at a colleague, wide-eyed and frazzled after her recent round of interrogation. "How do you do it?" she asked.

As Apple's iPhone business sputtered, Cook's questions about the state of the company's retail business intensified. He wanted to know why traffic to stores had fallen and what was being done to reverse it. People who worked with Ahrendts said that she wouldn't always have figures ready or have the deep understanding of granular numbers that Cook demanded.

Amid a series of heated meetings in early 2019, Cook and Ahrendts agreed to part ways. An abrupt announcement in February that she would leave ignited rumors that she had been fired. Apple's public relations team swung into action to suppress the rumors, pushing a story that the departure had been planned. Indeed, Ahrendts told friends, she was ready to leave. She had spent five years at the company and made $173 million. She was ready to exit an empire where an aloof CEO assaulted staff with interrogation.

To fill the vacancy, Cook turned to one of his longtime lieutenants from operations, Deirdre O'Brien. She had been at Apple in 1998 when Cook had arrived and had distinguished herself as a key member of his operations team by skillfully forecasting the demand for coming Apple products. She was the opposite of Ahrendts in many ways. She kept her dark hair cut short and dressed in bland blazers and dark denim. She reveled in numbers and details, an outgrowth of having risen through Cook's organization. She immediately went to work with Cook and CFO Luca Maestri to counter the iPhone slump, adjusting prices around the world.

Shortly after Ahrendts's departure was announced, Jimmy Iovine revealed that he, too, planned to part ways with the company. Apple Music, which he had helped build, was gaining subscribers at a steady clip, and the Hollywood content that he had pushed for was being developed by the former Sony executives Zack Van Amburg and Jamie Erlicht. In his estimation, Apple had become too big and too bureaucratic to keep pace with popular culture. It wasn't something he could change, so at sixty-five years old, five years after the $3 billion sale of Beats to Apple, he decided to retire.

Eddy Cue, who ran services, replaced Iovine with a longtime member of the Apple Music team. Oliver Schusser, who ran the company's international content divisions, was the opposite of his energetic Brooklyn-born predecessor. A low-key, efficient German, he had been with Apple for nearly fifteen years and brought a polished operations bent to the music organization.

In the span of a few months, two of Apple's most senior creatives had exited: one steeped in fashion, the other steeped in music. Much as Cook had remade Apple's board to the frustration of Jony Ive, he had begun to remake the company. With Apple mired in crisis, he turned to disciplined executives skilled at what he knew best: operations.

ON A BEAUTIFUL SPRING EVENING, stars descended on Apple Park. They had come for a victory celebration that had been years in the making.

Long before the latest iPhone crisis, Cook had been planning to pull back the curtains and share with the world the new product that he'd been working to create. By late March, he was ready to show off Apple's long-promised services.

That Sunday, Oprah Winfrey, Jennifer Aniston, and Reese Witherspoon circled their way around the massive headquarters on their
way to the cafeteria. Directors such as J. J. Abrams and Jon Chu
followed the same path. They were joined by entertainment agents
and producers, who filled the atrium of Caffè Macs with the kind of
boisterous, showy small talk customary at movie premieres and TV
screenings.

The gathering was so star-studded that everyone in the room
looked familiar. A walk to grab a bite of food might take you past
Jennifer Garner or Ewan McGregor. A trip to the bar might put you
side by side with *Breaking Bad* actor Aaron Paul or *Downton Abbey*
actor Michelle Dockery. Socializing with the famous was familiar to
most of the guests. What was unfamiliar was socializing alongside
celebrities on an isolated corporate campus a mile from a Residence
Inn and a Bed Bath & Beyond.

The surroundings evoked some mild irritation among the Hollywood crowd. They were accustomed to dictating how shows ran.
Instead, they had journeyed more than three hundred miles for a secretive event that Apple had refused to share any details about. Director M. Night Shyamalan had flown in from Pennsylvania, unsure
if he even had a role in the keynote session. To the irritation of an
industry that loved to gossip, Apple was keeping everything secret.

As the guests bantered, some marveled at the size of the four-
story-tall houseplants in the company cafeteria. Others joked about
the disconnect between their lavish $5 billion surroundings and a
drinks menu that included only wine and beer. The Hollywood set
favored light spirits, vodka sodas, or gin and tonics with lime. The
cheap-wedding vibe was another of Apple's mysteries, right alongside
its request that participants not publicize that they were there.

Before the event wrapped up, many of the actors and directors
gathered across campus at the Steve Jobs Theater. Apple brought in

the celebrity photographer Art Streiber, who often shot *Vanity Fair*'s multipage layouts of society's most important people. He arranged a curated group of the thirty-one celebrities across a series of platforms in the theater's circular lobby. Cook, who wore a high-collared black zip-up sweater and black pants, planted himself in the middle of the stars. He crossed his arms like an assembly-line supervisor and stared at the camera with a terse grin. A beaming Oprah sat to his right, a relaxed Abrams to his left. Aniston and Witherspoon were nearby. Cook looked confident and assured. He had paid Hollywood to come to him, testifying to the financial firepower of his global empire.

THE NEXT MORNING, perky greeters in Apple-logo T-shirts blurted out "Good morning!" and "Hello!" as the Hollywood guests ascended the small hill to the Steve Jobs Theater for a keynote speech. As more than five hundred members of the media passed by, agents and producers from Los Angeles snapped photographs of the company's spaceship headquarters that had hosted the prior night's reception. Though it was a year old, the ring of glass was still being treated like the eighth wonder of the world.

The Hollywood contingent joined an eclectic crowd inside the theater. The room filled with the tech reporters and analysts who regularly attended Apple events, as well as newcomers who wrote about entertainment, gaming, and credit cards. Everyone crammed together in front of a circular bar where staff poured coffee for guests. A cohort of bankers from Goldman Sachs trailed CEO David Solomon through the crowd. Agents and producers nibbled on breakfast bites and waited for the show to start.

Cook was downstairs in the greenroom, preparing to take the stage. It was a small, peaceful space with a sofa and a makeup area. Days before a previous event, he had asked that the subtle humming sound from a nearby generator be muted. Engineers and maintenance staff had mounted the giant generator on rubber pads. The 250-square-foot room had been quiet ever since.

Once the crowd slipped into the theater's $14,000 leather seats, Cook strolled onstage to boisterous applause and shouts from his own staff. "Thank you!" he said, waving. He clasped his hands prayerfully before his chin and assured the audience that day's event at Apple was going to be very different.

"For decades, Apple has been creating world-class hardware and world-class software," he said. "We've also been creating a growing collection of world-class services, and that is what today is all about." He paused. "So what is a service? Well, if you look it up in the dictionary—" As he spoke, the definition of the word *service* appeared in white on the black screen behind him:

SERVICE | NOUN: the action of helping or doing work for someone

Turning to the dictionary was an elementary school departure for the company. Jobs's presentations had enthralled people because they had been centered around it-just-works devices that were so intuitive that he had barely had to explain them. The magic had come from pulling them out of a box. But Cook was introducing a financial construct, not a product. He had started using the word *services* in 2014 after the company had changed a line item on its quarterly reports from "iTunes, software and services" to "Services." The category had been changed as iTunes sales had faded and Apple Music was being developed. After promising Wall Street two years earlier that Apple would double its "services" revenue by 2021, he had made the word a focal point of his update to investors. He had ticked off the number of subscribers who paid for apps each month and hyped the performance of Apple Music. It became the drumbeat of operations.

The strategy had been born partly out of necessity. Cook could see a future in which the App Store, the lifeblood of services, would dry up. Apple was coming under fire for the 30 percent toll it charged developers to sell their products to iPhone users. Two weeks earlier, Spotify had filed a complaint with the European Union that Apple was stifling competition by making it difficult for Spotify to market itself to iPhone users without giving Apple its cut. The subsequent antitrust investigation presaged a future when Epic Games, the maker of Fortnite, would take Apple to court with a similar complaint aimed at eliminating its fees. The pressure could cut into App Store sales, decimating its balance sheet and pummeling the company's share price. Cook knew that one way for the company to protect itself would be to introduce a suite of its own apps.

The strategy challenged the company's pedigree. Apple's strength had always been in making incredible devices with sophisticated software. Its track record in services was mixed. iTunes had been a runaway success that had transformed the music industry, but Apple Maps had been a bust. MobileMe, a 2008 online service for email, contacts, and calendar, hadn't worked, and Siri, Apple's novel voice assistant, had fallen behind its rivals in performance. In the absence of the next game-changing device, Cook was betting that he could persuade customers to stick with iPhones by getting them as tethered to Apple Music and other services as they had been to the devices themselves. He was digging a second moat.

FOLLOWING THE FORMULA of Apple's presentations, Cook transitioned the audience from service to service as he built to the show's most important unveiling. He started by introducing Apple News+, a service that cost $9.99 monthly and provided unlimited access to more than three hundred magazines, including *Vogue*, *The New Yorker*, and *National Geographic*. He introduced a credit card, the Apple Card, developed in collaboration with Goldman Sachs and

Mastercard. And he introduced Apple Arcade, a monthly video-game subscription service.

As Cook spoke, the Hollywood crowd grew restless and bored. Magazines, credit cards, and video games lacked the sizzle they were accustomed to from years of watching Jobs unveil world-changing devices.

Amid the growing apathy, Cook planted himself at the center of the stage and said that Apple planned to extend its work in entertainment beyond making Macs that directors used to edit films. He said that it was going to take a direct role in telling Hollywood stories. "We partnered with the most thoughtful, accomplished, and award-winning group of creative visionaries who have ever come together in one place to create a new service unlike anything that's been done before!"

Cook's voice rose to a shout, as it often did when he tried to convince the audience that Apple was doing something truly innovative. But the Hollywood set was unmoved. Their industry had been telling stories for more than a century. Beneath the raised voice and promise of something unique, they saw a tech company hell-bent on adding more TV shows and films to a world already awash in entertainment.

The screen behind Cook cemented their impression when it came alive with a billow of white clouds that gave way to an Apple logo and the word *TV+*. For much of the crowd, it looked and sounded like an intro to an HBO show.

Cook ceded the stage to a revolving door of hired talent who tried to sprinkle stardust over a company whose products had lost their sparkle. Former Sony executives Zack Van Amburg and Jamie Erlicht explained the mission of Apple TV+ by queuing up a video showing the directors Steven Spielberg, Sofia Coppola, Ron How-ard, and Damien Chazelle talking about how they told stories. When the video ended, Spielberg took the stage to a standing ovation and talked about his production company's show *Amazing Stories*, which would appear on Apple TV+. Afterward, Reese Witherspoon, Jennifer Aniston, and Steve Carell shared details of their Apple program, *The Morning Show*.

The script felt familiar. The entertainment industry compared it to the so-called up-front presentations TV networks do for advertisers each fall. At those events, executives and actors provide overviews of the upcoming season of shows in a bid to sell commercial time. Apple's service wouldn't have commercials. It would cost an as-yet-undetermined monthly fee, and it would debut at an as-yet-undetermined date. The lack of details frustrated the crowd.

But the audience's disappointment didn't stop Cook. When he returned to the stage, he said he had one more thing.

The room went dark as a video began to play with white words on a black screen. The video said that the broken world needed someone with a voice that could create connection; it needed a voice that had been missing.

When the lights rose, Oprah Winfrey stood onstage in a white blouse and black pants. The crowd erupted. People sprang to their feet to scream and applaud. Winfrey let their enthusiasm linger. "Okay," she finally said. "Hi." The crowd laughed as her friendly tone filled the room.

"We all crave connection," she said. "We search for common ground. We want to be heard, but we also need to listen, to open, be open, and contribute." She said that was why she had signed on to host a show on TV+: Apple would allow her to do what she had done for years in "a whole new way."

"Because"—she shrugged her shoulders and held up her hands in surrender, then she leaned forward as if to share a secret—"they are in a billion pockets, y'all," she said with a shake of the head. "A billion pockets."

As she finished, Cook walked onstage, applauding. He leaned down to hug her. "You're so incredible," he said softly.

Oprah's arm encircled his waist as he brushed a tear from the corner of his eye. Stunned by the emotional display, Cook's colleagues could only speculate that the small-town Alabamian was overwhelmed by the role he and Apple played in recruiting Oprah to be the foremost face of the TV programming they wanted to bring the world. She smiled and laughed. Bringing out people's emotional depth was her superpower. She had brought countless people to tears during her career. For her to unlock the emotions of a CEO who seldom betrayed his thoughts or feelings only added to her charm.

"I will never forget this," Cook said, laughing. He dabbed his eye again. "Sorry," he said to the crowd.

Behind him, a black-and-white image appeared of the previous night's photograph of all of Apple's stars but one. Apple had chosen a photograph Strieber had taken without Cook alongside the Hollywood storytellers. Instead, he stood onstage in full command of the ensemble that he had assembled.

"These are people who we admire for their great voices, incredible creativity and wonderfully diverse perspectives," Cook said. "They've impacted our culture, society, and we're so excited." His voice broke, and he stopped speaking for a moment. "And so humbled to be working with them."

As the crowd began to depart the theater, some were confused. The emissaries from Hollywood wanted to know more about the TV plan. The financial world scrambled for more details about the credit card. The publishing industry sought more information on the news app. Each faction was blinded by its tunnel vision. In the process, the revolution nearly slipped by unnoticed.

After years of being hounded by the same question—What's

the next new device?—Cook had finally delivered his answer: There isn't one.

His message hadn't been aimed at Main Street; it was for Wall Street. He wanted investors to see that Apple was making a major shift. Rather than its products creating glory, Cook outlined a future in which Apple basked in the glory of others. He didn't want to merely update the iPhone every year; he wanted people to pay Apple subscription fees for the movies they watched on that iPhone. He didn't want to enable digital payments; he wanted Apple to be the processor of every transaction. And he didn't want Apple to make the screen on which people read articles; he wanted to sell access to the magazines they read.

For years, Cook had seen new revenue opportunities in each of those businesses. He had plotted a path to get there, buying Beats in 2014, courting Hollywood agents and directors in the years that had followed, and forging strong ties with Goldman Sachs throughout that time. He saw in all of it a way to shed the burden of a device business that was running out of juice and enter a world of services that promised unlimited growth.

As Wall Street digested the strategy, Apple's share price soared. It nearly doubled by the end of the year. The longtime darling of Main Street had become a darling of Wall Street. Cook's conquest was complete.

Chapter 23

Yesterday

The stakeout began shortly after dawn. In the spring of 2019, a reporter with the technology news site The Information parked a black Nissan sedan in the shadow of San Francisco's hulking mansions. He stared across the street toward a two-story brick carriage house with a green garage. He watched, and he waited.

Rumors were once again sweeping across Silicon Valley that Jony Ive had disconnected from Apple. People close to the design team said he had stopped going into the office and shifted much of his work to a carriage-house studio a few blocks from his Pacific Heights home. The $3 million building had a one-bedroom apartment above a large garage where Ive had set up a glass conference table for product reviews. Apple staff began to come and go, irritating neighbors, who complained that a business was operating without a permit on their quiet residential street.

The reporter had come to find out if it was all true. Was Ive checked out of Apple but still holding on to power?

As the minutes passed, the street came to life. Construction crews working on nearby homes carried tools from recently parked trucks.

A cleaning woman arrived and walked into the carriage house. A deliveryman later dropped packages at the door.

The reporter watched it all, hoping to catch a glimpse of Ive or someone from Cupertino. Eventually, he climbed out of the car and walked down the street to get a closer look. He stared at the building to see if there were any security cameras and noticed that the lights in the apartment upstairs were on. Then someone inside closed the blinds.

THE INVITATIONS ARRIVED by email in early May. They featured a curving wheel of primary colors. Six lines of script followed, each written in one of the colors of Apple's original logo: blue, purple, red, orange, yellow, and green.

**Tim Cook, Jony Ive and Laurene Powell Jobs
invite you to a very special evening at Apple Park
in tribute to Steve.
Please join us for an evening of music,
food and celebration.**

May 17, 2019

Nearly fifteen years after Steve Jobs's earliest conversations about a new headquarters, three of the people who had been closest to him were uniting to celebrate the fulfillment of his vision with a grand opening of his final product.

In the weeks leading up to the event, a construction crew worked inside Apple Park to raise six semicircular aluminum frames in the colors of a rainbow. They formed arches over a stage that Ive had designed. The curved pieces of aluminum had been rolled individually

by a machinist over twelve days and transported on a custom cart
to Apple's campus. Each frame radiated with neon undertones. They
stacked up from small to large, creating a band shell reminiscent of
the famed Hollywood Bowl. The company named it Apple Stage.

In a message to staff about the stage, Ive said he had wanted
to create something that was immediately recognizable. "The idea
for the rainbow was one of those rare occasions where the earliest
thoughts worked on a number of different fronts," he said. "There is
the resonance with the rainbow logo that's been part of our identity
for many years. The rainbow is also a positive and joyful expression of
our inclusion values and I think that one of the primary reasons the
idea resonated so immediately and profoundly with us was . . . a semi-
circle relates so beautifully and naturally to the form of the ring."

Ive was excited about the event and planned to bring his wife,
Heather, and teenage boys, Harry and Charlie, to see it. But the day
before the celebration, he received terrible news. Mike Ive, his father,
had suffered a severe stroke. He had been rushed to a hospital near
his home in Somerset and then transported to London. Doctors were
concerned that he might die.

Ive immediately flew to England. He owed his life and career to
his father. From the hovercraft of Jony's childhood to teenage discussions about how things were made, Mike Ive had nurtured his
son's interest in the art of making. He had armed Jony with a deep
appreciation of the materials in products and helped him become a
skilled draftsman capable of bringing his imagination to life with the
stroke of a pen. Those tools had enabled him to graduate in the top
of his class from Newcastle Polytechnic and put him onto the radar
of Robert Brunner, who had brought him to Apple, where he had
teamed up with Jobs on the iMac, iPod, and iPhone.

Apple's cultish commitment to secrecy meant that Ive had shared
far less of his work with his father than he would have liked. Over the

years, friends had often asked Mike what product Apple might release next. Mike could only shrug. "I haven't a clue," he'd say. "Jony won't tell me." Instead, Ive often waited until his dad visited San Francisco. He would take his father to an Apple Store and they would browse through it together, discussing everything on display. It was a reversal from decades earlier, when father had taken son around stores in England, explaining how products were made. The onetime student had become the master.

Eventually, Ive would walk to the register and buy his father the latest iPod or iPhone. It was his way of scaling the wall between Apple and the outside world, a wall that extended even to his family.

As Ive's plane arrived in his homeland, he was filled with conflicting emotions. In California, he had left behind a joyous event celebrating the creative partner who had turned his dreams of making beautiful products into reality. In England, he arrived with the worry that he might lose the man who had given him the skills to pursue those dreams.

AT APPLE PARK, employees sprinted across a football field–size grass lawn toward the end of the Apple Stage rainbow. Word had spread that Lady Gaga would be performing that night to celebrate the official opening of the new campus, and staff members wanted to be as close as possible for the spectacle.

Tim Cook took the stage wearing black. Jumbotrons beamed a close-up image of him across the undulating park with its fruit trees and meditation pool. He welcomed everyone and directed their attention to the video screens. An image of Steve Jobs appeared. The voice of the late Apple cofounder filled the interior of the shimmering corporate coliseum.

"Man as a toolmaker has the ability to make a tool to amplify an inherent ability that he has," Jobs said. "And that's exactly what we're doing here. What we're doing here is building tools that amplify a human ability."

Soon the crowd heard Ive's familiar British accent. In the days before he had left for England, he had recorded a video to commemorate the special occasion. As he spoke, images flashed across the screen of the $5 billion monument to Apple's empire.

"You know sometimes when you get up in the morning and in that delicious semiconscious state, you just know it's going to be a rare day," Ive said. "A day that you won't forget. That's just how I felt this morning."

Few in the crowd knew that Ive wasn't there. They listened as he recounted walking through Hyde Park in London with Jobs in 2004. "Perhaps not surprisingly, we talked about walking and parks and trees," he said. "Our very earliest thoughts were of built spaces that enabled strong connections between one another and nature."

Ive said that concept had been the foundation of Apple Park. After many years of making products for others, he and Jobs had relished having a chance to finally make a product for themselves. The campus the crowd was standing in the middle of was an ambitious tool that had required years of designing, prototyping, engineering, and building.

"Something happened at the end of the project that I will never forget," he said. "Something seemingly trivial, but utterly significant." He recalled watching the sunset with his design team and how that moment had captured everything he and Jobs had wanted Apple Park to be: people in communion with one another and nature. Afterward, he said, he couldn't help but think how profoundly fortunate they were to work together in such a beautiful place.

AFTER THE SPEECHES, Lady Gaga paraded onstage wearing a chrome helmet. She was trailed by a dance troupe in skintight jumpsuits who shimmered beneath the stage lights like disco balls.

"Are you ready to celebrate Steve Jobs?" she shouted as she took the stage. "His brilliance! His kindness!"

The crowd roared as her band started into the synth-pop beats of her early hit "Poker Face." More songs and costume changes followed as she whipped Apple's engineers into a frenzy. Then she slowed things down and grew serious. "I'd like to say also thank you to Jony for designing such a beautiful place," she said. "Thank you to Tim Cook. To all of you."

She closed with "Shallow," a ballad she had cowritten for the film *A Star Is Born*. Seated at a piano and accompanied by acoustic guitar, she began to softly sing the lyrics about a couple weary with the burdens of life. They want to escape the weight of responsibility and yearn for the refuge of anonymity. Her voice erupted as she hit the song's bridge and soared to the safe place that they imagined:

> *Crash through the surface, where they can't hurt us*
> *We're far from the shallow now*

Across the Atlantic, Ive was holed up in a hotel room at Claridge's, around the corner from the London hospital caring for his father. He was up late and racked with disappointment and worry. He hated that he wasn't in Cupertino with colleagues celebrating the completion of Apple Park and worried that his father, who had survived the stroke, might have long-term physical impairments. Amid the push and pull of those conflicting emotions, his phone filled with messages and videos from colleagues of Lady Gaga's performance. He felt overwhelmed as he watched a video of her thanking him and grew sentimental as he watched her performing beneath the spotlight on the stage that he had designed.

The tension between work and family was something Ive had wrestled with for years. In 2008, he had been on the cusp of leaving

Apple so that he could spend more time with his parents and share
his homeland with his boys, but when Jobs's cancer had returned,
he had decided to stay at Apple and carry on their work together.
He had extended that commitment after Jobs had died, wanting to
be sure that his mentor's company survived. As the years passed, the
possibility of spending more time in the United Kingdom had evaporated and Ive's commitment to Apple had morphed from responsibility to chore. His father's stroke meant that whatever moments
he got with him in the future would never be the same as what they
would have shared if Ive had left California sooner. The watchmaker
of a new generation had been defeated by the very thing he had
wanted to redefine: time.

IVE RETURNED TO CALIFORNIA as a changed man, yet his responsibilities at Apple remained. To ensure the new campus fulfilled Jobs's
vision for collaboration, he turned his attention to bringing together
the software and industrial design teams that now shared the same
workspace.

On a Tuesday evening in late June, he convened his teams in San
Francisco at the special effects and production studio Industrial Light
& Magic. Founded by *Star Wars* creator George Lucas, it had created
the magic in movies such as *Jurassic Park* and *Jumanji*. Ive had reserved the venue's four-hundred-seat theater for a private screening
of the movie *Yesterday*.

The film imagined a world in which a singer-songwriter awakens
from an accident and discovers that he's the only person in the world
who remembers the Beatles. Ive was personal friends with its screen-
writer, Richard Curtis, who had written *Love Actually* and *Notting
Hill*. The concept behind the film resonated with Ive. Jobs, after all,
had named Apple partly in homage to the Beatles' label, Apple Records, and wanted the company to be like the band, a place where col-

leagues would come together to create something greater than they could create independently.

As the theater's lights dimmed, Ive and the designers watched as a struggling musician named Jack Malik played songs in pubs to small crowds that mostly ignored him. After a Y2K-like event wiped the Beatles from history, Malik scrambled to preserve their music. He made a demo of Beatles songs that caught the attention of a record label executive, who thought he was the most brilliant lyricist since Bob Dylan. Soon he found himself in Los Angeles being turned into a star, forced to deal with the label's commercial expectations that his album would set sales records. Meanwhile, he feared that someone would tell the world that he was a fraud who was playing someone else's songs.

The eternal conflict between art and commerce sat at the heart of the film. Malik wanted to be loyal to the artistic integrity of the Beatles and let the greatness of their music succeed. The record label wanted to guarantee profits by selling him as a rock genius. In one scene, Malik entered a conference room to find three dozen marketing executives staring at him. They were there to tell him what they wanted to name his album. The top marketer opened the meeting by dismissing some of Malik's ideas. *Sgt. Pepper's Lonely Hearts Club Band* had too many words; *The White Album* presented diversity issues; and *Abbey Road* was just a place where people drove on the wrong side of the street. The marketer said that they had chosen the perfect album title, *One Man Only*. Malik's face sank as he watched his ideas for honoring the Beatles get crushed by the label's commercial machinery.

Some designers in the theater saw echoes of Ive's own journey. After Ive had arrived at Apple, he had struggled to get the company to release the twentieth-anniversary Macintosh he had envisioned. He had then dealt with the uncertainty of the company's brush with bankruptcy. The hardships had made the string of hits he had developed with Jobs—iMac, iPod, iPhone, iPad—seem surreal. Colleagues said that Ive's perfectionism had left him with the traits of someone with nagging impostor syndrome. He seemed to have a latent concern that he would somehow be exposed as a fraud. Then, as Apple had ballooned under Cook and become the world's largest company, he had found himself in a high-pressure commercial atmosphere. The intimate meetings he had once had with Jobs had given way to gatherings in conference rooms full of executives, each with opinions about the products he was making. Layers of abstraction had arisen as the company had become bigger and bigger, creating a disconnect between his work making new products and the customers using them.

After Jobs's death, cracks had formed in Apple's version of the Beatles. A parade of talent exited the company. The losses included Bob Mansfield, its hardware leader, and Scott Forstall, its software sorcerer. Many members of their teams followed. Then the design team had begun to fray with the departures of Danny Coster and Chris Stringer followed by Daniele De Iuliis, Ive's carpool friend, and watch and AirPods leaders Rico Zorkendorfer and Julian Hönig. Over the course of two years, Ive had lost a third of the team that had spent more than a decade together. The band was breaking up.

When the movie ended, Ive stepped in front of the group to speak. He was clearly inspired by the film and felt compelled to share his thoughts about why he had wanted the team to see it. He explained that it was important for them to always work together and foster an environment inside Apple where art and creativity could flourish.

"Art needs the proper space and support to grow," he said. "When you're really big, that's especially important."

A DAY LATER, the designers got a note asking them to clear their calendars for a meeting with Ive. The request was unusual. No one could recall being asked to cancel meetings on such short notice, especially without being given a reason. En masse, they wiped their schedules.

The abrupt cancellations ricocheted across campus, angering engineers and operations staff who were scheduled to meet with members of the design team. The other divisions typically needed the studio's approval to advance their work, and they didn't have time for delay.

"I can't believe you canceled this meeting," an engineer wrote a designer.

"Sorry," a studio staffer replied. "It's a Jony thing."

THE DAY OF THE MEETING, June 27, 2019, Ive gathered the designers in an open area on the fourth floor near the newly unified software and industrial design studios. He watched as the crowd of a hundred-plus people packed in shoulder to shoulder around pockets of low-slung gray couches. The summer sunlight streamed through the building's glass and brightened the ceiling with a warm yellow glow.

Ive gazed out at a group that was multitudes larger than the one he had inherited. Through a combination of political force and operational savvy, he had gradually expanded his team to include material science experts, ergonomic researchers, and textile engineers. He had assumed oversight over a hundred-plus software designers responsible for the way Apple's icons looked and its devices behaved. The collective gave him more control over product than he ever had under Jobs but created more demands on his time. The bureaucratic responsibilities of the team he had built had become part of his undoing.

Ive told the group that he had completed his most important project, Apple Park, the last product Jobs had left behind. He said that the design team was positioned for future success but his role leading it was finished.

The faces before Ive turned ashen. Some people stared at him blankly. Others seemed gripped with fear, their silence suppressing a chorus of alarm that welled up inside them: *Holy fuck! This is happening! Holy fuck! This is happening!*

For many, Ive's words slowed time like a car skidding into a collision. Some began to weep as he explained that part of the reason he was leaving was that he'd grown weary of Apple's bureaucracy. Though they seldom acknowledged it to one another, they knew that the company's startup culture had faded. Without Jobs, some thought, Apple had become a machine with a heart of stone.

Cook had emboldened Apple's finance team. He had given accountants and operations people more voice in decision making. Their influence had been visible in the 2015 iPad that had never been made, the audits of longtime partners such as the photographer Andrew Zuckerman, and the rejected billings of the architect Foster + Partners. Jobs had held fast to the idea that lawyers and accountants were there to execute the decisions made by the people at the company's creative core. But over time, the company's bureaucratic caboose had become its engine.

Plus there were the meetings. Ive had taken to working from his studio in San Francisco partly to avoid having his calendar become clouds of blocked-off time. On campus, meetings had mushroomed in size, taking on the shape of the full conference room depicted in the film *Yesterday*. Decision making had slowed down, paralysis had set in, and Ive couldn't stand it.

"I don't want to go to any more fucking meetings," he told the team.

EVEN AS HE LAMENTED what Apple had become, Ive praised the team and implored them to keep Apple true to its identity. To be intentional. To be resolute. To continue to strive to surprise and delight the world. He was excited about the possibilities of what they could achieve now that the entire design team had been brought together at Apple Park. They had new resources and equipment. Their shared space should foster collaboration. And though he wouldn't be there every day, he said that he was going to set up an independent design firm with his friend Marc Newson that would continue to work with Apple.

Ive had named the firm in honor of Jobs. In the video that had played to open the Steve Jobs Theater, his longtime collaborator had said that products should express appreciation to humanity by being made with care and love. For his agency's name, Ive had condensed that tenet to two words that spoke to the essence of what he wanted to convey with every product he made: LoveFrom.

What Ive didn't say was that the firm's first client, Apple, had agreed to a more-than-$100-million exit package. The company struck a deal that would prevent Ive from going to work for a competitor and could be renewed annually to keep Ive and LoveFrom on hand to contribute to future projects. The designer's payout was on par with the golden parachutes many corporations might offer departing CEOs.

Late that afternoon, after the stock market closed, Apple issued a press release announcing Ive's departure. The release outlined a new reporting structure. After fifteen years of reporting directly to the CEO, Ive's old design team—the group of aesthetes once thought of as gods inside Apple—would report to Apple's chief operating officer, Jeff Williams, a mechanical engineer with an MBA.

Epilogue

The alchemy of Apple has long depended on visionary pairs. It was birthed by Steve Wozniak and Steve Jobs, revived by Jobs and Jony Ive, and sustained by Ive and Tim Cook.

In the months and years after Jobs's death, Silicon Valley anticipated that Apple's business would falter. Wall Street fretted about the road ahead. And loyal customers were concerned about the future of a beloved product innovator.

A decade later, its share price was at a record high. Its market value had risen more than eightfold to about $3 trillion, and its dominance of the global smartphone market continued unabated. It had become the darling of Wall Street, even as it had lost some of its glimmer as a disruptive innovator. Most important, it hadn't become the Sony, Hewlett-Packard, or Disney that Jobs had once feared.

Its endurance and financial success were a testament to the men Jobs had deputized to lead the company forward. Cook, the operator, showed artistry in expanding Apple's empire into China and services, while navigating the diplomatic headaches that confronted the corporate nation-state he built. Ive, the artist, showed operational savvy in leading the creation of the Apple Watch and completing Apple Park, the major new enterprises that had launched after Jobs's death.

In an email reflecting on their leadership, Laurene Powell Jobs said the company's endurance would not have been possible without the contributions of both men. They had played to each other's strengths, she said, while sustaining their "shared love of Steve and Apple."

Yet their success was darkened by the disappointment of their corporate divorce. The dissolution of their partnership was inevitable. The two men shared a love of Apple but little else. As Apple swelled in size alongside the iPhone's explosive growth, Cook began to change the fabric of the company out of necessity to manage its increased scale. At his direction, Apple expanded the number of products it made, scrutinized the money it spent, and shifted its focus from hardware to services. The strands connecting Cook and Ive frayed.

Cook's aloofness and unknowability made him an imperfect partner for an artist who wanted to bring empathy to every product. The CEO's colleagues said he showed limited interest in their advice on how to keep Ive happy and creatively fulfilled. Despite their repeated encouragement, he seldom went to the design studio to see Ive's team work. He failed to turn for counsel to people inside Apple with a track record of managing artists like Jimmy Iovine. When Ive first broached the idea of leaving the company in 2015, Cook focused on determining a succession plan. In the eyes of those who worked with him, his interest was in protecting the company more than protecting the individual. It was right for shareholders, even as colleagues found it difficult to witness.

Ive was not without blame. Weary after his decades of labor and grief-stricken after Jobs's death, the would-be keeper of Apple's flame flickered out. He made mistakes along the way. After Forstall's ouster, he assumed responsibility for software design and the management burdens that he soon came to disdain. He took on the Leica project while juggling the watch development and Apple Park. He burned himself out. His agreement to go part-time in 2015 spared the company a short-term drop in its share price but created an unhealthy arrangement for himself, his team, and the company he loved.

COOK'S POSITION as the company's leader reshaped it in ways that Jobs might never have imagined and Ive ultimately couldn't endure.

Jobs had admired Bob Dylan because the singer perpetually reinvented himself. The late CEO of Apple had brought that spirit to the company, reinventing its PC line with the iMac, transforming it with the iPod from a computer maker into a consumer electronics giant, and solidifying its preeminence with the iPhone. That hat trick of innovation had made him a latter-day Leonardo da Vinci.

No one expected Cook to replicate that, not even Cook himself. Instead of nourishing innovation, the industrial engineer played to his strengths, delivering one of the most lucrative business successions in history by squeezing more sales out of the business he had inherited. It was a triumph of method over magic, persistence over perfection, and improvement over revolution. Whereas Jobs had given Apple its identity by orchestrating leaps and upending industries, Cook focused on preserving the product he considered to be Jobs's greatest: Apple itself.

He played the role of a steward who made the company more reflective of his personality: cautious, collaborative, and tactical. He built an ecosystem of products and services around his predecessor's revolutionary inventions and sustained its reputation for releasing best-in-class updates to its line of hardware and software. Through his work, he helped the company generate enough cash—$66 billion after debt in fiscal 2021—to sustain it for many years, even if it pulled all its products off store shelves. In doing so, he kept alive the possibility that Apple would be able to surprise and delight the world again.

As long as iPhone sales hummed along, the faithful could wonder about the projects going on inside the secretive company: Would it ever make the car it continued to pursue? Would it release the augmented reality glasses that were under development? What about the noninvasive glucose-monitoring system? Would any of those products ever reach an Apple Store?

ON MAY 21, 2021, Cook arrived at a courthouse in Oakland to take the witness stand on the last day of an antitrust trial against Apple. Fortnite creator Epic Games had sued the tech giant, alleging that it unfairly prohibited competing app stores on iPhones and forced developers to give it a 30 percent cut of sales. The case struck at the heart of the services business Cook had engineered.

Cook had fulfilled his promise to double the size of services, which delivered $53 billion in sales in fiscal year 2020. Its revenue equaled that of Goldman Sachs and Caterpillar. He continued to introduce new Apple subscription services, including one for fitness, and the company benefited from a surge in App Store purchases by people who were stuck at home during the COVID-19 pandemic. Investors cheered the sales growth. They began to see Apple as more than a legacy hardware business that rose and fell depending on the popularity of each iPhone release. The company's price-to-earnings ratio jumped from its average of sixteen times earnings to more than thirty times earnings in 2020. It was a change so dramatic that some investors called it violent.

As Epic's suit advanced, Cook chafed at the threat it posed to what he had built. People at Apple dismissed it as a proxy suit being driven by the company's longtime nemesis, Microsoft. Cook was confident that the law was on Apple's side. It didn't have a dominant share of the U.S. smartphone market, and courts had been reluctant to rule that its dominance over iOS by itself made it a monopoly.

But he needed to nail his performance that day to ensure that the App Store would persevere.

Over a span of four hours, a federal judge and Epic lawyers peppered Cook with questions about Apple's business. The interrogator didn't handle being interrogated well. Cook argued that the prohibition of rival app stores and the fees it charged developers allowed the company to vet apps and protect users from security flaws. When Epic's lawyers asked about whether Apple had calculated its profit margin for the App Store, Cook insisted that the company avoided discussing profitability. The answer came as a surprise, given that the CEO held monthly meetings with a finance team to pore over services revenue. In an attempt to assess Cook's honesty, the lawyer asked about the claims that Google had paid Apple $8 billion to $12 billion to be the iPhone's default search engine.

"I don't remember the exact number," Cook said. When asked if it was upward of $10 billion, he developed a case of amnesia. Though he rose every morning to look at sales figures and expected staff to know about promotions at carrier stores all over the country, he said he didn't know about Google's payments.

The performance was written off even by some of Cook's longtime admirers as an embarrassment. Internal documents unearthed during the trial found Apple itself calculated that its operating margins for the App Store exceeded 75 percent in recent years, though Cook insisted that the document was from a "one-off presentation." At a moment of deep skepticism about Silicon Valley, he seemed to lean in to his position as a principal player in an era of tech oligopoly.

Yet the judge's verdict vindicated Apple's business practices. The company won a near total victory, with its App Store for video games deemed legal. The victory came with a caveat. The judge ruled that Apple could no longer block apps from directing customers to websites and other external payment systems where Apple wouldn't be able to collect its 30 percent cut. The ruling illustrated the challenges ahead for the App Store, which seems to be locked in a game of Jenga where regulators and developers are slowly pulling away blocks of a once-solid business. It appears to only be a matter of time before the App Store business is curtailed.

In addition to the Epic suit, Apple faced similar antitrust charges in Europe over allegations that it favored the distribution of Apple Music over Spotify. The Justice Department also launched an investigation that was expected to target the App Store. Between the trio of cases, it was clear that the services business that Cook had built would have to change. Indeed, the company began making concessions, lowering the fees it collected from some developers from 30 percent to 15 percent.

China cast a similar uncertainty over Apple's future. Under President Xi Jinping, the country's government had become more strident and nationalistic. The Communist Party had assumed control over Hong Kong and eliminated pro-democracy newspapers such as *Apple Daily*. In the remote region of Xianjing, the government had opened reeducation camps for Uyghurs, an ethnic minority group. Some Uyghurs had allegedly been forced to labor in factories, including those of seven suppliers connected to Apple. The company said it had found no evidence of forced labor in its supply chain, but the issue raised new questions about Cook's willingness to make compromises to maintain operations in the company's second largest market. For an executive who quoted Martin Luther King, Jr., and spoke at length about human rights and privacy in the United States, he took no such stands in China.

Taken together, Cook had engineered expansions of the services business and China operations that built incredible value for the company and its shareholders but grounded much of Apple's growth over the past decade on the shifting sands of an everyday business subject to the whims of regulators and autocrats.

WHEN CRITICISMS OR QUESTIONS came Cook's way, he could always point to the numbers. Since his promotion to CEO in August 2011, Apple's market capitalization had increased by more than $1.5 trillion and its total shareholder returns, including reinvested dividends, had been 867 percent, or about $500 billion. In late September of 2021, Apple's board of directors rewarded Cook with what amounted to a five-year employment extension from late 2020, awarding him a performance-based package of an additional 1 million shares by 2025. Over the course of a decade, he had vested the entirety of the 1.12 million performance-based shares he'd been granted since 2011 and become a billionaire.

The eye-popping figures were tucked into a dry corporate filing as discreet and understated as the man himself. They were a testament to his persistence.

AFTER HIS ANNOUNCEMENT, Jony Ive moved on. Following the final Apple keynote of his career in September 2019, he and the design team gathered for a party in San Francisco at Bix, a two-story restaurant off Jackson Square. It featured caviar, champagne, and a guest list that included NBA star Kobe Bryant. LCD Soundsystem front man James Murphy, a DJ, pumped out wide-ranging, high-energy music until well after midnight. It was the last Apple-themed party for its longtime artist.

Ive's departure became official two months later, when the company quietly removed his photograph and name from its leadership page. There was no fanfare, nor was there any commemoration of his contributions. In the most Apple way possible, he was there one day and gone the next.

In his wake, he left a legacy of products that was hard to fathom. He and Jobs had revived the company with the iMac and collaborated to deliver the product hits that had followed. Ive's aesthetic sensibilities heightened society's appreciation of design language. The company had indoctrinated the world in the principle of simplicity and the value of material in ways forefathers such as Dieter Rams could scarcely have imagined.

The Apple Watch and its conjoined twin, AirPods, had become valuable contributors to Apple's bottom line. In 2021, sales of the company's so-called wearables rose by 25 percent to $38.4 billion. The business's revenue was greater than the annual sales of Coca-Cola. Its troubled launch had humbled a company that had spent much of the decade coming to terms with the reality that it might never make another product as successful as the iPhone.

Yet Ive had built a sizable business for Apple on the spiritual rock of a new product category, just as Jobs had done in the previous decades. The wearables business was roughly half the size of the services business built by Cook, but it secured a commanding lead in its category and was expected to generate billions of dollars in uninterrupted sales for years to come.

IVE WOULD NEVER ARGUE that he had left Apple the right way. Some members of the design team remain frustrated at the failure of his succession planning and disappointed that the group's cohesion eroded. But leaving a job after more than twenty-five years isn't easy, especially not when you layer on the grief caused by losing a creative partner.

Frustrated by the growth of the company, the number of meetings, and the pressure from finance, Ive came to the realization that he could better help Apple externally than he could internally. He also wanted to find new ways to make a useful contribution to the world. He and Marc Newson assembled a team at LoveFrom made up of several longtime Apple colleagues, including software designer Chris Wilson, industrial designer Eugene Whang, and Foster + Partners architect James McGrath. The collection of creatives picked up clients that appealed to their interest. Airbnb, the home rental company, hired Ive to assist with redesigning its app and developing new products, and Ferrari tapped Ive and Newson to assist in designing its first electric vehicle and expanding its luxury apparel and luggage businesses. The firm also continued to advise Apple.

One of his biggest undertakings after leaving was a collaboration with Prince Charles on a sustainability initiative called the Terra Carta. Taking its name from the Magna Carta, which granted rights to the English people, the Prince Charles–led effort aims to address climate change by giving rights to nature. Ive developed a seal with an original typeface and swirling green fauna that is awarded to companies that distinguish themselves for their sustainability efforts.

He also continues to advise Apple on future projects, including renewed efforts to develop a car and the development of augmented reality devices, such as glasses. Like a rock star who boasts about his next album, he tells people that those future devices will be the best work of his career.

"The story of Jony Ive as a paradigm-shifting industrial designer is not done being written," a longtime colleague on the car project said.

The Apple design team has largely moved on. Core members of the team say that Ive and LoveFrom have minimal influence over their work. He's become a respected adviser rather than a controlling director. In his absence, they say that the group has become friendlier and more democratic in its collaboration, especially with colleagues in engineering and operations. The designers acknowledge that they are subject to more cost pressures now than when Ive was there to deflect those concerns. However, they say it's not so severe that they have been unable to do their work. And they insist that the work they're doing is the best work they've done.

With Ive's exit, it's unclear if design will ever regain its position as the dominant voice over product direction. It was Jobs who elevated Ive and the studio to a perch of power that allowed them to permeate the company with a devotion to design unrivaled in corporate America. In the near term, it would be foolish to expect a new sovereign to emerge with that much power and influence. After all, it took Ive nearly two decades to achieve such commanding clout.

After his team disbanded, they gradually took the measure of the company they had left behind. In reflecting on Jobs's legacy, they would often observe that he had made products that had changed the world. When asked how his successor, Cook, would be remembered—and by extension their final decade at Apple—some smiled. "For making a fuck ton of money," they said.

ON A LATE SUMMER AFTERNOON, Jony Ive was stretched out on an ivory-colored couch taking a midday nap. The top floor of his Pacific Heights home was a firmament of white, with a vaulted white ceiling towering over white walls. The color in the room came from an avant-garde bouquet of pink dahlias and garden roses overflowing with unruly vitality from a vase atop a small buffet table. The austerity of it all directed people's attention toward the floor-to-ceiling windows that looked north across the bay, where the blue-green water stretched from the towering red-painted steel of the Golden Gate Bridge to the ruins of the prison on Alcatraz Island, the landmarks of the city he loved.

As he slept, one of his personal assistants quietly climbed the stairs with a light lunch on a tray. Ive had a scheduled call in an hour's time and expected to be roused beforehand. The assistant set the tray down on a side table near the couch and whispered, "Jony, wake up." He didn't hear her and dozed on until he felt a soft tap on his shoulder. He slowly blinked open his eyes and gazed across the room toward the windows, where rays of midday light streamed into the room in narrow lines of yellow.

"Wow," he remarked. "The quality of the light coming in the room is beautiful."

The observation surprised the assistant. Turning to follow his gaze, she studied the light to try to see, as best she could, through his eyes. Later, thinking back to this quiet moment, she recalled that some people can see a hundred times as many colors as others do. Known as tetrachromats, they have a fourth type of color receptor in their retina that heightens their color perception.

"It is lovely," she told Ive.

She reminded him about his call and pointed to his lunch before slipping out of the room, leaving him alone, free from the responsibility of completing a campus, stressing over an aluminum computer part, or agonizing over the leather bands on the next Apple Watch. Liberated to make what he wanted next, if he wanted to make anything at all. Unburdened and at peace.

Acknowledgments

This book would not have been possible without the cooperation of current and former employees at Apple. To have this remarkable period of corporate history documented, they looked past the company's culture of omertà and shared stories of how they had made products that changed the world. I will be forever grateful for their generosity.

The journey of this book began with a cup of coffee and a suggestion to learn more about Jony Ive while covering Apple for the *Wall Street Journal*. When the time came to tell Ive's story, the paper's global tech editor, Jason Dean, emboldened my reporting, guided my writing, and worked wonders to sharpen the words on the page.

A cadre of other *Journal* editors steered the Apple coverage that was foundational to this book, including Brad Olson, Scott Austin, Brad Reagan, Scott Thurm, Jamie Heller, Matthew Rose, Tammy Audi, Jason Anders, and Matt Murray, who endorsed this endeavor.

Betsy Morris, my first editor at the *Journal,* pointed me toward the human story in Apple's resurgence, and John Helyar, whose *Barbarians at the Gate* is an inspiration, distilled the tale and offered continual guidance. They are part of an Atlanta work family who mentored me, including Valerie Bauerlein, Arian Campo-Flores, Mike Esterl, Betsy McKay, and Cameron McWhirter.

Many *Journal* colleagues supported the work on these pages, including Yoko Kubota, Yukari Kane, Joe Flint, Liz Hoffman, Jim Oberman, and Erich Schwartzel. To my coworkers in San Francisco, especially Tim Higgins, Aaron Tilley, and Georgia Wells—thank you.

Mauro DiPreta coaxed me into writing this book after reading the *Journal*'s coverage of Ive's departure in 2019. He saw the potential for a sweeping narrative about the world's largest company and brought that tale to life with thoughtful edits. My agent, Daniel Greenberg of Levine Greenberg Rostan Literary Agency, counseled me throughout the process, from proposal to final draft. The team at William Morrow brought it all together, including Vedika Khanna and Lynn Anderson.

Thomas French was my first reader, occasional therapist, and patient coach. He sees possibilities in the smallest details and thematic opportunities in every anecdote. I literally—and I mean that in the truest sense of the phrase—could not have done it without him.

Sean Lavery, my fact checker, was a pro's pro, scrutinizing words, verifying facts, and navigating complexities. John Bauernfeind, my research assistant, unearthed details about Jony Ive and Tim Cook that opened critical avenues of reporting.

An all-star support team generously fielded calls, provided edits, and kept me sane. Laura Stevens read and improved an entire draft in forty-eight hours; Eliot Brown helped set the structure and refine the manuscript; Justin Catanoso assisted with outlining; John Ourand strategized on whom to call; and Rob Copeland pushed my reporting. As brilliant as they are, they are first and foremost amazing friends.

The technology industry is complex, and learning about it requires guides. I was blessed with many, but none were more valuable than John Markoff and Talal Shamoon.

Special thanks and love to my family for their encouragement, including my in-laws, Sally and Mark, and the Coopers, Jennifer, Josh, Madelynn, and Nathaniel. My parents, Marilynn and Russ, supported my interest in journalism and armed me with skills I use daily, including listening with curiosity and appreciating the written word.

My deepest gratitude goes to Amanda Bell, who sat beside me on the optimism-despair roller coaster that culminated in this book. She will always be more than I deserve.

A Note on Sources

At Apple, current and former employees adhere to a strict code of silence. A corporate omertà.

Like the Italian Mafia that coined the term to protect itself, the iPhone syndicate is united in its commitment to guard the secrets of its operations. Staff are indoctrinated to bleed the company's original six-color logo and undergo an orientation where they are instructed not to discuss their work with anyone outside the confines of Cupertino.

Many companies have similar policies. At Apple, that policy pervades the culture. The company is structured to safeguard information. Initiatives are given code names; business strategies are limited to the most senior executives; subordinates are partitioned in ways that limit their knowledge of future products; and everyone buys into the idea that confidentiality prevents rivals from stealing ideas and preserves the mystery that helps the company secure hundreds of millions of dollars in free advertising from press coverage of its showy events.

There's a shared belief that success hangs on secrecy. Staff are convinced that anyone who speaks to the media disadvantages the company. Some who have spoken to reporters, even after leaving

Apple, have been ostracized by colleagues and friends. Others have been fired or sued.

This culture makes reporting on Apple a challenge. Employees can be tight-lipped with one another. Married couples in separate divisions at the company often go years without telling each other about their respective jobs. One couple told me that it wasn't until long after they had both retired that they had finally opened up with each other about what they had done. They said the conversation had taken courage.

Notes

Prologue

1   *He considered imitation:* Kia Makarechi, "Apple's Jonathan Ive in Conversation with Vanity Fair's Graydon Carter," *Vanity Fair,* October 16, 2014, https://www.vanityfair.com/news/daily-news/2014/10/jony-ive-graydon-carter-new-establishment-summit.

2   *The fifty-eight-year-old Cook:* Interview with Joe O'Sullivan, former Apple operations executive.

Chapter 1: One More Thing

5   *Jony Ive steeled himself:* ABC 7 Morning News, October 4, 2011; Lisa Brennan-Jobs, *Small Fry;* interviews with multiple Apple executives who visited Jobs at home after he stopped coming into the office.

7   *Staff nicknamed the phone:* Interviews with former Apple employees; Jim Carlton, *Apple;* Yukari Iwatani Kane and Geoffrey A. Fowler, "Steven Paul Jobs, 1955–2011: Apple Co-founder Transformed Technology, Media, Retailing and Built One of the World's Most Valuable Companies," *Wall Street Journal,* October 6, 2011, https://www.wsj.com/articles/SB10001424052702304447804576410753210811910; macessentials, "The Lost 1984 Video: Young Steve Jobs Introduces the Macintosh," YouTube, January 23, 2009, https://www.youtube.com/watch?v=2B-XwPjn9YY; Andrew Pollack, "Now, Sculley Goes It Alone," *New York Times,* September 22, 1985.

9   *Though Jobs hadn't attended rehearsals:* noddyrulezzz, "Apple iPhone 4S—Full Keynote—Apple Special Event on 4th October 2011," YouTube, October 6, 2011, https://www.youtube.com/watch?v=Nqol1AH_zeo; Geoffrey A. Fowler and John Letzing, "New iPhone Bows but Fails to Wow," *Wall Street Journal,* October 5, 2011, https://

www.wsj.com/articles/SB100014240529702045246045766109919 78907616.

10 *Staff saved an aisle seat:* Apple, "Apple Special Event, October 2011" (video), Apple Events, October 4, 2011, https://podcasts.apple.com/us /podcast/apple-special-event-october-2011/id275834665?i= 1000099827893.

12 *The following afternoon:* Nick Wingfield, "A Tough Balancing Act Remains Ahead for Apple," *New York Times,* October 5, 2011, https:// www.nytimes.com/2011/10/06/technology/for-apple-a-big-loss -requires-a-balancing-act.html.

12 *Less than fifteen miles away:* Jony Ive, "Jony Ive on What He Misses Most About Steve Jobs," *Wall Street Journal,* October 4, 2021, https://www.wsj.com/articles/jony-ive-steve-jobs-memories-10th -anniversary-11633354769?mod=hp_featst_pos3.

13 *Jobs had anticipated the pitfalls:* Walter Isaacson, *Steve Jobs;* James B. Stewart, *Disney War* (New York: Simon & Schuster, 2005); Michael G. Rukstad, David Collis, and Tyrrell Levine, "The Walt Disney Company: The Entertainment King," Harvard Business School, January 5, 2009, https://www.hbs.edu/faculty/Pages/item.aspx?num =27931; Brady MacDonald, "'The Imagineering Story': After Walt Disney's Death, Imagineering Wonders 'What Would Walt Do?,'" *Orange County Register,* November 4, 2019, https://www.ocregister .com/2019/11/04/the-imagineering-story-after-walt-disneys-death -imagineering-wonders-what-would-walt-do/; Christopher Bonanos, *Instant: The Story of Polaroid* (New York: Princeton Architectural Press, 2012); Christopher Bonanos, "Shaken like a Polaroid Picture," Slate, September 17, 2013, https://slate.com/technology/2013/09 /apple-and-polaroid-a-tale-of-two-declines.html; interview with Carl Johnson, former executive vice president of advertising, Polaroid; Chunka Mui, "What Steve Jobs Learned from Edwin Land of Polaroid," *Forbes,* October 26, 2011, https://www.forbes.com/sites /chunkamui/2011/10/26/what-steve-jobs-learned-from-edwin-land -of-polaroid/; John Nathan, "Sony CEO's Management Style Wasn't Made in Japan," *Wall Street Journal,* October 7, 1999, https://www .wsj.com/articles/SB939252647570595508; John Nathan, *Sony.*

16 *Jobs wanted Apple to defy:* Dialogue based on interviews with Apple staff; Brian X. Chen, "Simplifying the Bull: How Picasso Helps to Teach Apple's Style," *New York Times,* August 10, 2014, https:// www.nytimes.com/2014/08/11/technology/-inside-apples-internal -training-program-.html.

18   *Two weeks later:* Wylsacom, "A Celebration of Steve's Life (Apple, Cupertino, 10/19/2011) HD," YouTube, https://www.youtube.com /watch?v=ApnZTL-AspQ.

Chapter 2: The Artist

21   *Staff called it the holy of holies:* Walter Isaacson, *Steve Jobs:* description of the space, as well as interviews with Apple staff.

22   *Jony Ive grew up:* Interviews about the family with friends and work colleagues of Mike Ive, including John Chapman, Richard Tufnell, and Tim Longley.

24   *In the years that followed:* John Arlidge, "Jonathan Ive Designs Tomorrow," *Time,* March 17, 2014, https://time.com/jonathan-ive-apple -interview/; Rick Tetzeli, "Why Jony Ive Is Apple's Design Genius," *Smithsonian Magazine,* December 2017, https://www.smithsonian mag.com/innovation/jony-ive-apple-design-genius-180967232/; interviews with Mike Ive's former colleagues and friends Ralph Tabberer, Richard Tuffnel, John Cave, and Netta Cartwright; interview with Jony Ive's classmates Rob Chatfield, Stephen Palmer, and Dan Slee; Leander Kahney, *Jony Ive;* Rob Waugh, "How Did a British Polytechnic Graduate Become the Design Genius Behind $200 Billion Apple?," *Daily Mail,* March 19, 2013, https://www.dailymail .co.uk/home/moslive/article-1367481/Apples-Jonathan-Ive-How -did-British-polytechnic-graduate-design-genius.html.

27   *Ive's teenage art folder:* Interview with Walton design teacher Dave Whiting; Kahney, *Jony Ive;* NAAIDT HMI Mike Ive presentation 2001, http://archive.naaidt.org.uk/spd/record.html?Id=29&Adv=1&All =3; interview with Ive's former colleague Ralph Tabberer.

28   *In 1983, before Ive graduated:* Interview with Roberts Weaver managing director Phil Gray; interviews regarding Newcastle Polytechnic with 1988 graduate Craig Mounsey; 1989 classmates Steve Bailey, Sean Blair, and David Tonge; 1990 classmate Jim Dawton; professors John Elliott and Bob Young; and faculty member Mark Bailey.

30   *Newcastle offered Ive:* Interviews about Newcastle Polytechnic with Northumbria University Director of Innovation Design Mark Bailey, Ive's classmates Steve Bailey and Sean Blair, and professors John Elliott and Bob Young; tour of Squires Building; "Memphis Group: Awful or Awesome," The Design Museum, https://designmuseum.org /discover-design/all-stories/memphis-group-awful-or-awesome; Dieter Rams, *Less But Better;* "Tough on the sheets" tagline recalled by Sean Blair; Nick Carson, "If It Looks Over-Designed, It's Under-Designed,"

https://www.channel4.com/ten4, reprinted at https://ncarson.files.word press.com/2007/01/ten4-jonathanive.pdf.

32 *In 1987, Ive fulfilled his promise:* Luke Dormehl, *The Apple Revolution;* interviews about Roberts Weaver with Clive Grinyer, Peter Phillips, Jim Dawton, Phil Gray, and Barrie Weaver; interviews about hearing aid with classmate Jim Dawton and professor John Elliott; interview regarding Macintosh with Ann Irving; "Q&A with Jonathan Ive," The Design Museum, October 3, 2014, https://designmuseum.org /designers/jonathan-ive; Ian Parker, "The Shape of Things to Come: How an Industrial Designer Became Apple's Greatest Product," *The New Yorker,* February 16, 2015, https://www.newyorker.com/magazine /2015/02/23/shape-things-come.

34 *In order to graduate:* Interviews with professors John Elliott and Bob Young, as well as classmates Jim Dawton, Sean Blair, Craig Mounsey, and David Tonge; Melanie Andrews, "Jonathan Ive and the RSA's Student Design Awards," RSA, May 25, 2012, https://www.thersa.org /blog/2012/05/jonathan-ive-amp-the-rsas-student-design-awards; "Apple's Jonathan Ive in Conversation with Vanity Fair's Graydon Carter" (video), *Vanity Fair,* October 16, 2014, https://www.vanityfair .com/video/watch/the-new-establishment-summit-apples-jonathan -ive-in-conversation-with-vf-graydon-carter; "The First Phone Jony Ive Ever Designed" (video), *Vanity Fair,* October 28, 2014, https://www .youtube.com/watch?v=oF21m-6yV0U; interview with Clive Grinyer; Kahney, *Jony Ive;* Sheryl Garratt, "Interview: Jonathan Ive," *Times Magazine,* December 3, 2005.

36 *The model, which won:* Interview with Craig Mounsey, Newcastle Polytechnic graduate and head of concepts and innovation at Fisher & Paykel Technologies.

36 *Flush with prize money:* Interviews with Ive's friend David Tonge and Lunar Design cofounder Robert Brunner; Molly Wood, "We Love Stories About Silicon Valley Success, but What Is Its History?," Podchaser, July 10, 2019, https://www.podchaser.com/podcasts /marketplace-tech-50980/episodes/we-love-stories-about-silicon -41846275; Andrews, "Jonathan Ive and the RSA's Student Design Awards."

37 *Ive's return to Roberts Weaver:* Interviews with Roberts Weaver's managing director Phil Gray and lead partner Barrie Weaver; interviews with Tangerine's Peter Phillips, Clive Grinyer, Martin Darbyshire, and Jim Dawton; Kahney, *Jony Ive;* Parker, "The Shape of Things to

Come"; Waugh, "How Did a British Polytechnic Graduate Become the Genius Behind Apple Design?"; Peter Burrows, "Who Is Jonathan Ive?," *Bloomberg Businessweek*, September 24, 2006, https://www .bloomberg.com/news/articles/2006-09-24/who-is-jonathan-ive; "Q&A with Jonathan Ive," The Design Museum, October 3, 2014, https://designmuseum.org/designers/jonathan-ive; "The First Phone Jony Ive Ever Designed" (video), *Vanity Fair*, Oct. 28, 2014, https:// www.youtube.com/watch?v=oF21m-6yV0U.

39   *Desperately in need of a boost:* Interview with then Apple design chief Robert Brunner; interviews with Tangerine's Clive Grinyer, Peter Phillips, and Martin Darbyshire; interview with Steve Bailey; Burrows, "Who Is Jonathan Ive?"; Parker, "The Shape of Things to Come."

Chapter 3: The Operator

42   *On some days, Cook:* Steven Levy, "An Oral History of Apple's Infinite Loop," *Wired*, September 16, 2018, https://www.wired.com/story /apple-infinite-loop-oral-history/.

43   *Tim Cook grew up wanting:* Violla Young, "Tim Cook (CEO of Apple) Interview in Oxford," YouTube, July 18, 2018, https://www.youtube .com/watch?v=QPQ8qQP4zdk: "I saw my dad go to work and not love what he did. He worked for his family . . . but he never loved what he did. And so I wanted to get a job that I loved."

43   *Born in 1960:* Michael Finch II, "Tim Cook—Apple CEO and Robertsdale's Favorite Son—Still Finds Time to Return to His Baldwin County Roots," AL.com, February 24, 2014, updated January 14, 2019, https://www.al.com/live/2014/02/tim_cook_--_apple_ceo_and_robe .html.

43   *Dozier, a rural outpost:* Joe R. Sport, *History of Crenshaw County*.

44   *His family had arrived:* Ancestry.com research on Canie Dozier Cook, 1902–1985; Daniel Dozier Cook, 1867–1938; Alexander Hamilton Cook, 1818–1872; and William Cook, 1780–1820.

44   *His father supported the family:* 1930 and 1940 U.S. Federal Census Records on Ancestry.com.

44   *He would one day boast:* Tim Cook interview by Debbie Williams, WKRG TV, January 16, 2009.

44   *When Cook eventually became:* Interview with Linda Booker, Robertsdale resident and widow of one of Donald Cook's friends.

44   *Cook's father had:* John Underwood, "Living the Good Life," Gulf Coast Media, July 13, 2018.

45 *The Cooks, who were:* "Robert Quinley Services Held"; "Bay Minette Wreck Takes Three Lives," Ancestry.com.

45 *Donald said they had chosen:* Finch, "Tim Cook—Apple CEO and Robertsdale's Favorite Son—Still Finds Time to Return to His Baldwin County Roots."

45 *Most of its 2,300 residents:* Jack House, "Vanity Fair to Expand Its Robertsdale Plant," *Baldwin Times,* October 31, 1963.

45 *Children roamed the neighborhoods:* Interviews with local residents, including Barbara Davis, Fay Farris, and Rusty Aldridge, as well as articles from *Baldwin Times* in 1977 and 1978.

45 *Donald and Geraldine went:* Interviews with Robertsdale High teachers Fay Farris, Barbara Davis, and Eddie Page.

45 *They attended Robertsdale United Methodist Church:* Interview with Clem Bedwell, a classmate of Tim Cook and former church member.

46 *Donald spent his days there:* Underwood, "Living the Good Life"; "Industry Wage Survey: Shipbuilding and Repairing, September 1976," Bulletin no. 1968, Bureau of Labor Statistics, U.S. Department of Labor, 1977, https://fraser.stlouisfed.org/files/docs/publications/bls/bls _1968_1977.pdf.

46 *Cook's parents drilled into him:* Homecoming, "With Tim Cook," SEC Network, September 5, 2017.

46 *At Lee, he showed:* Interview with Jimmy Stapleton, Lee Drug Store pharmacist.

46 *His teachers compared him:* Interview with Cook's teacher Fay Farris.

46 *"He was one of those kids":* Interviews with Cook's teachers Eddie Page and Ken Brett.

47 *In 1971, Cook watched:* Homecoming, "With Tim Cook," SEC Network, September 5, 2017; Kirk McNair, "Remembering Alabama's 1971 Win over Auburn," 247sports.com, November 24, 2017, https://247sports .com/college/alabama/Board/116/Contents/As-this-year-1971 -Alabama-Auburn-game-had-major-ramifications-110969031/; Creg Stephenson, "Check Out Vintage Photos from 1972 'Punt Bama Punt' Iron Bowl," AL.com, November 24, 2015, updated January 13, 2019, https://www.al.com/sports/2015/11/check_out_vintage_photos _from.html.

47 *Within a year, Cook was telling:* Finch, "Tim Cook—Apple CEO and Robertsdale's Favorite Son—Still Finds Time to Return to His Baldwin County Roots."

47 *The community was unofficially:* Interviews with Wayne Ellis, Robertsdale High classmate, and others.

48 *In 1969, the local school:* Interviews with Wayne Ellis, Fay Farris, and others.

48 *In sixth or seventh grade:* Todd C. Frankel, "The Roots of Tim Cook's Activism Lie in Rural Alabama," *Washington Post,* March 7, 2016, https://www.washingtonpost.com/news/the-switch/wp/2016/03/07 /in-rural-alabama-the-activist-roots-of-apples-tim-cook/; Matt Richtel and Brian X. Chen, "Tim Cook, Making Apple His Own," *New York Times,* June 15, 2014, https://www.nytimes.com/2014/06/15/technology /tim-cook-making-apple-his-own.html.

48 *After he became Apple's CEO:* Auburn University, "Tim Cook Receiving the IQLA Lifetime Achievement Award," YouTube, December 14, 2013, https://www.youtube.com/watch?v=dNEafGCf-kw.

48 *After the speech:* In the *New York Times* article "Tim Cook, Making Apple His Own," reporters Matt Richtel and Brian X. Chen wrote that Apple "did confirm the details of the cross-burning story," while also noting that Cook declined to be interviewed.

48 *For years, former classmates:* Facebook Group, Robertsdale, Past and Present, "Discussion: 'Apple's CEO Tim Cook: An Alabama Day That Forever Changed His Life,' AL.com," Facebook, June 15, 2014, https://www.facebook.com/groups/263546476993149/permalink /863822150298909/.

49 *The old friends haven't:* Interview with Lisa Straka Cooper. A spokeswoman for Apple declined to comment.

49 *He found his social group:* Interviews with Mike Vivars and Eddie Page.

49 *Though he spent most of his time:* Interviews with classmates Rusty Aldridge, Johnny Little, Clem Bedwell, and Lisa Straka Cooper.

49 *"Tim was strange":* Interview with Johnny Little, former classmate, of Robertsdale, Alabama.

49 *Cook counted Cooper:* Interview with Lisa Straka Cooper.

50 *As business manager:* Interview with yearbook teacher Barbara Davis.

50 *Still intent on attending Auburn:* Trice Brown, "Apple CEO Tim Cook Was Robertsdale High School's Salutatorian in 1978, but Whatever Happened to the Valedictorian?," Lagniappe, July 1, 2020, https://lagniappemobile.com/apple-ceo-tim-cook-was-robertsdale -high-schools-salutatorian-in-1978-but-whatever-happened-to-the -valedictorian/.

51 *It was regarded by some:* "Letter About Elimination of Gays Disgusting," *Auburn Plainsman,* March 4, 1982, https://content.lib.auburn .edu/digital/collection/plainsman/id/2559/.

51 *In the small farming community:* Interviews with Fay Farris, Barbara Davis, Mike Vivar, and Johnny Little.

51 *The atmosphere made Cook:* Tim Cook on *The Late Show with Stephen Colbert,* September 15, 2015; interviews with Mike Vivar, Lisa Straka Cooper, and Rusty Aldridge.

52 *One day at school:* Interview with Fay Farris.

52 *As Cook finished high school:* Interview with Mike Vivar.

52 *Cook joined a group of eight:* Interview with Rusty Aldridge, a fellow resident.

53 *There were Funnel Fever:* Auburn University yearbook, 1982, https://content.lib.auburn.edu/digital/collection/gloms1980/id/17321/.

53 *"I have to say":* Homecoming, "With Tim Cook," SEC Network, September, 5, 2017.

53 *Cook got involved:* Ibid.

53 *It was a practical choice:* Interview with Fay Farris; Auburn University Bulletin 1978–1982, says that tuition was $200 to $240; Leslie Cardé, "Tim Cook," Inside New Orleans, Summer 2019, 48–49, https://issuu.com/in_magazine/docs/1907inoweb/49; Ray Garner, "Steve Jobs' World Man," *Business Alabama,* November 1999, 59–60.

54 *His classmates said:* Interviews with Pamela Palmer, Auburn, industrial engineering graduate, 1981; Mike Peeples, Auburn, industrial engineering graduate, 1981; and Paul Stumb, Auburn, industrial engineering graduate, 1982.

54 *They remember him:* Interview with Paul Stumb, Auburn, industrial engineering graduate, 1982.

54 *His professors admired:* "He could cut through all the junk and get down to the gist of the problem very quickly," Professor Robert Bulfin said. Yukari Kane, *Haunted Empire: Apple After Steve Jobs,* 98.

54 *In 1982, Cook was inducted:* Interview with Auburn University professor Sa'd Hamasha, who works with the honor society.

55 *When he returned to Auburn:* Kit Eaton, "Tim Cook, Apple CEO, Auburn University Commencement Speech 2010," Fast Company, August 26, 2011, https://www.fastcompany.com/1776338/tim-cook-apple-ceo-auburn-university-commencement-speech-2010.

55 *The resulting IBM:* Andrew Pollack, "Big I.B.M. Has Done It Again," *New York Times,* March 27, 1983, https://www.nytimes.com/1983/03/27/business/big-ibm-has-done-it-again.html.

55 *Executives wore starched white:* Michael W. Miller, "IBM Formally Picks Gerstner to Be Chairman and CEO—RJR Executive Doesn't Have a Turnaround Plan Yet for U.S. Computer Giant," *Wall Street*

Journal, March 29, 1993; interview with Richard L. Daugherty, former IBM vice president of worldwide PC manufacturing.

56 *The job put him at the forefront:* Tim Cook on *The David Rubenstein Show,* June 13, 2018; anunrelatedusername, "IBM Manufacturing Systems—Keyboard Assembly," YouTube, https://www.youtube.com /watch?v=mEN6Rry4ekk; Gene Bylinsky, "The Digital Factory," *Fortune,* November 14, 1994, https://archive.fortune.com/magazines /fortune/fortune_archive/1994/11/14/79947/index.htm; interview with Richard L. Daugherty.

56 *Cook made his mark:* Interview with Richard L. Daugherty; John Marcom, Jr., "Slimming Down: IBM Is Automating, Simplifying Products to Beat Asian Rivals," *Wall Street Journal,* April 14, 1986.

57 *Between Christmas and New Year's:* Interviews with Richard L. Daugherty and Gene Addesso, IBM plant manager.

57 *In marketing, he studied:* Bill Boulding, "What Tim Cook Told Me When I Became Dean of Duke University's Fuqua School of Business," Linkedin, December 10, 2015, https://www.linkedin.com /pulse/what-tim-cook-told-me-when-i-became-dean-duke-fuqua -school-boulding.

58 *Shortly afterward, Cook faced:* Andrew Gumbel, "Tim Cook: Out, Proud, Apple's New Leader Steps into the Limelight," *Guardian,* November 1, 2014, https://www.theguardian.com/theobserver/2014 /nov/02/tim-cook-apple-gay-coming-out.

58 *Around that time, he found:* Violla Young, "Tim Cook (CEO of Apple) Interview in Oxford."

58 *In 1992, four years later:* Interview with Dave Boucher, former IBM general manager.

59 *The Philadelphia-based business:* Interviews with Thomas Coffey, former finance chief at Intelligent Electronics; Gregory Pratt, former president of Intelligent Electronics.

59 *Margins were thin:* Interview with Thomas Coffey; Raju Nasiretti, "Extra Bites: Intelligent Electronics Made Much of Its Profit at Suppliers' Expense," *Wall Street Journal,* December 6, 1994; staff reporter, "Intelligent Electronics Agrees to Settle Class-Action Suits," *Wall Street Journal,* February 21, 1997, https://www.wsj.com/articles /SB856485760719766500; Leslie J. Nicholson, "Intelligent Electronics Pays $10 Million to Shareholders in Lawsuit," *Philadelphia Inquirer,* December 2, 1997; Intelligent Electronics Inc. Form 10-Q, Exton, Pennsylvania: Intelligent Electronics, September 16, 1997, https://www.sec .gov/Archives/edgar/data/814430/0000814430-97-000027.txt.

59   *The executives climbing:* Interview with Larry Deaton, former IBM executive and colleague of Tim Cook.

59   *He would report to a board:* Intelligent Electronics Inc., Form DEF 14A Proxy Statement, July 23, 1996, https://bit.ly/2XD4Hri.

60   *In his first year, he shuttered:* Kevin Merrill, "IE Beefs Up Memphis, Inacom Makes Addition on West Coast," *Computer Reseller News,* September 6, 1995.

60   *In 1996, Cook and Coffey:* Interview with Tom Coffey.

60   *The board of directors decided:* Raju Narisetti, "Intelligent Electronics Sale," *Wall Street Journal,* July 21, 1997; Raju Narisetti, "Xerox Agrees to Buy XLConnect and Parent Intelligent Electronics," *Wall Street Journal,* March 6, 1998, https://www.wsj.com/articles /SB889104642954787000.

60   *Eventually, he persuaded:* "Ingram Micro Will Buy Division," *Wall Street Journal,* May 1, 1997; interview with Tom Coffey.

60   *As the sales process played out:* Interview with Greg Petsch, former senior vice president of manufacturing and quality at Compaq Computer Corporation.

61   *In early 1998, Petsch got:* Interview with Greg Petsch.

62   *Cook paused to think:* Tim Cook on *The Charlie Rose Show,* September 12, 2014; Apple CEO Tim Cook on *The David Rubenstein Show,* June 13, 2018.

62   *The total was more than $1 million:* Interview with Rick Devine, executive recruiter who found Cook.

63   *"There's no way Apple":* Interview with executive recruiter Rick Devine, who recruited Cook for Steve Jobs.

Chapter 4: Keep Him

64   *Jony Ive's yellow Saab convertible:* Interviews with Robert Brunner and Clive Grinyer.

64   *San Francisco hadn't yet been:* "San Francisco in the 1990s [Decades Series]," Bay Area Television Archive, https://diva.sfsu.edu/collections /sfbatv/bundles/227905.

64   *Its Mission District had:* "Look Back: Pioneers of '90s Mission Arts Scene," San Francisco Museum of Modern Art, https://www.sfmoma .org/read/mission-school-1990s/; Stephanie Buck, "During the First San Francisco Dot-Com Boom, Techies and Ravers Got Together to Save the World," Quartz, August 7, 2017, https://qz.com/1045840 /during-the-first-san-francisco-dot-com-boom-techies-and-ravers -got-together-to-save-the-world/.

64   *He ditched his spiky hair:* Emma O'Kelly, "I've Arrived," Design Week, December 6, 1996, https://www.designweek.co.uk/issues/5-december -1996/ive-arrived/.

65   *The design team worked:* Interview with Robert Brunner.

65   *"It didn't offer a metaphor":* Paul Kunkel, *AppleDesign,* 253.

65   *In 1992, Apple's profits:* G. Pascal Zachary and Ken Yamada, "Apple Picks Spindler for Rough Days Ahead," *Wall Street Journal,* June 21, 1993.

66   *To reverse the declining sales:* Interviews with Robert Brunner and Tim Parsey.

66   *Ive's boss, Robert Brunner:* Emma O'Kelly, "I've Arrived," *Design Week,* December 6, 1996, https://www.designweek.co.uk/issues/5-december -1996/ive-arrived/; John Markoff, "At Home with: Jonathan Ive: Making Computers Cute Enough to Wear," *New York Times,* February 5, 1998, https://www.nytimes.com/1998/02/05/garden/at-home-with -jonathan-ive-making-computers-cute-enough-to-wear.html.

66   *"This is Jony":* Interview with Tim Parsey, former Apple design studio manager, 1991–1996.

67   *The resulting computer featured:* TheLegacyOfApple, "Jony Ive Intro- duces the 20th Anniversary iMac," YouTube, May 21, 2013, https:// www.youtube.com/watch?v=et6-hK-LA4A.

67   *Lee wanted to conduct:* Interview with Robert Brunner.

67   *It underforecast demand:* Jim Carlton, "Fading Shine: What's Eat- ing Apple? Computer Maker Hits Some Serious Snags—Talk Rises About Booting Spindler as Share Falls and Laptops Catch Fire—The Search for a Power Mac," *Wall Street Journal,* September 21, 1995.

67   *The CEO was eventually:* Jim Carlton, "Apple Ousts Spindler as Its Chief, Puts National Semi CEO at Helm, *Wall Street Journal,* Febru- ary 2, 1996, https://www.wsj.com/articles/SB868487469994949500.

67   *Financial analysts began:* Jim Carlton, *Apple.*

68   *Ive bristled:* O'Kelly, "I've Arrived."

68   *He considered quitting:* Interview with Clive Grinyer; O'Kelly, "I've Arrived."

68   *One day in July 1997:* Kahney, *Jony Ive.*

68   *The return of Jobs unsettled:* Isaacson, *Steve Jobs;* Brent Schendler and Rick Tetzeli, *Becoming Steve Jobs.*

68   *"What's wrong with this place?":* Isaacson, *Steve Jobs,* 317.

69   *They were all rattled:* Interview with Doug Satzger.

69   *He spoke with Richard Sapper:* Alyn Griffiths, "'Steve Jobs once wanted to hire me'—Richard Sapper," Dezeen, June 19, 2013, https://

www.dezeen.com/2013/06/19/steve-jobs-once-wanted-to-hire-me
-richard-sapper/.

69   *"Keep him":* Interview with Hartmut Esslinger.

69   *Ahead of Jobs's visit:* Ian Parker, "The Shape of Things to Come: How
an Industrial Designer Became Apple's Greatest Product," *The New
Yorker,* February 16, 2015, https://www.newyorker.com/magazine
/2015/02/23/shape-things-come.

69   *The team had tidied up:* Interview with Doug Satzger.

70   *"Fuck, you've not been":* Parker, "The Shape of Things to Come."

70   *"We were on the same wavelength":* Isaacson, *Steve Jobs,* 342.

70   *Ive prodded the group:* Karnjana Karnjanatawe, "Design Guru Says
Job Is to Create Products People Love," *Bangkok Post,* January 27,
1999.

71   *As they talked at meetings:* Interview with Doug Satzger; Leander
Kahney, *Jony Ive.*

71   *Ive liked the idea:* Karnjanatawe, "Design Guru Says Job Is to Create
Products People Love."

71   *"It has a sense":* Isaacson, *Steve Jobs,* 349.

71   *The team made models:* Leander Kahney, *Jony Ive.*

71   *The plastic shell:* Interview with Peter Phillips, who was working at
LG at the time.

71   *It cost $60 per unit:* Isaacson, *Steve Jobs,*

72   *In early May 1998:* Kahney, *Jony Ive.*

72   *"What the fuck is this?!?":* Isaacson, *Steve Jobs,* 352; interview with
Wayne Goodrich.

72   *"Steve, you're thinking":* Interview with Wayne Goodrich. Another
person who was present for this didn't recall Ive interacting with Jobs
but agreed that Ive had a calming effect on him.

73   *Apple was selling an iMac:* Karnjanatawe, "Design Guru Says Job Is to
Create Products People Love."

73   *On the day of the launch:* David Redhead, "Apple of Our Ive," *Design
Week,* Autumn 1998, 36-43.

73   *Much of the credit:* Jon Rubinstein, Ive's manager, led the development
of the iMac, making the critical choices of the components and firmware that powered the machine.

73   *In his home country:* John Ezard, "iMac Designer Who 'Touched Millions' Wins £25,000 Award," *Guardian,* June 3, 2003.

74   *Over the next three weeks:* Kahney, *Jony Ive;* interview with Doug
Satzger.

74   *In most places that decision:* Isaacson, *Steve Jobs.*

75 *In early 2001, Jobs moved:* Steven Levy, "An Oral History of Apple's Infinite Loop," *Wired,* September, 16, 2018, https://www.wired.com /story/apple-infinite-loop-oral-history.

76 *"Jony and I think up":* Isaacson, *Steve Jobs,* 342.

80 *a Bang & Olufsen phone:* Austin Carr, "Apple's Inspiration for the iPod? Bang & Olufsen, Not Braun," *Fast Company,* November 6, 2013, https://www.fastcompany.com/3016910/apples-inspiration-for -the-ipod-bang-olufsen-not-dieter-rams.

80 *They handed the ingredients:* Isaacson, *Steve Jobs;* Tony Fadell told Isaacson that Ive had been given the product to "skin," a turn of phrase meaning that he was asked to create its exterior.

80 *The design concept struck Ive:* Isaacson, *Steve Jobs.*

80 *The studio favored it:* Interview with Doug Satzger: "Color is hard. It alienates people. It makes some people happy and pisses some off. Black is too heavy. White is fresh and light."

81 *After Apple released the iPod:* Ron Adner, "From Walkman to iPod: What Music Teaches Us About Innovation," *The Atlantic,* March 5, 2012.

81 *Despite the triumph:* Kahney, *Jony Ive.*

82 *"There's no one who can":* Isaacson, *Steve Jobs,* 342.

82 *Designers said Thomas Meyerhoffer:* Interviews with members of the design team. A member of the team said that Meyerhoffer's exit was planned before Jobs's return and that an effort was made to persuade him to stay.

82 *During one such process:* Interview with Doug Satzger.

83 *"Being a British gentleman":* Interview with Tim Parsey.

84 *They obsessed over their hobbies:* Justin Housman, "Designer Rides: From Lamborghinis to Surfboards, Julian Hoenig Knows a Thing or Two About Design," Surfer, November 13, 2013, https://www.surfer .com/features/julian-hoenig/.

85 *The design team's power:* Brian Merchant, *The One Device.*

85 *"Imagine the back":* Brent Schendler and Rick Tetzeli, *Becoming Steve Jobs,* 310.

87 *A month later, Jobs unveiled:* Jonathan Turetta, "Steve Jobs iPhone 2007 Presentation (HD)," YouTube, May 13, 2013, https://www.youtube .com/watch?v=vN4U5FqrOdQ.

87 *Over the years:* Schendler and Tetzeli, *Becoming Steve Jobs,* 356–57.

87 *He bought a $3 million:* Simon Trump, "Designer of the iPod Tunes into Nature," *Telegraph,* May 24, 2008, https://www.telegraph.co.uk /news/uknews/2023212/Designer-of-the-iPod-tunes-into-nature.html.

88   *He told his longtime friend:* Interview with Clive Grinyer; Parker, "The Shape of Things to Come."

88   *In May 2009, Jony Ive arrived:* Isaacson, *Steve Jobs.* Representatives of Ive said he hadn't spoken with Isaacson about that important episode. Isaacson didn't provide Jobs's response or detail who had provided Ive's quotes in this exchange.

89   *Ive kick-started the evaluation:* Isaacson, *Steve Jobs.*

90   *Accompanied by Heather:* "Apple Design Chief Jonathan Ive Is Knighted" (video), BBC, May 23, 2012, https://www.bbc.com/news/uk -18171093; Yukari Kane, *Haunted Empire.*

91   *Later that day, Ive shed:* Interview and photographs provided by event organizer Tracy Breeze.

91   *He told friends:* Interview with Richard Tufnell, a friend of Mike Ive and former colleague at Middlesex Polytechnic: "Jony was his finest creation. He was incredibly proud. He pinches himself in some ways and thinks *Is he my son?*"

Chapter 5: Intense Determination

92   *After relocating from Texas:* Property records show that Cook lived in a 544-square-foot apartment.

93   *In his first days at Apple:* Interviews with former Apple vice president, Consumer Products & Asia Operations, Joe O'Sullivan and finance chief Fred Anderson.

93   *"I saw grown men cry":* Interview with Joe O'Sullivan.

94   *Cook called inventory:* Adam Lashinsky, "Tim Cook: The Genius Behind Steve," *Fortune,* November 23, 2008, https://fortune.com/2008/11 /24/apple-the-genius-behind-steve/; Adam Lashinsky, *Inside Apple.*

94   *Apple's operations team:* Interview with Joe O'Sullivan; Kane, *Haunted Empire.*

95   *The operations team's pursuit:* Interview with Joe O'Sullivan and other members of the team.

96   *Cook overruled him:* Interview with Joe O'Sullivan.

96   *He rallied his team:* Kane, *Haunted Empire.*

96   *A year later, he slashed:* Isaacson, *Steve Jobs.*

97   *While at Compaq:* Interviews with Joe O'Sullivan and other members of hardware operations.

97   *With $2,500 borrowed:* Jason Dean, "The Forbidden City of Terry Gou," *Wall Street Journal,* August 11, 2007, https://www.wsj.com/articles /SB118677584137994489.

97 *At Apple's request, Foxconn:* Background interviews with Apple leadership.

98 *Though his compensation topped:* Apple Form Def 14A, March 6, 2000.

99 *A problem could be:* Interview with Joe O'Sullivan.

100 *"This is really bad":* Lashinsky, "Tim Cook: The Genius Behind Steve."

100 *On occasion, Cook would get:* Interviews with former senior Apple officials; Brent Schlender and Rick Tetzeli, *Becoming Steve Jobs.*

100 *"He knew how important":* Schlender and Tetzeli, *Becoming Steve Jobs,* 393.

103 *On a flight to Japan:* Walter Isaacson, *Steve Jobs.*

103 *Jobs expected its forthcoming:* Interviews with senior Apple executives, who credit the vision for this supply-chain maneuver to Jobs, who asked for Cook to buy as much memory as possible. See also Lashinsky, "Tim Cook: The Genius Behind Steve"; "Apple Announces Long-Term Supply Agreements for Flash Memory," Apple, November 21, 2005, https://www.apple.com/newsroom/2005/11/21Apple-Announces-Long-Term-Supply-Agreements-for-Flash-Memory/; Leander Kahney, *Tim Cook.*

103 *The technique, which was used:* Leander Kahney, *Jony Ive.*

104 *Cook's lieutenant, Williams:* Corning Incorporated, "Apple & Corning Press Conference: Remarks from Apple COO Jeff Williams," YouTube, May 17, 2017, https://www.youtube.com/watch?v=AZgULosw6cY.

105 *The CEO, Wendell Weeks:* Isaacson, *Steve Jobs.*

105 *On an earnings call:* "Apple Inc., Q1 2009 Earnings Call," S&P Capital IQ, January 21, 2009, https://www.capitaliq.com/CIQDotNet/Transcripts/Detail.aspx?keyDevId=6156218&companyId=24937.

106 *Dubbed the "Cook Doctrine":* Adam Lashinsky, "The Cook Doctrine at Apple," *Fortune,* January 22, 2009, https://fortune.com/2009/01/22/the-cook-doctrine-at-apple/.

106 *On August 11, 2011:* Schlender and Tetzeli, *Becoming Steve Jobs,* 403–6.

106 *Jobs said he had studied:* z400racer37, "Apple CEO Tim Cook at D10 Full 100 Minute Video," YouTube, July 6, 2012, https://www.youtube.com/watch?v=eUAPHgiEniQ.

107 *The selection surprised some outsiders:* Walter Isaacson, *Steve Jobs.*

107 *"He's always been real smart":* Donna Riley-Lein, "Apple No. 2 Has Local Roots," *Independent,* December 25, 2008.

107 *Knowing that Cook was a bachelor:* Interview with Donna Riley-Lein.

108 *Cook was confident:* z400racer37, "Apple CEO Tim Cook at D10 Full 100 Minute Video," YouTube, July 6, 2012 https://www.youtube.com /watch?v=eUAPHgiEniQ.

109 *But it was in line:* Yukari Kane, *Haunted Empire.*

109 *The change immediately generated:* Jessica E. Vascellaro, "Apple in His Own Image," *Wall Street Journal,* November 2, 2011, https://www.wsj .com/articles/SB10001424052970204394804577012161036609728.

109 *Not everyone was reassured:* Tripp Mickle, "How Tim Cook Made Apple His Own," *Wall Street Journal,* August 7, 2020, https://www .wsj.com/articles/tim-cook-apple-steve-jobs-trump-china-iphone -ipad-apps-smartphone-11596833902.

109 *"I knew what I needed to do":* Homecoming, "With Tim Cook," SEC Network, September 5, 2017.

110 *The designers looked:* "We all looked at him and thought, 'He doesn't get it,'" a member of the design team said in an interview. "That's when we knew it was going to be different."

Chapter 6: Fragile Ideas

112 *Then he turned around:* A few members of Apple's design team remember the watch effort becoming official in this moment. Others recall a text exchange where the idea first circulated, and designer Julian Hönig building an initial model of a watch afterward.

112 *Oracle founder Larry Ellison:* Charlie Rose, "Oracle CEO Larry Ellison: Google CEO Did Evil Things, Apple Is Going Down" (video), CBS News, August 13, 2013, https://www.cbsnews.com/news/oracle -ceo-larry-ellison-google-ceo-did-evil-things-apple-is-going-down/.

113 *The designer believed:* Cambridge Union, "Sir Jony Ive | 2018 Hawking Fellow | Cambridge Union," YouTube, November 28, 2018, https://www.youtube.com/watch?v=KywJimWe_Ok.

113 *"It will have the simplest":* Walter Isaacson, *Steve Jobs,* 555.

114 *would be beyond its control:* People involved in the project said the TV effort failed to take flight because Apple executive Eddy Cue was unable to negotiate licensing agreements with the owners of TV stations, including Walt Disney Company and CBS Corp. Shalini Ramachandran and Daisuke Wakabayashi, "Apple's Hard-Charging Tactics Hurt TV Expansion," *Wall Street Journal,* July 28, 2016, https://www.wsj.com/articles/apples-hard-charging-tactics-hurt-tv -expansion-1469721330.

114 *At school, he excelled:* Adam Satariano, Peter Burrows, and Brad Stone, "Scott Forstall, the Sorcerer's Apprentice," *Bloomberg Businessweek,*

October 13, 2011; Computer History Museum, "CHM Live | Original iPhone Software Team Leader Scott Forstall (Part Two)," June 28, 2017, https://www.youtube.com/watch?v=IiuVggWNqSA; Code.org, "Code Break 9.0: Events with Macklemore & Scott Forstall," YouTube, May 20, 2020, https://youtu.be/-bcO-X9thds.

115 *Both attended Stanford University:* Computer History Museum, "CHM Live | Original iPhone Software Team Leader Scott Forstall (Part Two)."

115 *When sales of the computers:* Jim Carlton, *Apple.*

115 *The hiring process:* Interview with William Parkhurst, who designed the process, and other former NeXT engineers.

115 *Ten minutes into Forstall's interview:* Computer History Museum, "CHM Live | Original iPhone Software Team Leader Scott Forstall (Part Two)."

115 *Forstall worked on NeXT's:* Ibid.

116 *Forstall would spend hours:* Interview with Dan Grillo, NeXT colleague.

116 *In 2004, Forstall had:* Computer History Museum, "CHM Live | Original iPhone Software Team Leader Scott Forstall (Part Two)."

117 *They would have lunch:* Ibid.

118 *To create the iPhone:* Satariano, Burrows, and Stone, "Scott Forstall, the Sorcerer's Apprentice."

118 *Forstall and Fadell vied:* Tony Fadell left the company in 2008.

118 *Forstall also rankled:* Interview with Henri Lamiraux, former iOS engineering vice president.

118 *"Scott was very controlling":* Interview with Henri Lamiraux.

118 *Ive had wanted:* Peter Burrows and Connie Guglielmo, "Apple Worker Said to Tell Jobs IPhone Might Cut Calls," Bloomberg, July 15, 2010, https://www.bloomberg.com/news/articles/2010-07-15/apple-engineer-said-to-have-told-jobs-last-year-about-iphone-antenna-flaw.

119 *Jobs held a press conference:* Geoffrey A. Fowler, Ian Sherr, and Niraj Sheth, "A Defiant Steve Jobs Confronts 'Antennagate,'" *Wall Street Journal,* July 16, 2010, https://www.wsj.com/articles/SB10001424052748704913304575371131458273498.

119 *Ive's design team had obsessed:* "An Introduction to BEZIER Curves," presentation by Apple Industrial Design to Foster + Partners, circa 2014.

121 *It was an advantage:* Matt Hamblen, "Android Smartphone Sales Leap to Second Place, Gartner Says," Computerworld, February 9, 2011, https://www.computerworld.com/article/2512940/android-smartphone-sales-leap-to-second-place-in-2010--gartner-says.html.

121   *It called for Apple to create:* Hansen Hsu and Marc Weber, "Oral History of Kenneth Kocienda and Richard Williamson," Computer History Museum, October 12, 2017, https://archive.computerhistory .org/resources/access/text/2018/07/102740223-05-01-acc.pdf.

122   *Williamson approached Forstall and Schiller:* Ibid.

122   *In April 2012, charter buses:* Yukari Kane, *Haunted Empire.*

123   *The remainder of the world:* Hsu and Weber, "Oral History of Kenneth Kocienda and Richard Williamson."

123   *In June, Forstall stepped:* Apple, "Apple WWDC 2012 Keynote Address" (video), Apple Events, June 11, 2012, https://podcasts .apple.com/us/podcast/apple-wwdc-2012-keynote-address/id 275834665?i=1000117538651.

124   *Apple customers were reporting:* Juliette Garside, "Apple Maps Service Loses Train Stations, Shrinks Tower and Creates New Airport," *Guardian,* September 20, 2012, https://www.theguardian.com /technology/2012/sep/20/apple-maps-ios6-station-tower.

124   *In Dublin, users discovered:* Kilian Doyle, "Apple Gives Dublin a New 'Airfield,'" *Irish Times,* September 20, 2012, https://www.irishtimes .com/news/apple-gives-dublin-a-new-airfield-1.737796.

124   *In New York, the Brooklyn Bridge:* Nilay Patel, "Wrong Turn: Apple's Buggy iOS 6 Maps Leads to Widespread Complaints," Verge, September 20, 2012, https://www.theverge.com/2012/9/20/3363914 /wrong-turn-apple-ios-6-maps-phone-5-buggy-complaints.

125   *Cook drafted a letter:* Jordan Crook, "Tim Cook Apologizes for Apple Maps, Points to Competitive Alternatives," Techcrunch, September 28, 2012, https://techcrunch.com/2012/09/28/tim-cook-apologizes -for-apple-maps-points-to-competitive-alternatives/.

126   *To Cook, it was clear:* Tim Cook on *The Charlie Rose Show,* September 12, 2014.

Chapter 7: Possibilities

129   *watching through:* Ian Parker, "The Shape of Things to Come: How an Industrial Designer Became Apple's Greatest Product," *The New Yorker,* February 16, 2015, https://www.newyorker.com/magazine/2015 /02/23/shape-things-come.

129   *The sketchbooks had:* "Inside Apple," *60 Minutes,* CBS, December 20, 2015.

129   *There were only 150:* Banksy, *Monkey Queen,* MyArtBroker, https:// www.myartbroker.com/artist/banksy/monkey-queen-signed-print/; Banksy-Value.com, https://bit.ly/39gTqzk.

130 **Beside the print sat:** Good Fucking Design Advice, "Classic Advice Print," gfda.co, https://gfda.co/classic/.

131 **When Jobs had returned:** Joel M. Podolny and Morten T. Hansen, "How Apple Is Organized for Innovation," *Harvard Business Review*, November–December 2020, https://hbr.org/2020/11/how-apple-is-organized-for-innovation; Tony Fadell, "For the record, I fully believe . . . ," Twitter, October 23, 2000, https://twitter.com/tfadell/status/1319556633312268288.

131 **Steve Jobs had championed:** Klaus Göttling, "Skeumorphism Is Dead, Long Live Skeumorphism," Interaction Design Foundation, https://www.interaction-design.org/literature/article/skeuomorphism-is-dead-long-live-skeuomorphism.

132 **Apple's heads of operations:** St. Regis Lobby description provided by the hotel via email at author's request.

135 **Ive wanted to bring:** Erica Blust, "Apple Creative Director Alan Dye '97 to Speak Oct. 20," Syracuse University, https://news.syr.edu/blog/2010/10/18/alan-dye/; "Alan Dye," *Design Matters with Debbie Millman* (podcast), June 1, 2007, https://www.designmattersmedia.com/podcast/2007/Alan-Dye; "Bad Boys of Design III," *Design Matters with Debbie Millman* (podcast), May 5, 2006, https://www.designmattersmedia.com/podcast/2006/Bad-Boys-of-Design-III; Debbie millman, "Adobe & AIGA SF Presents Design Matters Live w Alan Dye," YouTube, https://www.youtube.com/watch?v=gBre88MsZZo.

136 **The corners of iOS 7:** "An Introduction to BEZIER Curves," presentation by Apple Industrial Design to Foster + Partners, circa 2014.

136 **A few months after taking over:** Interview with Bob Burrough, former Apple engineer, who attended the meeting.

139 **They learned how the British:** Interview with David Rooney, writer and former curator of timekeeping at the Royal Observatory, Greenwich, U.K.; David Belcher, "Wrist Watches: From Battlefield to Fashion Accessory," *New York Times*, October 23, 2013, https://www.nytimes.com/2013/10/23/fashion/wrist-watches-from-battlefield-to-fashion-accessory.html; Benjamin Clymer, "Apple, Influence, and Ive," *Hodinkee Magazine*, vol. 2, https://www.hodinkee.com/magazine/jony-ive-apple; Esti Chazanow, "9 Types of Uncommon Mechanical Watch Complications," LIV Swiss Watches, December 21, 2019, https://p51.livwatches.com/blogs/everything-about-watches/9-types-of-uncommon-mechanical-watch-complications; Jason Heaton, "In Defense of Quartz Watches," *Outside*, July 17, 2019, https://www.outsideonline.com/outdoor-gear/tools/defense-quartz-watches/.

140 *The most accurate heart rate:* Mark Sullivan, "What I Learned Working with Jony Ive's Team on the Apple Watch," *Fast Company,* August 15, 2016, https://www.fastcompany.com/3062576/what-i-learned-working -with-jony-ives-team-on-the-apple-watch.

141 *He had designed everything:* Catherine Keenan, "Rocket Man: Marc Newson," *Sydney Morning Herald,* July 30, 2009.

141 *Newson's in squiggles:* Jony Ive and Marc Newson on *The Charlie Rose Show,* November 21, 2013, https://charlierose.com/videos/17469.

142 *The drawings also included:* "Crown (Watchmaking)," Foundation High Horology, https://www.hautehorlogerie.org/en/watches-and-culture /encyclopaedia/glossary-of-watchmaking/.

142 *Ive would become:* Maria Konnikova, "Where Do Eureka Moments Come From?," *The New Yorker,* May 27, 2014, https://www.newyorker .com/science/maria-konnikova/where-do-eureka-moments-come -from.

Chapter 8: Can't Innovate

143 *Early in his tenure leading:* This anecdote is based on a firsthand source to whom Cook told the story. Apple has contested the anecdote, saying that it is inaccurate. Cook did not respond to repeated inquiries.

144 *Cook supported the rollout:* "Apple Fans Crowd New Downtown Palo Alto Store," Palo Alto Online, October 27, 2012, https://www.paloalto online.com/news/2012/10/27/apple-fans-crowd-new-palo-alto -store.

144 *In the new iPhone's first weekend:* "iPhone 5 First Weekend Sales Top Five Million," Apple, September 24, 2012, https://www.apple .com/newsroom/2012/09/24iPhone-5-First-Weekend-Sales -Top-Five-Million/; "iPhone 4S First Weekend Sales Top Four Million," Apple, October 17, 2011, https://www.apple.com/newsroom /2011/10/17iPhone-4S-First-Weekend-Sales-Top-Four-Million/; "iPhone 4 Sales Top 1.7 Million," Apple, June 28, 2010, https://www .apple.com/newsroom/2010/06/28iPhone-4-Sales-Top-1-7-Million/.

144 *The new model delivered:* Matt Burns, "Apple's Stock Price Crashes to Six Month Low and There's No Bottom in Sight," TechCrunch, November 15, 2012, https://techcrunch.com/2012/11/15/apples-stock -price-is-crashing-and-the-bottom-is-not-in-sight/. The market value fell from $656.34 billion on September 18, 2012, to $493.51 billion on November 15, 2012, per Macrotrends.

145 *The gaps between the iPhone:* Jon Russell, "IDC: Samsung Shipped Record 63.7m Smartphones in Q4 '12," TNW, January 25, 2013,

https://thenextweb.com/news/idc-samsung-shipped-record-63-7m
-smartphones-in-q4-12.

145 *Pendleton and his colleagues*: Interview with Scott Pendleton; Michal
     Lev-Ram, "Samsung's Road to Global Dominatation," *Fortune*,
     January 22, 2013, https://fortune.com/2013/01/22/samsungs-road
     -to-global-domination/. Brian X. Chen, "Samsung Saw Death of
     Apple's Jobs as a Time to Attack," *New York Times*, April 16, 2014,
     https://bits.blogs.nytimes.com/2014/04/16/samsung-saw-death-of
     -steve-jobs-as-a-time-to-attack/.

145 *But Samsung had ripped:* Ina Fried, "Apple Designer: We've Been
     Ripped Off," All Things Digital, July 31, 2012, https://allthingsd
     .com/20120731/apple-designer-weve-been-ripped-off/.

146 *"What if we had someone":* Interview with Scott Pendleton.

148 *Its ad had caught:* Scott Peters, "Rock Center: Apple CEO Tim Cook
     Interview," YouTube, January 20, 2013, https://www.youtube.com
     /watch?v=zz1GCpqd-0A.

149 *He so regularly shot down:* Peter Burrows and Adam Satariano, "Can
     Phil Schiller Keep Apple Cool?," Bloomberg, June 7, 2012, https://
     www.bloomberg.com/news/articles/2012-06-07/can-phil-schiller
     -keep-apple-cool.

149 *Tech reviewers panned:* Sean Hollister, "Apple's New Mac Ads Are
     Embarrassing," Verge, July 28, 2012.

150 *In late January 2013:* Ian Sherr and Evan Ramstad, "Has Apple Lost
     Its Cool to Samsung?," *Wall Street Journal,* January 28, 2013, https://
     www.wsj.com/articles/SB1000142412788732385490457826409007
     4879024.

150 *Schiller forwarded the article:* Jay Yarrow, "Phil Schiller Exploded on
     Apple's Ad Agency in an Email,"Business Insider,April 7,2014,https://
     www.yahoo.com/news/phil-schiller-exploded-apples-ad-163842747
     .html.

152 *Media Arts Lab had revived: Apple v. Samsung,* U.S. District Court,
     Northern District of California, C-12-00630, vol. 3, 498–756, April 4,
     2014.

153 *In its reply, Apple had left:* Elise J. Bean, *Financial Exposure: Carl
     Levin's Senate Investigations into Finance and Tax Abuse* (New York:
     Palgrave Macmillan, 2018), e-book; interview with Elise Bean.

154 *A favorable agreement:* Ibid.; interview with Elise Bean.

154 *He considered the rate unreasonable:* Offshore Profit Sharing and the
     U.S. Tax Code—Part 2 (Apple Inc.), Hearing Before the Permanent
     Subcommittee on Investigations of the Committee on Homeland

Security and Government Affairs, United States Senate, May 21, 2013, https://www.govinfo.gov/content/pkg/CHRG-113shrg81657 /pdf/CHRG-113shrg81657.pdf.

158 *He then sat down and listened:* Ibid., 9.

160 *The music flourished:* Apple, "Apple WWDC 2013 Keynote Address" (video), Apple Events, June 10, 2013, https://podcasts.apple .com/us/podcast/apple-wwdc-2013-keynote-address/id275 834665?i=1000160871947.

162 *"I'm really glad you liked that":* Ibid.

163 *The New York Times' David Pogue:* David Pogue, "Yes, There's a New iPhone. But That's Not the Big News," *New York Times,* September 17, 2013, https://pogue.blogs.nytimes.com/2013/09/17/yes-theres -a-new-iphone-but-thats-not-the-big-news/; Darrell Etherington, "Apple iOS 7 Review: A Major Makeover That Delivers, but Takes Some Getting Used To," TechCrunch, September 18, 2013, https:// techcrunch.com/2013/09/17/ios-7-review-apple/.

164 *The brand campaign:* TouchGameplay, "Official Designed by Apple in California Trailer," YouTube, June 10, 2013, https://www.youtube .com/watch?v=0xD569Io7kE.

164 *Slate pilloried it:* Seth Stevenson, "Designed by Doofuses in California," Slate, August 26, 2013, https://slate.com/business/2013/08 /designed-by-apple-in-california-ad-campaign-why-its-so-terrible .html.

164 *One of the original corporate raiders:* Cara Lombardo, "Carl Icahn Is Nearing Another Landmark Deal. This Time It's with His Son," *Wall Street Journal,* October 19, 2019, https://www.wsj.com/articles/carl -icahn-is-nearing-another-landmark-deal-this-time-its-with-his -son-11571457602; interview with Carl Icahn.

167 *Ahrendts had tripled:* Jeff Chu, "Can Apple's Angela Ahrendts Spark a Retail Revolution?," *Fast Company,* January 6, 2014, https://www .fastcompany.com/3023591/angela-ahrendts-a-new-season-at-apple.

167 *"We have thousands of techies":* Nicole Nguyen, "Meet the Woman Who Wants to Change the Way You Buy Your iPhone," BuzzFeed News, October 25, 2017, https://www.buzzfeednews.com/article /nicolenguyen/meet-the-woman-who-wants-to-change-the-way -you-buy-your.

167 *She was perceived:* Forty thousand retail employees: see Apple Inc.; Form 10-K, United States Securities and Exchange Commission, September 28, 2013, https://www.sec.gov/Archives/edgar /data/320193/000119312513416534/d590790d10k.htm.

Chapter 9: The Crown

170 *Ive and Newson had come to the project:* Paul Goldberger, "Designing Men," *Vanity Fair,* October 10, 2013, https://www.vanityfair.com /news/business/2013/11/jony-ive-marc-newson-design-auction#~o.

170 *A single product:* Ibid.

170 *a new camera that stripped away:* Jony Ive and Marc Newson, *The Charlie Rose Show,* November 21, 2013, https://charlierose.com /videos/17469.

171 *The camera design took more:* Goldberger, "Designing Men."

173 *In weekly meetings:* "Apple Unveils Apple Watch—Apple's Most Personal Device Ever," Apple, September 9, 2014, https://www.apple .com/newsroom/2014/09/09Apple-Unveils-Apple-Watch-Apples -Most-Personal-Device-Ever/.

174 *Ultimately, he chose:* Apple Watch marketing site, April 30, 2015, via Wayback Machine—Internet Archive, https://web.archive.org /web/20150430052623/http://www.apple.com/watch/apple-watch/.

174 *A similar process played out:* The Apptionary, "Full March 9, 2015, Apple Keynote Apple Watch, Macbook 2015," YouTube, March 9, 2015, https://www.youtube.com/watch?v=U2wJsHWSafc; Benjamin Clymer, "Apple, Influence, and Ive," *Hodinkee Magazine,* vol. 2, https:// www.hodinkee.com/magazine/jony-ive-apple.

174 *Comparable care was given:* Ariel Adams, "10 Interesting Facts about Marc Newson's Watch Design Work at Ikepod," A Blog to Watch, September 9, 2014, https://www.ablogtowatch.com/10-interesting-facts -marc-newson-watch-design-work-ikepod/.

175 *To support and broaden:* Jim Dallke, "Inside the Small Evanston Company Whose Tech Was Acquired by Apple and Used by SpaceX," CHICAGOINNO, February 15, 2017, https://www.bizjournals.com /chicago/inno/stories/inno-insights/2017/02/15/inside-the-small -evanston-company-whose-tech-was.html; "Charlie Kuehmann, VP at SpaceX and Tesla Motors, Is Visiting Georgia Tech!," Georgia Institute of Technology, https://materials.gatech.edu/event/charlie -kuehmann-vp-spacex-and-tesla-motors-visiting-georgia-tech.

175 *The work fell to:* Kim Peterson, "Did Apple Invent a New Gold for Its Luxury Watch?," Moneywatch, CBS News, March 10, 2015, https://www.cbsnews.com/news/did-apple-invent-a-new-gold-for -its-luxury-watch/; "Crystalline Gold Alloys with Improved Hardness," patent no. WO 2015038636A1, March 19, 2015, https:// patentimages.storage.googleapis.com/59/52/60/086e50f497e052 /WO2015038636A1.pdf; Apple Videos, "Apple Watch Edition—

Gold," YouTube, August 13, 2015, https://www.youtube.com/watch?v=
S-aEW0vWdT4.

176 *It reflected his philosophy:* Walter Isaacson, *Steve Jobs.*

176 *With the watch:* Anick Jesdanun, "Pick Your Apple Watch: 54 Combinations of case, band, size," Associated Press, April 9, 2015, https://apnews.com/0cf0112b699a407e9fcc8286946949ff.

177 *In 2004, he had gone:* Christina Passariello, "How Jony Ive Masterminded Apple's New Headquarters," *Wall Street Journal Magazine,* July 26, 2017, https://www.wsj.com/articles/how-jony-ive-masterminded-apples-new-headquarters-1501063201.

179 *With prototypes of the watch:* David Pierce, "iPhone Killer: The Secret History of the Apple Watch," *Wired,* May 1, 2015, https://www.wired.com/2015/04/the-apple-watch/.

181 *Williams had earned the label:* Apple Inc. Definitive Proxy Statement, Schedule 14A, United States Securities and Exchange Commission, January 7, 2013, https://www.sec.gov/Archives/edgar/data/320193/000119312513005529/d450591ddef14a.htm.

183 *Operating off the knowledge:* "Monitor Your Heart Rate with Apple Watch," Apple, https://support.apple.com/en-us/HT204666.

184 *Samsung was rising in power:* Jon Russell, "IDC: Smartphone Shipments Hit 1B for the First Time in 2013, Samsung 'Clear Leader' with 32% Share," TNW, January 27, 2014, https://thenextweb.com/news/idc-smartphone-shipments-passed-1b-first-time-2013-samsung-remains-clear-leader.

187 *They came up with ideas:* Mark Gurman, "Apple Store Revamp for Apple Watch Revealed: 'Magical' Display Tables, Demo Loops, Sales Process," 9to5Mac, March 29, 2015, https://9to5mac.com/2015/03/29/apple-store-revamp-for-apple-watch-revealed-magical-tables-demo-loops-sales-process/.

190 *Wintour was mesmerized:* Interview with Anna Wintour.

Chapter 10: Deals

192 *Brand-conscious consumers:* Ian Johnson, "China's Great Uprooting: Moving 250 Million into Cities," *New York Times,* June 15, 2013, https://www.nytimes.com/2013/06/16/world/asia/chinas-great-uprooting-moving-250-million-into-cities.html; Rui Zhu, "Understanding Chinese Consumers," *Harvard Business Review,* November 14, 2013, https://hbr.org/2013/11/understanding-chinese-consumers.

192 *The effort required navigating:* WikiLeaks, "Cablegate: Apple Iphone Facing Licensing Issues in China," Scoop Independent News, June 12,

2009, https://www.scoop.co.nz/stories/WL0906/S00516/cablegate
-apple-iphone-facing-licensing-issues-in-china.htm?from-mobile
=bottom-link-01.

193 *He was especially close:* Zheng Jun, "Interview with Cook: Hope That
the Mainland Will Become the First Batch of New Apple Products
to Be Launched," Sina Technology (translated) January 10, 2013;
John Underwood, "Living the Good Life," Gulf Coast Media, July 13,
2018, https://www.gulfcoastnewstoday.com/stories/living-the-good
-life,64626.

193 *It posted its lowest:* Apple Inc., Form 10-Q for the fiscal quarter ended
December 27, 2013, Securities and Exchange Commission, https://
www.sec.gov/Archives/edgar/data/320193/000119312515259935
/d927922d10q.htm.

194 *Publicly, Cook called the theory:* "Apple Inc. Presents at Goldman Sachs
Technology & Internet Conference 2013," S&P Capital IQ, Feb-
ruary 12, 2013, https://www.capitaliq.com/CIQDotNet/Transcripts
/Detail.aspx?keyDevId=227981668&companyId=24937.

195 *"Mr. Xi, will you now use":* "CNBC Exclusive: CNBC Transcript: Apple CEO Tim Cook and China Mobile Chairman Xi Guohua Speak
with CNBC's Eunice Yoon Today," CNBC, January 15, 2014, https://
www.cnbc.com/2014/01/15/cnbc-exclusive-cnbc-transcript-apple
-ceo-tim-cook-and-china-mobile-chairman-xi-guohua-speak-with
-cnbcs-eunice-yoon-today.html.

196 *Cook and Xi later headed:* "CEO Tim Cook Visits Beijing," Getty
Images, January 17, 2014, https://www.gettyimages.com/detail/news
-photo/tim-cook-chief-executive-officer-of-apple-inc-visits-a
-news-photo/463193469; Dhara Ranasinghe, "Apple Takes a Fresh
Bite into China's Market," CNBC, January 17, 2014, https://www
.cnbc.com/2014/01/16/apple-takes-a-fresh-bite-into-chinas-market
.html; Mark Gurman, "Apple CEO Cook Hands Out Autographed
iPhones at China Mobile Launch, Says 'Great Things' Coming,"
9to5Mac, January 16, 2014, https://9to5mac.com/2014/01/16/tim
-cook-hands-out-autographed-iphones-at-china-mobile-launch-says
-great-things-in-product-pipeline/.

199 *He liked to say:* Marco della Cava, "For Iovine and Reznor, Beats
Music Is 'Personal,'" *USA Today,* January 11, 2014, https://www.usa
today.com/story/life/music/2014/01/11/beats-music-interview
-jimmy-iovine-trent-reznor/4401019/.

200 *Pained by the thought:* Tripp Mickle, "Jobs, Cook, Ive—Blevins? The
Rise of Apple's Cost Cutter," *Wall Street Journal,* January 23, 2020,

https://www.wsj.com/articles/jobs-cook-iveblevins-the-rise-of
-apples-cost-cutter-11579803981.

201 *Seele, the German manufacturer:* Sydney Franklin, "How the World's
Largest Curved Windows Were Forged for Apple HQ," Architizer,
https://architizer.com/blog/inspiration/stories/architectural-details
-apple-park-windows/.

202 *they would marvel:* "Steel-and-Glass Design with Curved Glass
for LACMA," Seele, https://seele.com/references/los-angeles-county
-museum-of-arts-usa.

202 *"Could this be smaller?":* The architects at Foster + Partners worked to
reduce the strip of steel from one inch to less than a half inch, according to people familiar with the event and the project.

203 *At the time, Tesla was:* Mike Ramsey, "Tesla Motors Nearly
Doubled Staff in 2014," *Wall Street Journal,* February 27, 2015,
https://www.wsj.com/articles/tesla-motors-nearly-doubled-staff
-in-2014-1425072207; Daisuke Wakabayashi and Mike Ramsey,
"Apple Gears Up to Challenge Tesla in Electric Cars," *Wall Street Journal,* February 13, 2015, https://www.wsj.com/articles/apples-titan-car
-project-to-challenge-tesla-1423868072.

203 *The largest options:* "2015 Global Health Care Outlook: Common
Goals, Competing Priorities," Deloitte, https://www2.deloitte.com
/content/dam/Deloitte/global/Documents/Life-Sciences-Health
-Care/gx-lshc-2015-health-care-outlook-global.pdf; "The World's
Automotive Industry," International Organisation of Motor Vehicles Manufacturers, November 29, 2006, https://www.oica.net/wp
-content/uploads/2007/06/oica-depliant-final.pdf.

203 *he had thought they made:* Tom Relihan, "Steve Jobs Talks Consultants, Hiring, and Leaving Apple in Unearthed 1992 Talk," MIT
Sloan School of Management, May 10, 2018, https://mitsloan.mit
.edu/ideas-made-to-matter/steve-jobs-talks-consultants-hiring-and
-leaving-apple-unearthed-1992-talk.

204 *One night after work:* "Tim Cook," Charlie Rose, September 12, 2014,
https://charlierose.com/videos/18663.

204 *As each one had arrived:* Ben Fritz and Tripp Mickle, "Apple's iTunes
Falls Short in Battle for Video Viewers," *Wall Street Journal,* July 9, 2017,
https://www.wsj.com/articles/apples-itunes-falls-short-in-battle
-for-video-viewers-1499601601.

206 *Dre had a history:* Tom Connick, "Dr. Dre Discusses History of Abuse
Towards Women: 'I Was Out of My Fucking Mind,'" NME, July 11, 2017,
https://www.nme.com/news/music/dr-dre-discusses-abuse-women

-fucking-mind-2108142; Joe Coscarelli, "Dr. Dre Apologizes to the 'Women I've Hurt,'" *New York Times,* August 21, 2015, https://www.nytimes.com/2015/08/22/arts/music/dr-dre-apologizes-to-the-women-ive-hurt.html.

207 *"Remember that scene":* Wall Street Journal, "Behind the Deal—The Weekend That Nearly Blew the $3 Billion Apple Beats Deal," YouTube, July 13, 2017, https://www.youtube.com/watch?v=A0md3ok60g8.

Chapter 11: Blowout

210 *The sacrifices:* "Jony Ive: The Future of Design," Hirshhorn Museum, November 29, 2017, podcast posted to Soundcloud.com by Fuste, https://soundcloud.com/user-175082292/jony-ive-the-future-of-design; Ian Parker, "The Shape of Things to Come: How an Industrial Designer Became Apple's Greatest Product," *The New Yorker,* February 16, 2015, https://www.newyorker.com/magazine/2015/02/23/shape-things-come.

210 *It stood two stories high:* Justin Sullivan, "Apple Unveils iPhone 6," Getty Images, September 9, 2014, https://www.gettyimages.com/detail/news-photo/the-new-iphone-6-is-displayed-during-an-apple-special-event-news-photo/455054182; Karl Mondon, "Final Preparations Are Made Monday Morning, September 8, 2014, for Tomorrow's Big Apple Media Event," Getty Images, September 8, 2014, https://www.gettyimages.in/detail/news-photo/final-preparations-are-made-monday-morning-sept-8-for-news-photo/1172329286; Karl Mondon, "Different Models of the New Apple Watch Are on Display," Getty Images, September 9, 2014, https://www.gettyimages.com/detail/news-photo/different-models-of-the-new-apple-watch-are-on-display-for-news-photo/1172329258.

211 *Some three thousand miles away:* Don Emmert/AFP, "People Wait in Line on Chairs September 9, 2014 Outside the Apple Store on 5th Avenue," Getty Images, September 9, 2014, https://www.gettyimages.com/detail/news-photo/people-wait-in-line-on-chairs-september-9-2014-outside-the-news-photo/455039230.

211 *A New Yorker writer at work:* Parker, "The Shape of Things to Come," *The New Yorker.*

212 *"Everything's great":* "Apple Special Event, September 2014" (video), Apple Events, September 9, 2014, https://podcasts.apple.com/us/podcast/apple-special-event-september-2014/id275834665?i=1000430692664.

217 *"It shows innovation":* "Apple Watch: Will It Revolutionize the Personal Device?," *Nightline,* ABC, September 9, 2014, https://abc news.go.com/Nightline/video/apple-watch-revolutionize-personal -device-25396956.

217 *"I still don't know":* Suzy Menkes, "A First Look at the Apple Watch," *Vogue,* September 9, 2014, https://www.vogue.co.uk/article/suzy-menkes -apple-iwatch-review.

218 *Eager to quell the unrest:* Chris Welch, "Apple Releases One-Click Tool to Delete the U2 Album You Didn't Want," Verge, September 15, 2014, https://www.theverge.com/2014/9/15/6153165/apple-u2 -songs-of-innocence-removal-tool; Robert Booth, "U2's Bono Issues Apology for Automatic Apple iTunes Download," *Guardian,* October 15, 2014, https://www.theguardian.com/music/2014/oct/15/u2-bono-issues -apology-for-apple-itunes-album-download.

219 *Early one morning:* Colette Paris, "Apple Watch at Colette Paris," Facebook, October 1, 2014, https://www.facebook.com/www.colette .fr/photos/a.10152694538705266/10152694539145266.

220 *Nearby, Newson spoke:* Miles Socha, "Apple Unveils Watch at Colette," *Women's Wear Daily,* September 30, 2014, https://wwd.com/fashion -news/fashion-scoops/apple-unveils-watch-at-colette-7959364/.

221 *Lagerfeld dismissed Alaïa:* Emilia Petrarca, "Karl Lagerfeld Talks Death and His Enemies in a Wild New Interview," *New York,* April 13, 2018, https://www.thecut.com/2018/04/karl-lagerfeld-numero -interview-azzedine-alaia-virgil-abloh.html#_ga=2.218658718 .629632365.1631210806-1193973995.1631210803; Ella Alexander, "Full of Faults," *Vogue,* June 23, 2011, https://www.vogue.co.uk/article /alaia-criticises-karl-lagerfeld-and-anna-wintour.

221 *Ive, who clung:* "Apple Azzedine Alaia Party with Lenny Kravitz, Marc Newson, Jonathan Ive for Apple Watch," AudreyWorldNews, November 11, 2014, http://www.audreyworldnews.com/2014/11/apple -azzedine-alaia-party.html; Vanessa Friedman, "The Star of the Show Is Strapped on a Wrist," *New York Times,* October 1, 2014, https:// www.nytimes.com/2014/10/02/fashion/apple-watch-azzedine-alaia -paris-fashion-week.html.

Chapter 12: Pride

222 *Rising in California:* "Apple Inc., Q4 2014 Earnings Call, Oct 20, 2014," S&P Capital IQ, October 20, 2014, https://www.capitaliq .com/CIQDotNet/Transcripts/Detail.aspx?keyDevId=273702454& companyId=24937.

222 *The company's daily sales figures:* Apple Inc., Form 10-Q for the fiscal quarter ended December 27, 2014, United States Securities and Exchange Commission, https://www.sec.gov/Archives/edgar/data /320193/000119312515023697/d835533d10q.htm.

222 *On average, five hundred iPhones:* Walt Mossberg, "The Watcher of the Apple Watch: Jeff Williams at Code 2015 (Video)," Vox, June 18, 2015, https://www.vox.com/2015/6/18/11563672/the-watcher-of-the -apple-watch-jeff-williams-at-code-2015-video.

223 *"Demand for the new iPhones":* "Apple Inc., Q4 2014 Earnings Call, Oct 20, 2014," S&P Capital IQ, October 20, 2014, https:// www.capitaliq.com/CIQDotNet/Transcripts/Detail.aspx?keyDev Id=273702454&companyId=24937.

223 *In the fall of 2014:* Ryan Phillips, "Tim Cook, Nick Saban Among Newest Members of Alabama Academy of Honor," *Birmingham Business Journal,* October 27, 2014, https://www.bizjournals.com/birmingham /morning_call/2014/10/tim-cook-nick-saban-among-newest -members-of.html.

223 *Cook had penned an editorial:* Tim Cook, "Workplace Equality Is Good for Business," *Wall Street Journal,* November 3, 2013, https://www.wsj .com/articles/SB10001424052702304527504579172302377638002.

224 *Two years earlier:* Jena McGregor, "Anderson Cooper was Tim Cook's Guide for Coming Out as Gay," *Washington Post,* August 15, 2016, https://www.washingtonpost.com/news/on-leadership/wp /2016/08/15/why-tim-cook-talked-with-anderson-cooper-before -publicly-coming-out-as-gay/.

224 *He told Cooper that:* Anderson Cooper on *The Howard Stern Show,* May 12, 2020, https://www.howardstern.com/show/2020/05/12 /robin-quivers-struggles-turning-down-houseguests-amidst-global -pandemic/.

224 *Cook called Tyrangiel:* Bloomberg Surveillance, "Apple CEO Tim Cook: I'm Proud to Be Gay" (video), Bloomberg, October 30, 2014, https://www.bloomberg.com/news/videos/2014-10-30/apple-ceo -tim-cook-im-proud-to-be-gay.

225 *"Throughout my professional life":* Tim Cook, "Tim Cook Speaks Up," Bloomberg, October 30, 2014, https://www.bloomberg.com/news /articles/2014-10-30/tim-cook-speaks-up.

226 *Acceptance of gay and lesbian:* "LGBT Rights," Gallup, https://news .gallup.com/poll/1651/gay-lesbian-rights.aspx.

226 *It was a view:* "The History of the Castro," KQED, 2009, https:// www.kqed.org/w/hood/castro/castroHistory.html.

226   *It had amended:* "Apple Gives Benefits to Domestic Partners," *San Francisco Chronicle,* July 25, 1992.

226   *A 2008 profile:* Adam Lashinsky, "Tim Cook: The Genius Behind Steve," *Fortune,* November 23, 2008, https://fortune.com/2008/11/24 /apple-the-genius-behind-steve/; Owen Thomas, "Is Apple COO Tim Cook Gay?," Gawker, November 10, 2008, https://www.gawker .com/5082473/is-apple-coo-tim-cook-gay.

227   *In 2011,* **Out** *magazine:* Nicholas Jackson, "To Be the Most Powerful Gay Man in Tech, Cook Needs to Come Out," *The Atlantic,* August 25, 2011, https://www.theatlantic.com/technology/archive/2011/08 /to-be-the-most-powerful-gay-man-in-tech-cook-needs-to-come out/244083/.

227   *A Gawker article:* Ryan Tate, "Tim Cook: Apple's New CEO and the Most Powerful Gay Man in America," Gawker, August 24, 2011, https://www.gawker.com/5834158/tim-cook-apples-new-ceo-and -the-most-powerful-gay-man-in-america; interviews with Ben Ling and friends of Ben Ling, who said that Ling and Cook never dated.

228   *He placed an iPad:* Erin Edgemon, "Apple CEO Tim Cook Criticizes Alabama for Not Offering Equality to LGBT Community," AL.com, October 27, 2014, updated January 13, 2020, https://www.al.com /news/montgomery/2014/10/apple_ceo_tim_cook_criticizes.html; WKRG, "Apple's Tim Cook Honored, Slams Alabama Education System," YouTube, November 12, 2014, https://www.youtube .com/watch?v=P6xZSCyPWmA.

228   *"We're all familiar with":* Ismail Hossain, "Apple CEO Tim Cook Speaks at Alabama Academy of Honor Induction," YouTube, January 3, 2015, https://www.youtube.com/watch?v=frpvn_0bxQs.

229   *A prominent conservative news outlet:* Ryan Boggus, "Sims Unloads on Apple CEO for 'Swooping In' to 'Lecture Alabama on How We Should Live,'" Yellowhammer News, October 28, 2014, https://yellow hammernews.com/sims-unloads-apple-ceo-swooping-lecture -alabama-live/.

229   *"I kept it to my small circle":* "Exclusive: Amanpour Speaks with Apple CEO Tim Cook" (video), CNN, October 25, 2018, https://www.cnn .com/videos/business/2018/10/25/tim-cook-amanpour-full.cnn.

230   *Headlined "Tim Cook Speaks Up":* Cook, "Tim Cook Speaks Up."

231   *People in the gay community:* Marc Hurel, "Tim Cook of Apple: Being Gay in Corporate America (letter)," *New York Times,* October 31, 2014, https://www.nytimes.com/2014/11/01/opinion/tim-cook-of-apple -being-gay-in-corporate-america.html; James B. Stewart, "The Com-

ing Out of Apple's Tim Cook: 'This Will Resonate,'" *New York Times*, October 30, 2014.

Chapter 13: Out of Fashion

233 *Not long after the meeting:* Flight records for N586GV; Ian Parker, "The Shape of Things to Come," *The New Yorker*, February 16, 2015, https://www.newyorker.com/magazine/2015/02/23/shape-things-come.

233 *By 2015, Ive was being chauffeured:* Parker, "The Shape of Things to Come"; Jake Holmes, "2014 Bentley Mulsanne Adds Pillows, Privacy Curtains and Wi-Fi," Motortrend, January 23, 2013, https://www.motortrend.com/news/2014-bentley-mulsanne-adds-pillows-privacy-curtains-and-wi-fi-199127/.

234 *"Year after year, they've kept":* "Cramer: Own Apple, Don't Trade It" (video), *Mad Money with Jim Cramer*, CNBC, January 28, 2015, https://www.cnbc.com/video/2015/01/28/cramer-own-apple-dont-trade-it.html.

234 *Cramer praised Cook:* "Cook Calls Cramer: Happy 10th Anniversary!" (video), *Mad Money with Jim Cramer*, CNBC, March 12, 2015, https://www.cnbc.com/video/2015/03/12/cook-calls-cramer-happy-10th-anniversary.html.

235 *In an article headlined:* Vanessa Friedman, "This Emperor Needs New Clothes," *New York Times*, October 15, 2014, https://www.nytimes.com/2014/10/16/fashion/for-tim-cook-of-apple-the-fashion-of-no-fashion.html.

235 *In contrast, Ive reached:* Parker, "The Shape of Things to Come."

236 *"Now we've got":* "Apple Special Event, March 2015" (video), Apple Events, March 9, 2015, https://podcasts.apple.com/us/podcast/apple-special-event-march-2015/id275834665?i=1000430692662.

236 *The aluminum Apple:* Press Release, "Apple Watch Available in Nine Countries on April 24," Apple, March 9, 2015, https://www.apple.com/newsroom/2015/03/09Apple-Watch-Available-in-Nine-Countries-on-April-24/.

236 *The company's previous:* Apple Inc., 2011 Form 10-K for the year ended September 24, 2011, (filed October 26, 2011), p. 30, SEC, https://www.sec.gov/Archives/edgar/data/320193/000119312511282113/d220209d10k.htm.

236 *On CNBC, anchors asked:* Jay Yarow, "There's 'Lackluster Interest' in Apple Watch, Says UBS," Business Insider, May 1, 2015, https://www.businessinsider.com/ubs-on-the-apple-watch-2015-5; "Can Apple

Watch Move the Needle?" (video), CNBC, March 13, 2015, https://www
.cnbc.com/video/2015/03/10/can-apple-watch-move-the-needle.html.

238 *paid about $2 an hour:* Karen Turner, "As Apple's Profits Decline,
iPhone Factory Workers Suffer, a New Report Claims," *Washington Post,* September 1, 2016, https://www.washingtonpost.com
/news/the-switch/wp/2016/09/01/as-apples-profits-decline-iphone
-factory-workers-suffer-a-new-report-claims/.

238 *Late in the assembly process:* Daisuke Wakabayashi and Lorraine
Luk, "Apple Watch: Faulty Taptic Engine Slows Rollout," *Wall Street
Journal,* April 29, 2015, https://www.wsj.com/articles/apple-watch
-faulty-taptic-engine-slows-roll-out-1430339460.

240 *At Infinite Loop:* Interview with Patrick Pruniaux, who joined
Deneve's team from the watchmaker Tag Heuer.

240 *He knotted a black satin tie:* Alan F. "Rich and Famous in Milan Get
Free Apple Watch," PhoneArena.com, April 17, 2015, https://www
.phonearena.com/news/Rich-and-famous-in-Milan-get-free-Apple
-Watch-Apple-Watch-Band-and-more_id68390.

241 *Ive reveled in:* Nick Compton, "Road-Testing the Apple Watch at
Salone del Mobile 2015," *Wallpaper,* April 13, 2015, https://www
.wallpaper.com/watches-and-jewellery/the-big-reveal-road-testing
-the-apple-watch-at-salone-del-mobile-2015.

241 *Moments after the designers:* Micah Singleton, "Jony Ive: It's Not
Our Intent to Compete with Luxury Goods" (video), Verge, April
24, 2015, https://www.theverge.com/2015/4/24/8491265/jony-ive
-interview-apple-watch-luxury-goods; Scarlett Kilcooley-O'Halloran,
"Apple Explains Its Grand Plan to Suzy Menkes" (video), *Vogue,* April
22, 2015, http://web.archive.org/web/20150425201744/https://www
.vogue.co.uk/news/2015/04/22/the-new-luxury-landscape.

242 *He had pursued the iPad:* Imran Chaudhri, "So the Real Story Is
That Steve's Brief," Twitter, December 16, 2019, https://twitter.com
/imranchaudhri/status/1206785636855758855?lang=en.

243 *In the United States:* Associated Press, "Shoppers Get to Know Apple
Watch on First Day of Sales," CTV News, April 10, 2015, https://
www.ctvnews.ca/sci-tech/shoppers-get-to-know-apple-watch-on
-first-day-of-sales-1.2320387.

243 *Instead, the line outside:* Tim Higgins, Jing Ceo, and Amy Thomson,
"Apple Watch Debut Marks a New Retail Strategy for Apple," Bloomberg,
April 24, 2015, https://www.bloomberg.com/news/articles/2015-04-24
/apple-watch-debut-marks-a-new-retail-strategy-for-apple.

244 *"Neither felt like a luxury":* Sam Byford, Amar Toor, and Tom Warren, "We Went Shopping for an Apple Watch in Tokyo, Paris, and London," Verge, April 10, 2015, https://www.theverge.com/2015/4/10/8380993 /apple-watch-tokyo-paris-london-shopping.

244 *It was among the first:* Nilay Patel, "Apple Watch Review," Verge, April 8, 2015, https://www.theverge.com/a/apple-watch-review; Nicole Phelps, "Apple Watch: A Nine-Day Road Test," *Vogue,* April 8, 2015, https://www.vogue.com/article/apple-watch-test-drive.

244 *Their views were best captured:* Joshua Topolsky, "Apple Watch Review: You'll Want One, but You Don't Need One," Bloomberg, April 8, 2015, https://www.bloomberg.com/news/features/2015-04-08/apple -watch-review-you-ll-want-one-but-you-don-t-need-one.

245 *The customer apathy led:* Jay Yarow, "There's 'Lackluster Interest' in Apple Watch, Says UBS," Business Insider, May 1, 2015, https:// www.businessinsider.com/ubs-on-the-apple-watch-2015-5; sfgoldberg, "Long Sync Times, Delayed Notifications, and Other Issues— Explained!," Apple, May 12, 2015, https://discussions.apple.com /thread/7039051.

245 *They warned him:* Interview with Patrick Pruniaux.

246 *He grew sick:* Parker, "The Shape of Things to Come."

248 *Some observers speculated:* "Fortune 500," Fortune, 2015, https://fortune .com/fortune500/2015/search/.

248 *In advance of the change:* Stephen Fry, "When Stephen Fry Met Jony Ive: The Self-Confessed Tech Geek Talks to Apple's Newly Promoted Chief Design Officer," *Telegraph,* May 26, 2015, https:// www.telegraph.co.uk/technology/apple/11628710/When-Stephen -Fry-met-Jony-Ive-the-self-confessed-fanboi-meets-Apples-newly -promoted-chief-design-officer.html.

Chapter 14: Fuse

251 *the iPhone business:* Apple Inc., 2015 Form 10-K for the year ended September 26, 2015, (filed October 28, 2011), p. 30, SEC, https:// www.sec.gov/Archives/edgar/data/320193/000119312515356351 /d17062d10k.htm.

251 *addressed recent news reports:* Daisuke Wakabayashi and Mike Ramsey, "Apple Gears Up to Challenge Tesla in Electric Cars," *Wall Street Journal,* February 13, 2015, https://www.wsj.com/articles/apples-titan -car-project-to-challenge-tesla-1423868072; Tim Bradshaw and Andy Sharman, "Apple Hiring Automotive Experts to Work in Secret

Research Lab," *Financial Times,* February 13, 2015, https://www.ft
.com/content/84906352-b3a5-11e4-9449-00144feab7de.

252 *The tradition dated back:* Nik Rawlinson, "History of Apple: The Story
of Steve Jobs and the Company He Founded," *Macworld,* April 25, 2017,
https://www.macworld.co.uk/feature/history-of-apple-steve-jobs
-mac-3606104/.

253 *Jeff Robbin, Apple's vice president:* Evan Minsker, "Trent Reznor Talks
Apple Music: What His Involvement Is, What Sets It Apart," Pitchfork,
July 1, 2015, https://pitchfork.com/news/60190-trent-reznor-talks
-apple-music-what-his-involvement-is-what-sets-it-apart/.

254 *The figure was a hundredfold increase:* Todd Wasserman, "Report:
Beats Music Had Only 111,000 Subscribers in March," Mashable,
May 13, 2014.

256 *The negotiations were challenged:* Josh Duboff, "Taylor Swift: Apple
Crusader, #GirlSquad Captain, and the Most Influential 25-Year-Old
in America," *Vanity Fair,* August 11, 2015, https://www.vanityfair
.com/style/2015/08/taylor-swift-cover-mario-testino-apple-music.

256 *Tim Cook led the faithful:* Apple, "Apple—WWDC 2015," YouTube,
June 15, 2015, https://www.youtube.com/watch?v=_p8AsQhaVKI.

257 *A decade earlier:* "Steve Jobs to Kick Off Apple's Worldwide Developers Conference 2003," Apple, May 8, 2003, https://www
.apple.com/newsroom/2003/05/08Steve-Jobs-to-Kick-Off-Apples
-Worldwide-Developers-Conference-2003/; "Apple Launches the
iTunes Music Store," Apple, April 28, 2003, https://www.apple.com
/newsroom/2003/04/28Apple-Launches-the-iTunes-Music-Store/;
Apple Novinky, "Steve Jobs Introduces iTunes Music Store—Apple
Special Event 2003," YouTube, April 3, 2018, https://www.youtube
.com/watch?v=NF9o46zK5Jo.

258 *The pop star Taylor Swift:* Duboff, "Taylor Swift: Apple Crusader,
#GirlSquad Captain, and the Most Influential 25-Year-Old in Amer-
ica"; interview with Scott Borchetta.

259 *To Apple, Love Taylor:* Peter Helman, "Read Taylor Swift's Open
Letter to Apple Music," Stereogum, June 21, 2015, https://www
.stereogum.com/1810310/read-taylor-swifts-open-letter-to-apple
-music/news/.

259 *That Father's Day morning:* "HBO's Richard Plepler and Jimmy Iovine
on Dreaming and Streaming—FULL CONVERSATION," *Vanity
Fair,* October 8, 2015, https://www.vanityfair.com/video/watch/hbo
-richard-plepler-jimmy-iovine-dreaming-streaming.

259 *Swift's independent record label:* Duboff, "Taylor Swift: Apple Cru-

sader, #GirlSquad Captain, and the Most Influential 25-Year-Old in America"; Fortune Magazine, "How Technology Is Changing the Music Industry," YouTube, July 17, 2015, https://www.youtube.com/watch?v=5ZdVA-_deYE.

260 *"What is this?":* Interview with Scott Borchetta.

260 *"This is a drag":* Jim Famurewa, "Jimmy Iovine Interview: Producer Talks Apple Music, Zane Lowe, and Taylor Swift's Wrath," *Evening Standard,* August 6, 2015, https://www.standard.co.uk/tech/jimmy-iovine-interview-producer-talks-apple-music-zane-lowe-and-taylor-swift-s-wrath-10442663.html.

260 *"Here's the good news":* Fortune Magazine, "How Technology Is Changing the Music Industry"; interview with Scott Borchetta.

260 *"What's the right rate?":* Interview with Scott Borchetta.

260 *At the time, Spotify was:* Tim Ingham, "Pandora: Our $0.001 per Stream Payout Is 'Very Fair' on Artists. And Besides, Now We Can Help Them Sell Tickets," MusicBusiness Worldwide, February 22, 2015, https://www.musicbusinessworldwide.com/pandora-our-0-001-per-stream-payout-is-very-fair/.

261 *"Taylor," Cue said:* Interview with Scott Borchetta.

261 *It later signed with Borchetta's:* Anne Steele, "Apple Music Reveals How Much It Pays When You Stream a Song," *Wall Street Journal,* April 16, 2021, https://www.wsj.com/articles/apple-music-reveals-how-much-it-pays-when-you-stream-a-song-11618579800.

261 *"People think it's too good":* Interview with Scott Borchetta.

262 *Riccio's team lurched:* Taylor Soper, "Amazon Echo Sales Reach 5M in Two Years, Research Firm Says, as Google Competitor Enters Market," GeekWire, November 21, 2016, https://www.geekwire.com/2016/amazon-echo-sales-reach-5m-two-years-research-firm-says-google-competitor-enters-market/.

263 *The system, which they had developed:* Sean Hollister, "Microsoft Releases Xbox One Cheat Sheet: Here's What You Can Tell Kinect to Do," Verge, November 25, 2013, https://www.theverge.com/2013/11/25/5146066/microsoft-releases-xbox-one-cheat-sheet-heres-what-you-can-tell; Liz Gaines, "Apple Aiming at PrimeSense Acquisition, but Deal Is Not Yet Done," All Things D, November 17, 2013, https://allthingsd.com/20131117/apple-aiming-at-primesense-acquisition-but-deal-is-not-yet-done.

263 *The cost increase would offset:* Linda Sui, "Apple iPhone Shipments by Model: Q2 2007 to Q2 2018," Strategy Analytics, February 11, 2019, https://www.strategyanalytics.com/access-services/devices/mobile

-phones/handset-country-share/market-data/report-detail/apple
-iphone-shipments-by-model-q2-2007-to-q4-2018.

264   *The* **Wall Street Journal***'s:* Joanna Stern, "Apple Music Review: Behind
a Messy Interface Is Music's Next Big Leap," *Wall Street Journal,* July
7, 2015, https://www.wsj.com/articles/apple-music-review-behind-a
-messy-interface-is-musics-next-big-leap-1436300486; Brian X. Chen,
"Apple Music Is Strong on Design, Weak on Networking," *New York
Times,* July 1, 2015, https://www.nytimes.com/2015/07/02/technology
/personaltech/apple-music-is-strong-on-design-weak-on-social
-networking.html; Micah Singleton, "Apple Music Review," Verge,
July 8, 2015, https://www.theverge.com/2015/7/8/8911731/apple
-music-review; Walt Mossberg, "Apple Music First Look: Rich,
Robust—but Confusing," Recode, June 30, 2015, https://www.vox
.com/2015/6/30/11563978/apple-music-first-look-rich-fluid-but
-somewhat-confusing.

264   *It made the app busier:* Susie Ochs, "Turning Off Connect Makes
Apple Music Better," *Macworld,* July 1, 2015, https://www.macworld
.com/article/225829/turning-off-connect-makes-apple-music-better
.html.

265   *It had 10 million paid:* Matthew Garrahan and Tim Bradshaw,
"Apple's Music Streaming Subscribers Top 10M," *Financial Times,*
January 10, 2016, https://www.ft.com/content/742955d2-b79b-11e5
-bf7e-8a339b6f2164.

Chapter 15: Accountants

266   *The family had been using:* Walter Isaacson, *Steve Jobs.*

266   *Jobs had spent more:* Walter Isaacson, *Steve Jobs,* 366.

266   *For Ive, who had consulted:* Brad Stone and Adam Satariano, "Tim
Cook Interview: The iPhone 6, the Apple Watch, and Remaking a
Company's Culture," Bloomberg, September 18, 2014, https://www
.bloomberg.com/news/articles/2014-09-18/tim-cook-interview-the
-iphone-6-the-apple-watch-and-being-nice.

268   *In his youth, he had worked:* Buster Hein, "These Are the Fabulous
Rides of Sir Jony Ive," Cult of Mac, February 27, 2014, https://www
.cultofmac.com/254380/jony-ives-cars/.

269   *His vision differed from:* Daisuke Wakabayashi, "Apple Scales Back
Its Ambitions for a Self-Driving Car," *New York Times,* August 22,
2017, https://www.nytimes.com/2017/08/22/technology/apple-self
-driving-car.html.

269   *As the debate simmered:* Jack Nicas, "Apple, Spurned by Others, Signs

Deal with Volkswagen for Driverless Car," *New York Times,* May 23, 2018, https://www.nytimes.com/2018/05/23/technology/apple-bmw-mercedes-volkswagen-driverless-cars.html.

270 ***Outside, an actor performed as Siri:*** Aaron Tilley and Wayne Ma, "Before Departure, Apple's Ive Faded from View," The Information, June 27, 2019, https://www.theinformation.com/articles/before-departure-apples-jony-ive-faded-from-view.

271 ***For the seats:*** Foster + Partners, "The Steve Jobs Theater at Apple Park," fosterandpartners.com, September 15, 2017, https://www.fosterandpartners.com/news/archive/2017/09/the-steve-jobs-theater-at-apple-park/; Gordon Sorlini, "Full Leather Trim," The Official Ferrari Magazine, March 29, 2021, https://www.ferrari.com/en-GM/magazine/articles/full-leather-trim-poltrona-frau-dashboards; Seung Lee, "Apple's New Steve Jobs Theater Is Expected to Be a Major Reveal of Its Own," *Mercury News,* September 11, 2017, https://www.mercurynews.com/2017/09/11/apples-new-steve-jobs-theater-is-expected-to-be-a-major-reveal-of-its-own/.

273 ***Reverend Jesse Jackson:*** Dawn Chmielewski, "Rev. Jesse Jackson Lauds Apple's Diversity Efforts, but Says March Not Over," Recode, March 10, 2015, https://www.vox.com/2015/3/10/11560038/rev-jesse-jackson-lauds-apples-diversity-efforts-but-says-march-not.

274 ***The film was based on:*** Stephen Galloway, "A Widow's Threats, High-Powered Spats and the Sony Hack: The Strange Saga of 'Steve Jobs,'" *Hollywood Reporter,* October 7, 2015, https://www.hollywoodreporter.com/movies/movie-features/a-widows-threats-high-powered-829925/.

275 ***Just days after the anniversary:*** "Jony Ive, J. J. Abrams, and Brian Grazer on Inventing Worlds in a Changing One—FULL CONVERSATION" (video), *Vanity Fair,* October 9, 2015, https://www.vanityfair.com/video/watch/the-new-establishment-summit-jony-ive-j-j-abrams-and-brian-grazer-on-inventing-worlds-in-a-changing-one-2015-10-09.

275 ***Ive told The New Yorker:*** Ian Parker, "The Shape of Things to Come: How an Industrial Designer Became Apple's Greatest Product," *The New Yorker,* February 16, 2015, https://www.newyorker.com/magazine/2015/02/23/shape-things-come.

276 ***It struck Andrew Bolton:*** Interview with Andrew Bolton; Guy Trebay, "At the Met, Andrew Bolton Is the Storyteller in Chief," *New York Times,* April 29, 2015, https://www.nytimes.com/2015/04/30/fashion/mens-style/at-the-met-andrew-bolton-is-the-storyteller-in-chief.html.

277 *The woman, who looked:* Christina Binkley, "Karl Lagerfeld Runway Show Features Pregnant Model in Neoprene Gown," *Wall Street Journal,* July 9, 2014, https://www.wsj.com/articles/BL-SEB-82150.

277 *She called Ive to see:* Interview with Anna Wintour.

277 *He approached Cook:* Interview with Anna Wintour; Maghan McDowell, "Yahoo's $3 Million Met Ball Sponsorship Comes Under Fire," *Women's Wear Daily,* December 16, 2015, https://wwd .com/fashion-news/fashion-scoops/yahoos-3-million-met-ball -sponsorship-comes-under-fire-10299361/.

278 *The partnership had been born:* Christina Passariello, "Apple's First Foray into Luxury with Hermès Watch Breaks Tradition," *Wall Street Journal,* September 11, 2015, https://www.wsj.com/articles/apple-breaks -traditions-with-first-foray-into-luxury-1441944061; interview with Andrew Bolton.

Chapter 16: Security

280 *Early one December morning:* Rick Braziel, Frank Straub, George Watson, and Rod Hoops, *Bringing Calm to Chaos: A Critical Incident Review of the San Bernardino Public Safety Response to the December 2, 2015, Terrorist Shooting Incident at the Inland Regional Center,* Office of Community Oriented Policing Services, U.S. Department of Justice, 2016, https://www.justice.gov/usao-cdca/file/891996/download.

283 *Though Apple wouldn't unlock:* Apple, "Legal Process Guidelines: Government & Law Enforcement Within the United States," https:// www.apple.com/legal/privacy/law-enforcement-guidelines-us.pdf.

283 *The FBI reached:* Lev Grossman, "Inside Apple CEO Tim Cook's Fight with the FBI," *Time,* March 17, 2016, https://time.com/4262480/tim -cook-apple-fbi-2/; *The Encryption Tightrope: Balancing Americans' Security and Privacy, Hearing Before the Committee on the Judiciary, House of Representatives,* March 1, 2016, https://docs.house.gov/meetings /JU/JU00/20160301/104573/HHRG-114-JU00-Transcript -20160301.pdf.

283 *It also discovered:* Kim Zetter, "New Documents Solve a Few Mysteries in the Apple-FBI Saga," *Wired,* March 11, 2016, https://www .wired.com/2016/03/new-documents-solve-mysteries-apple-fbi -saga/.

283 *In early January:* John Shinal, "War on Terror Comes to Silicon Valley," *USA Today,* February 25, 2016, https://www.usatoday.com/story/tech /columnist/2016/02/25/war-terror-comes-silicon-valley/80918106/.

283 *The Obama administration representatives:* Ellen Nakashima, "Obama's Top National Security Officials to Meet with Silicon Valley CEOs," *Washington Post,* January 7, 2016, https://www.washingtonpost.com /world/national-security/obamas-top-national-security-officials-to -meet-with-silicon-valley-ceos/2016/01/07/178d95ca-b586-11e5 -a842-0feb51d1d124_story.html.

283 *The relationship between the government:* Glenn Greenwald, "NSA Prism Program Taps In to User Data of Apple, Google and Others," *Guardian,* June 7, 2013, https://www.theguardian.com/world/2013 /jun/06/us-tech-giants-nsa-data.

284 *He called on the government:* Jena McLaughlin, "Apple's Tim Cook Lashes Out at White House Officials for Being Wishy-Washy on Encryption," The Intercept, January 12, 2016, https://theintercept .com/2016/01/12/apples-tim-cook-lashes-out-at-white-house -officials-for-being-wishy-washy-on-encryption/.

284 *Comey expanded on the government's:* Daisuke Wakabayashi and Devlin Barrett, "Apple, FBI Wage War of Words," *Wall Street Journal,* February 22, 2016, https://www.wsj.com/articles/apple-fbi-wage -war-of-words-1456188800.

285 *"Director Comey, what's the risk":* Current and Projected National Security Threats to the United States, Hearing Before the Select Committee on Intelligence of the United States Senate, February 9, 2016, https:// www.govinfo.gov/content/pkg/CHRG-114shrg20544/pdf/CHRG -114shrg20544.pdf, 43–44; C-SPAN, "Global Threats" (video), c-span.org, February 9, 2016, https://www.c-span.org/video/?404 387-1/hearing-global-terrorism-threats.

286 *Newspaper reporters and newscasters:* Dustin Volz and Mark Hosenball, "FBI Director Says Investigators Unable to Unlock San Bernardino Killer's Phone Content," Reuters, February 9, 2016, https://www.reuters.com/article/california-shooting-encryption /fbi-director-says-investigators-unable-to-unlock-san-bernardino -killers-phone-content-idUSL2N15O246.

286 *They drafted an application:* Orin Kerr, "Opinion: Preliminary Thoughts on the Apple iPhone Order in the San Bernardino Case: Part 2, the All Writs Act," *Washington Post,* February 19, 2016, https://www.washingtonpost.com/news/volokh-conspiracy/wp /2016/02/19/preliminary-thoughts-on-the-apple-iphone-order-in -the-san-bernardino-case-part-2-the-all-writs-act/; Alison Frankel, "How a N.Y. Judge Inspired Apple's Encryption Fight: Frankel," Reuters,

February 17, 2016, https://www.reuters.com/article/apple-encryption
-column/refile-how-a-n-y-judge-inspired-apples-encryption-fight
-frankel-idUSL2N15W2HZ.

286   *On February 16: Attorneys for the Applicant United States of America. In
the Matter of the Search of an Apple iPhone Seized During the Execution
of a Search Warrant on a Black Lexus IS300, California License Plate
35KGD203,* ED No. 15-0451M, Government's *Ex Parte* Application,
U.S. District Court, Central District of California, February 16, 2016,
https://www.justice.gov/usao-cdca/page/file/1066141/download.

287   *Not to mention:* Issie Lapowsky, "Apple Takes a Swipe at Google in
Open Letter on Privacy," *Wired,* September 18, 2014, https://www
.wired.com/2014/09/apple-privacy-policy/.

288   *Federighi dissected the FBI's request: Attorneys for Apple Inc. Apple Inc's
Motion to Vacate Order Compelling Apple Inc to Assist Agents in Search
and Opposition to Government's Motion to Compel Assistance,* ED No.
CM 16-10 (SP), United States District Court, the Central District of
California, Eastern Division, March 22, 2016, https://epic.org/amicus
/crypto/apple/In-re-Apple-Motion-to-Vacate.pdf.

290   *It became a presidential campaign:* Scott Bixby, "Trump Calls for Apple
Boycott amid FBI Feud—Then Sends Tweets from iPhone," *Guardian,* February 19, 2016, https://www.theguardian.com/us-news/2016
/feb/19/donald-trump-apple-boycott-fbi-san-bernardino.

290   *Public opinion was split:* Devlin Barrett, "Americans Divided over Apple's Phone Privacy Fight, WSJ/NBC Poll Shows," *Wall Street Journal,*
March 9, 2016, https://www.wsj.com/articles/americans-divided-over
-apples-phone-privacy-fight-wsj-nbc-poll-shows-1457499601.

290   *On February 25, a week after:* ABC News, "Exclusive: Apple CEO
Tim Cook Sits down with David Muir (Extended Interview),"YouTube,
February 25, 2016, https://www.youtube.com/watch?v=tGqLTFv7v7c.

292   *The Justice Department escalated:* Eric Lichtblau and Matt Apuzzo,
"Justice Department Calls Apple's Refusal to Unlock iPhone a 'Marketing Strategy,'" *New York Times,* February 19, 2016, https://www
.nytimes.com/2016/02/20/business/justice-department-calls-apples
-refusal-to-unlock-iphone-a-marketing-strategy.html.

293   *It pointed to Cook's recent letter:* Matthew Panzarino, "Apple's Tim
Cook Delivers Blistering Speech on Encryption, Privacy," Tech-
Crunch, June 2, 2015, https://techcrunch.com/2015/06/02/apples
-tim-cook-delivers-blistering-speech-on-encryption-privacy/.

293   *But in that country:* Jack Nicas, Raymond Zhong, and Daisuke Wakabayashi, "Censorship, Surveillance and Profits: A Hard Bargain

for Apple in China," *New York Times,* May 17, 2021, https://www
.nytimes.com/2021/05/17/technology/apple-china-censorship-data
.html; Reed Albergotti, "Apple Puts CEO Tim Cook on the Stand to
Fight the Maker of 'Fortnite,'" *Washington Post,* May 21, 2021, https://
www.washingtonpost.com/technology/2021/05/21/apple-tim-cook
-epic-fortnite-trial/.

293 *"Every time I hear this":* The Encryption Tightrope: Balancing Americans' Security and Privacy, Hearing Before the Committee on the Judiciary,
House of Representatives.

294 *A few years later:* Michael Simon, "Apple's iPhone Privacy Billboard
Is a Clever CES Troll, but It's Also Inaccurate," *Macworld,* January
6, 2019, https://www.macworld.com/article/232305/apple-privacy
-billboard.html.

295 *The government had cracked:* Mark Hosenball, "FBI Paid Under $1
Million to Unlock San Bernardino iPhone: Sources," Reuters, April 28,
2016, https://www.reuters.com/article/us-apple-encryption/fbi-paid
-under-1-million-to-unlock-san-bernardino-iphone-sources-idU
SKCN0XQ032; Ellen Nakashima and Reed Albergotti, "The FBI
Wanted to Unlock the San Bernardino Shooter's iPhone. It Turned
to a Little-Known Australian Firm," *Washington Post,* April 14, 2021,
https://www.washingtonpost.com/technology/2021/04/14/azimuth
-san-bernardino-apple-iphone-fbi/.

295 *The agency, which didn't have:* "A Special Inquiry Regarding the Ac-
curacy of FBI Statements Concerning Its Capabilities to Exploit an
iPhone Seized During the San Bernardino Terror Attack Investigation,"
Office of the Inspector General, U.S. Department of Justice, March 2018,
https://www.oversight.gov/sites/default/files/oig-reports/o1803.pdf.

295 *For the first time:* Apple Press Release, "Apple Reports Second
Quarter Results," Apple, April 26, 2016, https://www.apple.com
/newsroom/2016/04/26Apple-Reports-Second-Quarter-Results/.

296 *As he spoke:* Daisuke Wakabayashi, "Apple Sinks on iPhone Stumble,"
Wall Street Journal, April 26, 2016.

Chapter 17: Hawaii Days

301 *"When Anna and Andrew":* Video of Jony Ive's speech at the Metropolitan Museum of Art obtained during reporting; Dan Howarth, "'Fewer Designers Seem to Be Interested in How Something
Is Actually Made' says Jonathan Ive," Dezeen, May 3, 2016, https://
www.dezeen.com/2016/05/03/fewer-designers-interested-in-how
-something-is-made-jonathan-ive-apple-manus-x-machina/.

302 *That evening, Ive and Cook:* Jim Shi, "See How Tech and Fashion Mixed at the Met Gala," Bizbash, May 10, 2016, https://www.bizbash .com/catering-design/event-design-decor/media-gallery/13481625 /see-how-tech-and-fashion-mixed-at-the-met-gala.

302 *Ive and Cook would arrive:* Patricia Garcia, "Watch the Weeknd and Nat Perform at the 2016 Met Gala," *Vogue,* May 3, 2016, https://www .vogue.com/article/the-weeknd-nas-met-gala-performance.

305 *"What am I going to do":* Tripp Mickle, "Jony Ive Is Leaving Apple, but His Departure Started Long Ago," *Wall Street Journal,* June 30, 2019, https://www.wsj.com/articles/jony-ive-is-departing-apple-but-he -started-leaving-years-ago-11561943376?mod=article_relatedinline.

306 *Ive and Newson were:* Alice Morby, "Jony Ive and Marc Newson Create Room-Size Interpretation of a Christmas Tree," Dezeen, November 21, 2016, https://www.dezeen.com/2016/11/21/jony-ive -marc-newson-immersive-christmas-tree-claridges-hotel-london/; Jessica Klingelfuss, "First Look at Sir Jony Ive and Marc Newson's Immersive Festive Installation for Claridge's," *Wallpaper,* November 19, 2016, https://www.wallpaper.com/design/first-look-jony-ive-marc -newson-festive-installation-claridges.

Chapter 18: Smoke

307 *The star of the show:* Jonathan Cheng, "Samsung Adds Iris Scanner to New Galaxy Note Smartphone," *Wall Street Journal,* August 2, 2016, https://www.wsj.com/articles/samsung-adds-iris-scanner-to-new -galaxy-note-smartphone-1470150004; "Gartner Says Worldwide Sales of Smartphones Grew 7 Percent in the Fourth Quarter of 2016," Gartner, February 15, 2017, https://www.gartner.com/en /newsroom/press-releases/2017-02-15-gartner-says-worldwide-sales -of-smartphones-grew-7-percent-in-the-fourth-quarter-of-2016.

308 *A marketer based in Marion:* Interview with Joni Barwick; Olivia So- lon, "Samsung Owners Furious as Company Resists Paying Up for Note 7 Fire Damage," *Guardian,* October 19, 2016, https://www .theguardian.com/technology/2016/oct/19/samsung-galaxy-note -7-fire-damage-owners-angry; "Samsung Exploding Phone Law- suits May Be Derailed by Fine Print," CBS News, February 3, 2017, https://www.cbsnews.com/news/samsung-galaxy-note-7-fine-print -class-action-waiver-lawsuits/; Joanna Stern, "Samsung Galaxy Note 7 Review: Best New Android Phone," *Wall Street Journal,* August 16, 2016, https://www.wsj.com/articles/samsung-galaxy-note-7-review -its-all-about-the-stylus-1471352401.

308 **U.S. consumer protection authorities received:** "Samsung Recalls Galaxy Note7 Smartphones Due to Serious Fire and Burn Hazards," United States Consumer Product Safety Commission, September 15, 2016, https://www.cpsc.gov/Recalls/2016/Samsung-Recalls-Galaxy-Note7-Smartphones/.

309 **They explained that Samsung:** Sijia Jiang, "China's ATL to Become Main Battery Supplier for Samsung's Galaxy Note 7: Source," Reuters, September 13, 2016, https://www.reuters.com/article/us-atl-samsung-battery/chinas-atl-to-become-main-battery-supplier-for-samsungs-galaxy-note-7-source-idUSKCN11J1EL; Sherisse Pham, "Samsung Blames Batteries for Galaxy Fires," CNN, January 23, 2017, https://money.cnn.com/2017/01/22/technology/samsung-galaxy-note-7-fires-investigation-batteries/.

310 **A giant Apple logo:** Tim Cook, Twitter, September 7, 2016, https://twitter.com/tim_cook/status/773530595284529152.

310 **When the CEO:** Apple, "Apple Special Event, October 2016" (video) Apple Events, September 7, 2016, https://podcasts.apple.com/us/podcast/apple-special-event-october-2016/id275834665?i=1000430692673.

310 **The company had sold:** Daisuke Wakabayashi, "Apple's Watch Outpaced the iPhone in First Year," *Wall Street Journal,* April 24, 2016, https://www.wsj.com/articles/apple-watch-with-sizable-sales-cant-shake-its-critics-1461524901; Apple Press Release, "Apple Reports Fourth Quarter Results," Apple (with consolidated financial statements), October 25, 2016, https://www.apple.com/newsroom/2016/10/apple-reports-fourth-quarter-results/.

311 **The second camera:** Apple Press Release, "Portrait Mode Now Available on iPhone 7 Plus with iOS 10.1," Apple, October 24, 2016, https://www.apple.com/newsroom/2016/10/portrait-mode-now-available-on-iphone-7-plus-with-ios-101/.

313 **A year before his death:** "Steve Jobs in 2010, at D8," Apple Podcasts, https://podcasts.apple.com/us/podcast/steve-jobs-in-2010-at-d8/id529997900?i=1000116189688.

313 **The comedy site CollegeHumor:** CollegeHumor, "The New iPhone Is Just Worse," YouTube, September 8, 2016, https://www.youtube.com/watch?v=RgBDdDdSqNE.

314 **The comedian Conan O'Brien:** "Apple's New AirPods Ad | Conan on TBS," YouTube, September 14, 2016, https://www.youtube.com/watch?v=z_wImaGRkNY.

314 **Days after the launch:** Paul Blake, "Exclusive: Apple CEO Tim Cook

Dispels Fears That AirPods Will Fall out of Ears," ABC News, September 13, 2016, https://abcnews.go.com/Technology/exclusive-apple-ceo-tim-cook-dispels-fears-airpods/story?id=42054658.

314 *Its software and hardware teams:* Interview with Chris Deaver, former human resources official, who did a white paper on the problem and developed a solution he called "Collaboration by Design."

315 *The loss led:* Interview with Chris Deaver, former senior human resources business partner; Chris Deaver, "From Think Different to Different Together: The Best Work of My Life at Apple," LinkedIn, August 29, 2019, https://www.linkedin.com/pulse/think-different-together-best-work-my-life-apple-chris-deaver/.

315 *The public blowback:* Jonathan Cheng and John D. McKinnon, "The Fatal Mistake that Doomed Samsung's Galaxy Note," *Wall Street Journal,* October 23, 2016, https://www.wsj.com/articles/the-fatal-mistake-that-doomed-samsungs-galaxy-note-1477248978.

316 *Cook had moderated customers' expectations:* Neil Mawston, "SA: Apple iPhone 7 Was World's Best-Selling Smartphone Model in Q1 2017," Strategy Analytics, May 10, 2017, https://www.strategyanalytics.com/strategy-analytics/news/strategy-analytics-press-releases/strategy-analytics-press-release/2017/05/10/strategy-analytics-apple-iphone-7-was-world%27s-best-selling-smartphone-model-in-q1-2017.

316 *Ted Weschler, an investment manager:* Mark Böschen, "Berkshire Hathaway Manager Establishes Apple Investment," *Manager Magazin,* October 28, 2016; Anupreeta Das, "Warren Buffett's Heirs Bet on Apple," *Wall Street Journal,* May 16, 2016, https://www.wsj.com/articles/buffetts-berkshire-takes-1-billion-position-in-apple-1463400389; Hannah Roberts, "Warren Buffett's Berkshire Hathaway Has More than Doubled Its Stake in Apple," Business Insider, February 27, 2017, https://www.businessinsider.com/warren-buffetts-berkshire-hathaway-has-more-than-doubled-its-stake-in-apple-2017-2; Becky Quick and Lauren Feiner, "Watch Apple CEO Tim Cook's Full Interview from the Berkshire Hathaway Shareholder Meeting," CNBC, May 6, 2019, https://www.cnbc.com/2019/05/06/apple-ceo-tim-cook-interview-from-berkshire-hathaway-meeting.html.

317 *During his Sunday trips:* Emily Bary, "What Warren Buffett Learned About the iPhone at Dairy Queen," *Barron's,* February 27, 2017, https://www.barrons.com/articles/what-warren-buffett-learned-about-the-iphone-at-dairy-queen-1488216174.

318 *In a bid to restart:* Daisuke Wakabayashi, "Apple Taps Bob Mansfield to Oversee Car Project," *Wall Street Journal,* July 25, 2016, https://

www.wsj.com/articles/apple-taps-bob-mansfield-to-oversee-car
-project-1469458580; Daisuke Wakabayashi and Brian X. Chen, "Apple Is Said to Be Rethinking Strategy on Self-Driving Cars," *New York Times,* September 9, 2016, https://www.nytimes.com/2016/09/10 /technology/apple-is-said-to-be-rethinking-strategy-on-self-driving -cars.html.

320 **Shortly after the iTunes shutdown:** Paul Mozur and Jane Perlez, "Apple Services Shut Down in China in Startling About-Face," *New York Times,* April 21, 2016, https://www.nytimes.com/2016/04/22/technology /apple-no-longer-immune-to-chinas-scrutiny-of-us-tech-firms.html.

321 **"We're going to get Apple":** Liberty University, "Donald Trump— Liberty University Convocation," YouTube, January 18, 2016, https:// www.youtube.com/watch?v=xSAyOlQuVX4.

321 **Polls all week:** Josh Katz, "Who Will Be President?," *New York Times,* November 8, 2016, https://www.nytimes.com/interactive/2016/upshot /presidential-polls-forecast.html; Gregory Krieg, "The Day That Changed Everything: Election 2016, as It Happened," CNN, November 8, 2017, https://www.cnn.com/2017/11/08/politics/inside -election-day-2016-as-it-happened/index.html.

Chapter 19: The Jony 50

323 **They exited into:** "About the Battery," The Battery, https://www.the batterysf.com/about.

325 **Stern had joined the company:** Shalini Ramachandran, "Apple Hires Former Time Warner Cable Executive Peter Stern," *Wall Street Journal,* September 14, 2016, https://www.wsj.com/articles/apple-hires -former-time-warner-cable-executive-peter-stern-1473887487.

326 **Such cost-conscious decisions:** Fred Imbert, "GoPro Hires Designer Away from Apple; Shares Spike," CNBC, April 13, 2016, https:// www.cnbc.com/2016/04/13/gopro-hires-apple-designer-daniel -coster-shares-jump.html; Paul Kunkel, *AppleDesign.*

327 **The company had begun:** Mike Murphy, "Apple Shares Just Closed at Their Highest Price Ever," Quartz, February 13, 2017, https:// qz.com/909729/how-much-are-apple-aapl-shares-worth-more -than-ever/.

327 **He had spent years:** Jay Peters, "One of the Apple Watch's Original Designers Tweeted a Behind-the-Scenes Look at Its Development," Verge, April 24, 2020, https://www.theverge.com/tldr/2020 /4/24/21235090/apple-watch-designer-imran-chaudhri-development -tweetstorm.

328 *Following common practice:* Apple grants staff shares of equity as part of total compensation. Staff could convert the entirety of those grants, known as restricted share units, after a vesting period of about four years. At Apple, the shares typically vest in fall and spring and a wave of employees depart or retire. In some instances, employees have received grants that vest early in a new year.

330 *The host in the Cotswolds:* Charlotte Edwardes, "Meet the Glamorous New Tribes Shaking Up the Cotswolds," *Evening Standard,* July 20, 2017, https://www.standard.co.uk/lifestyle/esmagazine/new-wolds -order-how-glamorous-new-arrivals-are-shaking-things-up-in-the -cotswolds-a3590711.html; Suzanna Andrews, "Untangling Rebekah Brooks," *Vanity Fair,* January 9, 2012, https://www.vanityfair.com /news/business/2012/02/rebekah-brooks-201202.

333 *He recalled a moment:* Bono, The Edge, Adam Clayton, Larry Mullen, Jr., with Neil McCormick, *U2 by U2* (London: itbooks, 2006), 270–75.

334 *The organ kicked in:* Bono, Adam Clayton, The Edge, Larry Mullen, Jr., "One," *Achtung Baby,* 1992, https://genius.com/U2-one-lyrics.

Chapter 20: Power Moves

336 *He vowed to bring:* "'America First': Full Transcript and Video of Donald Trump's Inaugural Address," *Wall Street Journal,* January 20, 2017, https://www.wsj.com/articles/BL-WB-67322.

337 *Apple was on track:* Apple Press Release, "Apple Reports Fourth Quarter Results (Consolidated Financial Statements)," Apple, November 2, 2017, https://www.apple.com/newsroom/2017/11/apple-reports -fourth-quarter-results/; Apple Press Release, "Apple Reports Fourth Quarter Results (Consolidated Financial Statements)," Apple, October 27, 2015, https://www.apple.com/newsroom/2015/10/27Apple -Reports-Record-Fourth-Quarter-Results/.

338 *Apple took a 30 percent cut:* Tripp Mickle, "Apple's Pressing Challenge: Build Its Services Business," *Wall Street Journal,* January 10, 2019, https://www.wsj.com/articles/apples-pressing-challenge-build-its -services-business-11547121605.

338 *Some 80 percent of that:* Tim Higgins and Brent Kendall, "Epic vs. Apple Trial Features Battle over How to Define Digital Markets," *Wall Street Journal,* May 2, 2021, https://www.wsj.com/articles /epic-vs-apple-trial-features-battle-over-how-to-define-digital -markets-11619964001.

339 *After detailing the strong:* "Apple Inc., Q1 2017 Earnings Call, Jan 31, 2017," S&P Capital IQ, https://www.capitaliq.com/CIQDotNet /Transcripts/Detail.aspx?keyDevId=415202390&companyId=24937.

339 *The App Store accounted for:* Mickle, "Apple's Pressing Challenge: Build Its Services Business."

340 *The exercise marked:* Nick Wingfield, "'The Mobile Industry's Never Seen Anything like This': An Interview with Steve Jobs at the App Store's Launch," *Wall Street Journal,* July 25, 2018, https://www.wsj .com/articles/the-mobile-industrys-never-seen-anything-like-this -an-interview-with-steve-jobs-at-the-app-stores-launch-1532527201.

341 *In the days after moving:* Timothy B. Lee, "Trump Claims 1.5 Million People Came to His Inauguration. Here's What the Evidence Shows," Vox, January 23, 2017, https://www.vox.com/policy-and -politics/2017/1/21/14347298/trump-inauguration-crowd-size; Abby Phillip and Mike DeBonis, "Without Evidence, Trump Tells Lawmakers 3 Million to 5 Million Illegal Ballots Cost Him the Popular Vote," *Washington Post,* January 23, 2017, https://www .washingtonpost.com/news/post-politics/wp/2017/01/23/at-white -house-trump-tells-congressional-leaders-3-5-million-illegal-ballots -cost-him-the-popular-vote/; Akane Otani and Shane Shifflett, "Think a Negative Tweet from Trump Crushes a Stock? Think Again," *Wall Street Journal,* February 23, 2017, https://www.wsj.com/graphics /trump-market-tweets/.

341 *Steve Jobs had been antipolitical:* G. Pascal Zachary, "In the Politics of Innovation, Steve Jobs Shows Less Is More," IEEE Spectrum, December 15, 2010, https://spectrum.ieee.org/in-the-politics-of-innovation -steve-jobs-shows-less-is-more.

341 *When Laurene Powell Jobs had tried:* Walter Isaacson, *Steve Jobs.*

342 *Cook's inbox overflowed:* Interview with Tim Cook.

342 *"Apple is open":* Edward Moyer, "Apple's Cook Takes Aim at Trump's Immigration Ban," CNET, January 28, 2017, https://www.cnet.com /news/tim-cook-trump-immigration-apple-memo-executive-order/.

343 *Cook assured Apple's staff:* Interview with Tim Cook.

344 *In late May:* Lizzy Gurdus, "Exclusive: Apple Just Promised to Give U.S. Manufacturing a $1 Billion Boost" (video), CNBC, May 3, 2017, https://www.cnbc.com/2017/05/03/exclusive-apple-just-promised -to-give-us-manufacturing-a-1-billion-boost.html.

344 *Cramer didn't point out:* Tripp Mickle and Yoko Kubota, "Tim Cook and Apple Bet Everything on China. Then Coronavirus Hit,"

Wall Street Journal, March 3, 2020, https://www.wsj.com/articles /tim-cook-and-apple-bet-everything-on-china-then-coronavirus -hit-11583172087; Glenn Leibowitz, "Apple CEO Tim Cook: This Is the No. 1 Reason We Make iPhones in China (It's Not What You Think)," Inc., December 21, 2017, https://www.inc.com/glenn -leibowitz/apple-ceo-tim-cook-this-is-number-1-reason-we-make -iphones-in-china-its-not-what-you-think.html.

344 *Apple's breakneck growth:* Apple Inc. Form 10-K 2017, Cupertino, CA: Apple Inc, 2017, https://www.sec.gov/Archives/edgar /data/320193/000032019317000070/a10-k20179302017.htm; Apple Inc. Form 10-K 2011, Cupertino, CA: Apple Inc, 2011, https:// www.sec.gov/Archives/edgar/data/320193/000119312511282113 /d220209d10k.htm; Apple Inc. Definitive Proxy Statement 2018, Cupertino, CA: Apple Inc., December 15, 2017, https://www.sec.gov /Archives/edgar/data/320193/000119312517380130/d400278dd ef14a.htm.

345 *Lest anyone think that Cook:* Jonathan Swan, "What Apple's Tim Cook Will Tell Trump," Axios, June 18, 2017, https://www.axios .com/what-apples-tim-cook-will-tell-trump-1513303073-74d6db9f -d6c2-46c7-8e24-a291325d88e9.html.

345 *"I hope you will put more":* David McCabe, "Tim Cook to Trump: Put 'More Heart' in Immigration Debate," Axios, June 20, 2017, https://www.axios.com/tim-cook-to-trump-put-more-heart-in -immigration-debate-1513303104-f5799556-4f78-4c80-aca3-d7b48 864a917.html.

346 *During a later interview:* "Excerpts: Donald Trump's Interview with the Wall Street Journal," *Wall Street Journal,* July 25, 2017, https:// www.wsj.com/articles/donald-trumps-interview-with-the-wall -street-journal-edited-transcript-1501023617?tesla=y; Tripp Mickle and Peter Nicholas, "Trump Says Apple CEO Has Promised to Build Three Manufacturing Plants in U.S.," *Wall Street Journal,* July 25, 2017, https://www.wsj.com/articles/trump-says-apple-ceo-has-promised -to-build-three-manufacturing-plants-in-u-s-1501012372.

347 *When Cook called Trump:* "Remarks by President Trump to the World Economic Forum," The White House, January 26, 2018, https://trumpwhitehouse.archives.gov/briefings-statements/remarks -president-trump-world-economic-forum/.

348 *In the spring of 2018:* Bob Davis and Lingling Wei, *Superpower Showdown.*

348   *The stock market shuddered:* Apple share price on March 20, 2018, was $42.33; on March 23, 2018, it was $39.84.

348   *If the Chinese retaliated:* Jack Nicas and Paul Mozur, "In China Trade War, Apple Worries It Will Be Collateral Damage," *New York Times,* June 18, 2018, https://www.nytimes.com/2018/06/18/technology/apple -tim-cook-china.html; Norihiko Shirouzu and Michael Martina, "Red Light: Ford Facing Hold-ups at China Ports amid Trade Friction," Reuters, May 9, 2018, https://www.reuters.com/article/us -usa-trade-china-ford/red-light-ford-facing-hold-ups-at-china-ports -amid-trade-friction-sources-idUKKBN1IA1O1; Eun-Young Jeong, "South Korea's Companies Eager for End to Costly Spat with China," *Wall Street Journal,* November 1, 2017, https://www.wsj.com/articles /south-koreas-companies-eager-for-end-to-costly-spat-with -china-1509544012.

349   *"Countries that embrace openness":* Yoko Kubota, "Apple's Cook to Trump: Embrace Open Trade," *Wall Street Journal,* March 24, 2018, https://www.wsj.com/articles/apples-cook-to-trump-embrace-open -trade-1521880744; "Apple CEO Calls for Countries to Embrace Openness, Trade and Diversity at China Development Forum" CCTV, March 24, 2018; "China to Continue Pushing Forward Opening Up and Reform: Li Keqiang," China Plus, March 27, 2018, http:// chinaplus.cri.cn/news/china/9/20180327/108308.html.

350   *The two-year-old service:* Caroline Cakebread, "With 60 Million Subscribers, Spotify Is Dominating Apple Music," Yahoo! Finance, August 1, 2017, https://finance.yahoo.com/news/60-million-subscribers-spotify -dominating-195250485.html.

350   *It was a departure:* Tripp Mickle and Joe Flint, "No Sex Please, We're Apple: iPhone Giant Seeks TV Success on Its Own Terms," *Wall Street Journal,* September 22, 2018, https://www.wsj.com/articles/ no-sex-please-were-apple-iphone-giant-seeks-tv-success-on-its -own-terms-1537588880; Margaret Lyons, "'Madam Secretary' Proved TV Didn't Have to Be Hip to Be Great," *New York Times,* December 8, 2019, https://www.nytimes.com/2019/12/08/arts/television /madam-secretary-finale.html.

351   *When the show debuted:* Maureen Ryan, "TV Review: Apple's 'Planet of the Apps,'" *Variety,* June 6, 2017, https://variety.com/2017/tv /reviews/planet-of-the-apps-apple-gwyneth-paltrow-jessica -alba-1202456477/; Jake Nevins, "Planet of the Apps Review— Celebrity Panel Can't Save Apple's Dull First TV Show," *Guardian,*

June 8, 2017, https://www.theguardian.com/tv-and-radio/2017/jun /08/planet-of-the-apps-review-apple-first-tv-show.

353 *When the duo exited:* Tripp Mickle and Joe Flint, "Apple Poaches Sony TV Executives to Lead Push into Original Content," *Wall Street Journal,* June 16, 2017, https://www.wsj.com/articles/apple -poaches-sony-tv-executives-to-lead-push-into-original-content -1497616203.

353 *Within a few months:* Joe Flint, "Jennifer Aniston, Reese Witherspoon Drama Series Headed to Apple," *Wall Street Journal,* November 8, 2017, https://www.wsj.com/articles/jennifer-aniston-reese-witherspoon -drama-series-headed-to-apple-1510167626.

354 *More than a year after Trump's arrival:* Interview with Larry Kudlow.

354 *In the days leading up to his visit:* Hanna Sender, William Mauldin, and Josh Ulick, "Chart: All the Goods Targeted in the Trade Spat," *Wall Street Journal,* April 5, 2018, https://www.wsj.com/articles/a-look-at -which-goods-are-under-fire-in-trade-spat-1522939292; Bob Davis, "Trump Weighs Tariffs on $100 Billion More of Chinese Goods," *Wall Street Journal,* April 5, 2018, https://www.wsj.com/articles/u-s -to-consider-another-100-billion-in-new-china-tariffs-1522970476.

355 *Cook strode through:* Aaron Steckelberg, "Inside Trump's West Wing," *Washington Post,* May 3, 2017, https://www.washingtonpost.com /graphics/politics/100-days-west-wing/.

355 *The country had recently passed:* Jack Nicas, Raymond Zhong, and Daisuke Wakabayashi, "Censorship, Surveillance and Profits: A Hard Bargain for Apple in China," *New York Times,* June 17, 2021, https:// www.nytimes.com/2021/05/17/technology/apple-china-censorship -data.html.

356 *When a reporter later asked:* "Remarks by President Trump Before Marine One Departure," The White House, August 21, 2019, https://trumpwhitehouse.archives.gov/briefings-statements/remarks -president-trump-marine-one-departure-011221/.

357 *On August 2, 2018, it became:* Noel Randewich, "Apple Breaches $1 Trillion Stock Market Valuation," Reuters, August 2, 2018, https:// www.reuters.com/article/us-apple-stocks-trillion/apple-breaches-1 -trillion-stock-market-valuation-idUSKBN1KN2BE.

357 *When the Trump administration issued:* Tripp Mickle and Jay Greene, "Apple Says China Tariffs Would Hit Watch, AirPods," *Wall Street Journal,* September 7, 2018, https://www.wsj.com/articles/apple-says -china-tariffs-would-hit-watch-airpods-1536353245; Tripp Mickle, "How Tim Cook Won Donald Trump's Ear," *Wall Street Journal,*

October 5, 2019, https://www.wsj.com/articles/how-tim-cook-won
-donald-trumps-ear-11570248040.

Chapter 21: Not Working

358 *"There's lots of ways":* Apple, "Apple Special Event, September
2017" (video) Apple Events, September 14, 2017, https://podcasts
.apple.com/us/podcast/apple-special-event-september-2017
/id275834665?i=1000430692674.

359 *In 1997, Jobs had rejected:* Walter Isaacson, *Steve Jobs.*

359 *He led Abloh across:* Nick Compton, "In the Loop: Jony Ive on Apple's
New HQ and the Disappearing iPhone," *Wallpaper,* December 2017,
https://www.wallpaper.com/design/jony-ive-apple-park.

360 *It looked like a supersize MacBook Air:* Tripp Mickle and Eliot Brown,
"Apple's New Headquarters Is a Sign of Tech's Boom, Bravado," *Wall
Street Journal,* May 14, 2017, https://www.wsj.com/articles/apples
-new-headquarters-is-a-sign-of-techs-boom-bravado-1494759606.

360 *As he stopped:* Interview with Steve Wozniak, September 14, 2017.

361 *Ive settled into:* Apple, "Apple Special Event, September 2017" (video),
Apple Events, September 14, 2017, https://podcasts.apple.com/us
/podcast/apple-special-event-september-2017/id275834665?i
=1000430692674

363 *The company had also run into:* Yoko Kubota, "Apple iPhone X Production Woe Sparked by Juliet and Her Romeo," *Wall Street Journal,*
September 27, 2017, https://www.wsj.com/articles/apple-iphone-x
-production-woe-sparked-by-juliet-and-her-romeo-1506510189.

363 *During a call with Wall Street:* "Apple Inc., Q4 2017 Earnings Call,
Nov 02, 2017," S&P Capital IQ, November 2, 2017, https://www
.capitaliq.com/CIQDotNet/Transcripts/Detail.aspx?keyDevId
=540777466&companyId=24937.

363 *Under Jobs and later Cook:* Apple Inc. Definitive Proxy Statement
2018, Cupertino, CA: Apple Inc., December 15, 2017, https://
www.sec.gov/Archives/edgar/data/320193/000119312517380130
/d400278ddef14a.htm.

364 *After the photographer Andrew Zuckerman:* Zuckerman's documentary work was abandoned. Apple has not said whether it will release
any of his work.

365 *The company monitors: Apple Inc v. Gerard Williams III, Williams Cross-
Complaint Against Apple Inc.,* Superior Court of the State of California, County of Santa Clara, November 6, 2019.

366 *In late 2017, he flew:* "Jony Ive: The Future of Design," November 29,

2017, https://hirshhorn.si.edu/event/jony-ive-future-design/; fuste, "Jony Ive: The Future of Design" (audio recording), Soundcloud, 2018, https://soundcloud.com/user-175082292/jony-ive-the-future-of -design.

368 *Over the next few weeks:* "Apple Park: Transcript of 911 Calls About Injuries from Walking into Glass," *San Francisco Chronicle,* March 2, 2018, https://www.sfchronicle.com/business/article/Apple-Park -Transcript-of-911-calls-about-12723602.php.

373 *In the midst:* Vanessa Friedman, "Is the Fashion Wearables Love Affair Over?," *New York Times,* January 12, 2018, https://www.nytimes .com/2018/01/12/fashion/ces-wearables-fashion-technology.html.

Chapter 22: A Billion Pockets

376 *In a remote corner of Utah:* Becca Hensley, "Review: Amangiri," *Condé Nast Traveler,* https://www.cntraveler.com/hotels/united-states /canyon-point/amangiri-canyon-point.

376 *Nature inspired and motivated Cook:* Michael Roberts, "Tim Cook Pivots to Fitness," *Outside,* February 10, 2021, https://www.outside online.com/health/wellness/tim-cook-apple-fitness-wellness-future/; "Tim Cook on Health and Fitness" (podcast), *Outside,* December 9, 2020, https://www.outsideonline.com/podcast/tim-cook-health -fitness-podcast/.

377 *As 2018 drew to a close:* Yoko Kubota, "The iPhone that's Failing Apple: iPhone XR," *Wall Street Journal,* January 6, 2019, https:// www.wsj.com/articles/the-phone-thats-failing-apple-iphone -xr-11546779603.

378 *Huawei, the country's largest:* "Compare Apple iPhone XR vs Huawei P20," Gadgets Now, https://www.gadgetsnow.com/compare-mobile -phones/Apple-iPhone-XR-vs-Huawei-P20.

378 *Once the market leader:* Yoko Kubota, "Apple iPhone Loses Ground to China's Homegrown Rivals," *Wall Street Journal,* January 3, 2019, https:// www.wsj.com/articles/apple-loses-ground-to-chinas-homegrown -rivals-11546524491.

378 *As boxes of iPhone XRs:* Debby Wu, "Apple iPhone Supplier Foxconn Planning Deep Cost Cuts," Bloomberg, November 21, 2018, https:// www.bloomberg.com/news/articles/2018-11-21/apple-s-biggest -iphone-assembler-is-said-to-plan-deep-cost-cuts.

379 *it developed a trade-in program:* Hayley Tsukayama, "Apple Launches iPhone Trade-in Program," *Washington Post,* August 30, 2013, https://www.washingtonpost.com/business/technology/apple

-launches-trade-in-program/2013/08/30/35c360a0-1183-11e3-b4cb -fd7ce041d814_story.html.

379 *The marketing team sweetened:* Mark Gurman, "Apple Resorts to Promo Deals, Trade-ins to Boost iPhone Sales," Bloomberg, December 4, 2018, https://www.bloomberg.com/news/articles/2018-12-04 /apple-is-said-to-reassign-marketing-staff-to-boost-iphone-sales.

379 *Apple lost more than $300 billion:* Apple's market value on September 4, 2018, was $1.131 trillion; on December 24, 2018, it was $695 billion. Source: Macrotrends.net.

379 *Its nearly decade-long run:* Jay Greene, "How Microsoft Quietly Became the World's Most Valuable Company," *Wall Street Journal,* December 1, 2018, https://www.wsj.com/articles/how-microsoft -quietly-became-the-worlds-most-valuable-company-1543665600.

380 *Shortly after the markets closed:* "Letter from Tim Cook to Apple Investors," Apple, January 2, 2019, https://www.apple.com/newsroom /2019/01/letter-from-tim-cook-to-apple-investors/.

380 *That afternoon, Cook sat down:* "CNBC Exclusive: CNBC Transcript: Apple CEO Tim Cook Speaks with CNBC's Josh Lipton Today," CNBC, January 2, 2019, https://www.cnbc.com/2019/01/02/cnbc -exclusive-cnbc-transcript-apple-ceo-tim-cook-speaks-with-cnbcs -josh-lipton-today.html.

381 *The following day:* Sophie Caronello, "Apple's Market Cap Plunge Must Be Seen in Context," Bloomberg, January 4, 2019, https://www .bloomberg.com/news/articles/2019-01-04/apple-s-market-cap -plunge-must-be-seen-in-context.

382 *She had spent five years:* Apple Inc. Definitive Proxy Statement, 2014, Schedule 14A, United States Securities and Exchange Commission, https://www.sec.gov/Archives/edgar/data/320193/00011931 2514008074/d648739ddef14a.htm; Apple Inc. Definitive Proxy Statement, 2017, Schedule 14A, United States Securities and Exchange Commission, https://www.sec.gov/Archives/edgar/data /320193/000119312517003753/d257185ddef14a.htm.

384 *Before the event wrapped up:* John Koblin, "Hollywood Had Questions. Apple Didn't Answer Them," *New York Times,* March 26, 2019, https://www.nytimes.com/2019/03/26/business/media/apple-tv -plus-hollywood.html.

386 *Once the crowd slipped:* Apple, "Apple Special Event, March 2019" (video), Apple Events, March 25, 2019, https://podcasts .apple.com/us/podcast/apple-special-event-march-2019/id275834665?i =1000433397233.

387 *Two weeks earlier, Spotify:* Valentina Pop and Sam Schechner, "Spotify Accuses Apple of Stifling Competition in EU Complaint," *Wall Street Journal,* March 13, 2019, https://www.wsj.com/articles/spotify -files-eu-antitrust-complaint-over-apples-app-store-11552472861.

391 *It nearly doubled:* Kevin Kelleher, "Apple's Stock Soared 89% in 2019, Highlighting the Company's Resilience," *Fortune,* December 31, 2019, https://fortune.com/2019/12/31/apple-stock-soared-in-2019/.

Chapter 23: Yesterday

392 *The $3 million building had:* "Pac Heights Carriage House in Contract at $4K Per Square Foot," SocketSite, June 10, 2015, https://socketsite .com/archives/2015/06/3-5m-carriage-house-in-contract-at-4k-per -square-foot.html.

393 *Then someone inside:* Aaron Tilley and Wayne May, "Before Departure, Apple's Jony Ive Faded from View," The Information, June 27, 2019, https://www.theinformation.com/articles/before-departure-apples -jony-ive-faded-from-view.

393 *The invitations arrived:* Copy of the invitation.

394 *In a message to staff:* Lewis Wallace, "How (and Why) Jony Ive Built the Mysterious Rainbow Apple Stage," Cult of Mac, May 9, 2019, https://www.cultofmac.com/624572/apple-stage-rainbow/.

394 *Ive was excited about:* Interview with Camille Crawford, former personal assistant to Jony Ive.

394 *Over the years:* Interview with John Cave, longtime friend of Mike Ive and colleague at Middlesex Polytechnic.

397 *"Are you ready to celebrate":* Sina Digital, "Apple Park, Apple's New Headquarters, Opens Lady Gaga Rainbow Stage Singing (translated)," Sina, May 19, 2019, https://tech.sina.com.cn/mobile/n/n /2019-05-19/doc-ihvhiqax9739760.shtml

397 *"I'd like to say":* Monster Nation, PAWS UP, "Lady Gaga Live at the Apple Park" (video), Facebook, May 17, 2019, https://www.facebook .com/MonsterNationPawsUp/videos/671078713305170/.

397 *They want to escape:* Peggy Truong, "The Real Meaning of 'Shallow' from 'A Star Is Born,' Explained," *Cosmopolitan,* February 25, 2019, https://www.cosmopolitan.com/entertainment/music/a26444189 /shallow-lady-gaga-lyrics-meaning/; Lady Gaga, Mark Ronson, Anthony Rossomando, and Andrew Wyatt, "Shallow," *A Star Is Born,* 2018, https://genius.com/Lady-gaga-and-bradley-cooper-shallow-lyrics.

398 *a place where colleagues:* Paulinosdepido, "Steve Jobs My Model in

Business Is the Beatles," YouTube, December 13, 2011, https://www
.youtube.com/watch?v=1QfK9UokAIo.

403 *The designer's payout:* "10 of the Largest Golden Parachutes CEOs
Ever Received," *Town & Country,* December 6, 2013.

403 *After fifteen years:* Tripp Mickle, "Jony Ive's Long Drift from Apple—
The Design Chief's Departure Comes After Years of Growing Distance and Frustration," *Wall Street Journal,* July 1, 2019.

Epilogue

404 *In the months and years:* Tripp Mickle, "How Tim Cook Made Apple
His Own," *Wall Street Journal,* August 7, 2020, https://www.wsj.com
/articles/tim-cook-apple-steve-jobs-trump-china-iphone-ipad-apps
-smartphone-11596833902.

404 *In an email:* Email from Laurene Powell Jobs, March 25, 2021.

406 *Through his work:* Apple Inc., "Apple Return of Capital and Net Cash
Position," Cupertino, CA, Apple Inc, 2021, https://s2.q4cdn.com
/470004039/files/doc_financials/2021/q3/Q3'21-Return-of-Capital
-Timeline.pdf.

407 *Its revenue equaled:* "Fortune 500," *Fortune,* 2020, https://fortune
.com/fortune500/2020/.

407 *The company's price-to-earnings ratio:* "Apple Inc.," FactSet, https://
www.factset.com.

407 *It was a change:* Tripp Mickle, "Apple Was Headed for a Slump. Then
It Had One of the Biggest Rallies Ever," *Wall Street Journal,* January 26. 2020, https://www.wsj.com/articles/apple-was-headed-for-a
-slump-then-it-had-one-of-the-biggest-rallies-ever-11580034601.

408 *When Epic's lawyers asked:* Tim Higgins, "Apple's Tim Cook Faces
Pointed Questions from Judge on App Store Competition," *Wall Street
Journal,* May 21, 2021, https://www.wsj.com/articles/apples-tim-cook
-expected-to-take-witness-stand-in-antitrust-fight-11621589408.

408 *Internal documents unearthed:* Tim Higgins, "Apple Doesn't Make
Videogames. But It's the Hottest Player in Gaming," *Wall Street Journal,*
October 2, 2021, https://www.wsj.com/articles/apple-doesnt-make
-videogames-but-its-the-hottest-player-in-gaming-11633147211.

408 *The company won:* Ben Thompson, "The *Apple v. Epic* Decision,"
Stratechery, September 13, 2021, https://stratechery.com/2021/the
-apple-v-epic-decision/.

409 *Some Uyghurs had allegedly been:* Wayne Ma, "Seven Apple Suppliers Accused of Using Forced Labor from Xinjiang," The Information,

May 10, 2021, https://www.theinformation.com/articles/seven-apple
-suppliers-accused-of-using-forced-labor-from-xinjiang.

410 *Since his promotion to CEO:* "Apple Inc. Notice of 2021 Annual Meeting of Shareholders and Proxy Statement," Apple, January 5, 2021, https://www.sec.gov/Archives/edgar/data/320193/000119312521001987/d767770ddef14a.htm.

410 *Apple's board of directors:* Anders Melin and Tom Metcalf, "Tim Cook Hits Billionaire Status with Apple Nearing $2 Trillion," Bloomberg, August 10, 2020, https://www.bloomberg.com/news/articles/2020-08-10/apple-s-cook-becomes-billionaire-via-the-less-traveled-ceo-route; Mark Gurman, "Apple Gives Tim Cook Up to a Million Shares That Vest Through 2025," Bloomberg, September 29, 2020, https://www.bloomberg.com/news/articles/2020-09-29/apple-gives-cook-up-to-a-million-shares-that-vest-through-2025.

411 *In 2021, sales of the company's:* Apple Inc. Form 10-K 2020. Cupertino, CA: Apple Inc, 2020, https://s2.q4cdn.com/470004039/files/doc_financials/2020/q4/_10-K-2020-(As-Filed).pdf.

412 *The collection of creatives:* Dave Lee, "Airbnb Brings in Jony Ive to Oversee Design," *Financial Times,* October 21, 2020, https://www.ft.com/content/8bc63067-4f58-4c84-beb1-f516409c9838; Tim Bradshaw, "Jony Ive Teams Up with Ferrari to Develop Electric Car," *Financial Times,* September 27, 2021, https://www.ft.com/content/c2436fb5-d857-4aff-b81e-30141879711c; Ferrari N.V. Press Release, "Exor, Ferrari and LoveFrom Announce Creative Partnership," Ferrari, September 27, 2021, https://corporate.ferrari.com/en/exor-ferrari-and-lovefrom-announce-creative-partnership; Nergess Banks, "This Is Ferrari and Superstar Designer Marc Newson's Tailored Luggage Line," *Forbes,* May 6, 2020, https://www.forbes.com/sites/nargessbanks/2020/05/05/ferrari-marc-newsons-luggage-collection/?sh=248a8e762d11.

413 *The color in the room came:* Interview with Camille Crawford.

414 *Turning to follow his gaze:* Interview with Camille Crawford; Alexa Tsoulis-Reay, "What It's Like to See 100 Million Colors," *New York,* February 26, 2015, https://www.thecut.com/2015/02/what-like-see-a-hundred-million-colors.html.

Bibliography

Books

Austin, Rob, and Lee Devin. *Artful Making: What Managers Need to Know About How Artists Work.* New York: FT Prentice Hall, 2003.

Brennan-Jobs, Lisa. *Small Fry: A Memoir.* New York: Grove Press, 2018.

Brunner, Robert, and Stewart Emery with Russ Hall. *Do You Matter? How Great Design Will Make People Love Your Company.* Upper Saddle River, NJ: FT Press, 2009.

Cain, Geoffrey. *Samsung Rising: The Inside Story of the South Korean Giant That Set Out to Beat Apple and Conquer Tech.* New York: Currency, 2020.

Carlton, Jim. *Apple: The Inside Story of Intrigue, Egomania, and Business Blunders.* New York: Times Books, 2017.

Davis, Bob, and Lingling Wei. *Superpower Showdown: How the Battle Between Trump and Xi Threatens a New Cold War.* New York: Harper Business, 2020.

Dormehl, Luke. *The Apple Revolution: Steve Jobs, the Counter Culture and How the Crazy Ones Took Over the World.* London: Virgin Books, 2012.

Esslinger, Hartmut. *Keep It Simple: The Early Design Years at Apple.* Stuttgart, Germany: Arnoldsche Art Publishers, 2013.

Higgs, Antonia. *Tangerine: 25 Insights into Extraordinary Innovation & Design.* London: Goodman, 2014.

Iger, Robert. *The Ride of a Lifetime: Lessons Learned from 15 Years as CEO of the Walt Disney Company.* New York: Random House, 2019.

Isaacson, Walter. *Steve Jobs*. New York: Simon & Schuster, 2011.

Ive, Jony, Andrew Zuckerman, and Apple Inc. *Designed by Apple in California*. Cupertino, CA: Apple, 2016.

Kahney, Leander. *Jony Ive: The Genius Behind Apple's Greatest Products*. New York: Portfolio/Penguin, 2013.

——. *Tim Cook: The Genius Who Took Apple to the Next Level*. New York: Portfolio/Penguin, 2019.

Kane, Yukari Iwatani. *Haunted Empire: Apple After Steve Jobs*. New York: Harper Business, 2014.

Kocienda, Ken. *Creative Selection: Inside Apple's Design Process During the Golden Age of Steve Jobs*. New York: St. Martin's Press, 2018.

Kunkel, Paul. *AppleDesign: The Work of the Apple Industrial Design Group*. Cupertino, CA: Apple, 1997.

Lashinsky, Adam. *Inside Apple: How America's Most Admired—and Secretive—Company Really Works*. New York: Business Plus, 2013.

Levy, Steven. *Insanely Great: The Life and Times of Macintosh, the Computer That Changed Everything*. New York: Penguin, 1994.

Merchant, Brian. *The One Device: The Secret History of the iPhone*. New York: Back Bay Books, 2017.

Moritz, Michael. *Return to the Little Kingdom: How Apple & Steve Jobs Changed the World*. New York: Overlook Press, 1984, 2009.

Nathan, John. *Sony: The Private Life*. Boston: Houghton Mifflin Harcourt, 1999.

Rams, Dieter. *Less but Better*. Berlin: Gestalten, 1995.

Schendler, Brent, and Rick Tetzeli. *Becoming Steve Jobs: The Evolution of a Reckless Upstart into a Visionary Leader*. New York: Crown Business, 2015.

Shenk, Joshua Wolf. *Powers of Two: How Relationships Drive Creativity*. New York: First Mariner Books, 2014.

Stalk, George, Jr., and Thomas M. Hout. *Competing Against Time: How Time-Based Competition Is Reshaping Global Markets.* New York: Free Press, 1990.

Vogelstein, Fred. *Dogfight: How Apple and Google Went to War and Started a Revolution.* New York: Sarah Crichton Books, 2013.

Film

First Monday in May. Andrew Rossi. Magnolia Pictures, 2016.

Objectified. Gary Hustwit. Plexi Productions, 2009.

September Issue. R. J. Cutler. A&E Indie Films, 2009.

The Defiant Ones. Allen Hughes. Alcon Entertainment, 2017.

Yesterday. Danny Boyle. Decibel Films, 2019.

Index

Abloh, Virgil, 359–60
Abrams, J. J., 275
Ahrendts, Angela
  Apple Watch and, 239
  basic Apple facts about, xiii, 381
  characteristics, 168
  new headquarters, 272
  reputation of, 167
  resignation of, 382
  traveling Apple Stores in
    China, 272, 381–82
AirPods, 312–15, 411
Alaïa, Azzedine, 220–21
All Writs Act (1789), 286
Anderson, Fred, 68, 93
Andre, Bart, 76–77, 170
Android system, 120–21
Aniston, Jennifer, 353, 389
"Antennagate," 119
App Store
  Cook and, 386–87
  creation of, 117
  Epic lawsuit and, 407, 408–9
  under Jobs, 340
  revenue, 255, 310, 338, 339
  Schiller and, 279
Apple Arcade, 388

Apple Card, 387–88
Apple Inc.
  under Anderson, 68
  balance of art with engineering
    under Jobs, 371
  Buffett and, 317–19
  CEOs between Jobs's tenures,
    8–9
  China and, 101–2, 319–21, 409
  consumer demand for
    transformational devices,
    337
  early days, 8
  Epic Games antitrust trial
    against, 387, 407–9
  growth of, 234, 344
  hardware versus services profits,
    325
  headquarters, 6, 42–43
  integration of work of divisions,
    314–15
  inventory, 93–95, 96, 315–16
  Jobs's illness and future of, 88
  Manus × Machina exhibit and,
    299–303
  market value of, 338, 357,
    379–81, 404, 410

Apple Inc. (*cont.*)
 meeting of employees after
  death of Jobs, 18–20
 memorial service for Jobs,
  18–20
 new headquarters, 177–78, 200,
  201–2, 267, 270–72, 367–74
 new manufacturing technique
  for lighter laptops, 103–4
 NeXT and, 9, 68
 pay packages, 363–65
 principles central to identity of,
  18, 358
 production outsourced, 97–98
 profits, 6, 74, 81
 rehiring of former employees,
  181–82
 rejection of legitimate billings,
  365–66
 sales, 9, 66, 97–98, 108, 163,
  182, 193–94, 295–96
 under Sculley, 65–66
 as services company, 337–38,
  339–41, 383–91, 407–9
 share price, 11, 59, 143, 144,
  164–65
 similarities to Walt Disney
  Company, 14
 social causes and, 108–9
 under Spindler, 66, 67
 structure of, 14, 16–17, 130–31
 as target takeover, 67–68
 taxes, 153–54, 157–59, 347
 television and, 350–54, 388–90
 Trump and, 336, 342–43
 work-hard, play-hard culture,
  84–85
 See also Cook, Timothy Donald,
  as CEO *and specific products*
Apple Maps, 196–97, 387

Apple Music
 artists' and labels' compensation,
  256, 258–61
 Beats and, 197, 198–99, 204,
  252–53
 criticisms of, 264–65
 development of, 254–56
 introduction of, 257–58
 Spotify as competitor, 199, 204,
  255, 260, 350, 387
 success of, 265, 310, 383
 Tidal as competitor, 204
Apple News+, 387
Apple Park, 177–78, 200, 201–2,
  267, 270–72, 367–74
Apple Pay, 35, 212
Apple TV+, 388–90
Apple University, 16
Apple Watch
 as Apple's first new product
  category after Jobs's death,
  188
 concerns about, 112–13, 243,
  244
 development of, 133–35, 138–
  42, 173–74, 175–77, 179–80,
  181, 182–86, 190–91, 210
 as fashion accessory, 186–87,
  189–90, 219–21, 235, 237,
  239–44, 245–46, 249–50,
  277–78, 279, 373
 introduction of, 211, 213–18
 marketing, 186–90, 240–44, 279
 production, 237–38
 reintroduction of, 236
 sales, 236–37, 244–45, 278,
  310–11, 337, 411
 Series 2, 311
Asai, Hiroki, 155, 156–57
Auburn University, 47, 52–54, 57

autonomous cars
  developmental problems, 297–98
  Mansfield and, 318–19
  Project Titan, 203–4, 251–52,
    267–70, 298
Avolonte Health, 139, 182

Bailey, Christopher, 167, 306
Beats Music and Beats Electronics
  Cook and, 197, 198–99, 205,
    206–9
  Cue and, 197, 198
  fusion with Apple team
    members, 252–53, 255–56
Bell, James, 273, 274
Berkshire Hathaway, 318–19
Blevins, Tony (the Blevinator),
  200–201
Bloomberg Businessweek, 117,
  224–26, 229, 230
"blue sky" projects, 34–36
board of directors, xvi, 274
  Wagner, Susan, 273
Bolton, Andrew, 276–78, 299–300
Bono, 215–16, 218–19, 333–34
Borchetta, Scott, 259–60, 261
Browett, John, 127, 167, 199
Brown, Gordon, 91
Brunner, Robert
  Beats and, 198
  departure from Apple, 67
  design team under Spindler as
    CEO, 66
  Ive promoted under, 67
  Juggernaut project, 39–40
  at Lunar Design, 37
Buffett, Warren, 316–19

Campbell, Bill, 273
Campbell, Naomi, 374–75

Carell, Steve, 353, 389
Chambers, Tony, 374–75
Chaudhri, Imran, xiv, 85, 179,
  327–28, 366
China
  anti-democratic moves by, 409
  Apple problems in, 319–21
  dependency on Apple, 101–2
  Foxconn, 97–98, 101, 103,
    237–38, 378
  as growth market, 192
  importance of, to Apple, 336,
    348
  iPhones in, 194–96, 222, 251,
    378
  privacy and Apple in, 293
  trade with U.S., 348, 349,
    354–56
  traveling Apple Stores in, 272,
    381–82
  Trump attacks on, 336, 343
China Mobile, 194–96
Christie, Greg, 85, 120, 132
Clow, Lee, 159
Coffey, Thomas, 59, 60
Cohen, Sacha Baron, 332
CollegeHumor, 313–14
Comey, James, 283–86, 293
Compaq, 60–61, 63, 65
computer-controlled (CNC)
  machines, 104
Cook, Donald Dozier (father of
  Tim), xvi, 43–46, 107
Cook, Gerald (brother of Tim),
  44–45
Cook, Geraldine Majors (mother of
  Tim), xvi, 43, 44, 100, 107
Cook, Timothy Donald
  Alabama Academy of Honor, 223
  appearance of, 2

Cook, Timothy Donald (*cont.*)
  at Auburn, 52–54
  background, xvi, 43–46
  basic Apple facts about, xiii, 2,
    62–63, 92
  boyhood of, 46–49
  as CEO
    abilities of, 336–37, 344, 345–46
    AirPods, 314
    Apple as services company,
      383–91, 407–9
    Apple Music and, 254–55, 257,
      260, 261
    Apple Watch and, 134, 188–89,
      213, 214–15, 216, 236, 244,
      245–46, 249–50, 279
    Apple's move into television,
      350–52, 353–54
    Apple's tax payments testimony
      of, 154, 157–59
    basic Apple facts about, 405,
      406
    Beats Electronics and, 198–99,
      205, 206–9
    Berkshire Hathaway and, 319
    board of directors appointments,
      273
    China and, 192–93, 320–21,
      349, 381–82
    clothes of, 235
    compensation package, 410
    concerns about, 112, 337
    custom-made Leica camera and,
      172
    debut as Apple spokesman, 6–7,
      9–11
    Deneve and, 279
    design team concerns about, 110
    en masse resignation of senior
      engineers, 202–3
    endurance of Apple after death
      of Jobs, 404–5
    Epic Games antitrust trial and,
      407–8
    financial doctrine of, 199
    focus of, 109
    growth of Apple under, 2, 344
    Icahn and, 165–66
    Iovine and, 350–51
    iPhone access for law
      enforcement and, 283–84,
      287–92, 293, 294–95
    iPhone mapping system failures,
      124–27
    iPhone X series and, 377–79
    iPhones sales and Apple market
      value, 380–81
    Ive and, 247–48, 274, 366
    on jobs Apple created in U.S.,
      344
    Jobs's death and, 18–19, 42–43
    Jobs's elevation of, to, 6, 106–7
    Maestri and, 199, 274
    management changes, 127–28
    marketing and, 186, 188–89
    Media Arts and, 159–61, 162
    new headquarters and, 200, 202,
      369–70, 371
    political influence and Apple,
      342, 355, 356
    privacy position of, 293
    public announcement of sexual
      orientation, 224–27, 229–31
    Samsung and, 147–48
    Schiller and, 152
    security detail for, 143–44
    share price under, 144
    similarities to Jeff Williams,
      180–81
    smart speakers and, 262

social causes and, 108–9

stage presence of, 6, 162, 212–13, 310

Steve Jobs Theater opening, 358–59, 361–62

stock buybacks, 165, 166

treatment of subordinates, 2

tribute to Jobs, 395

Trump and, 321, 346–47, 354–56, 357

Jeff Williams as consigliere of, 274

Jeff Williams as Cook's number two, 99

Compaq and, 61, 63

glass iPhone screens, 105

health scare, 58

high school years, 49–52

homosexuality of, 51–52

house purchase by, 99

at IBM, 54, 56–57, 58–59

importance of making difference, 62

at Intelligent Electronics, 58–61

inventory and, 93–95, 96

Ive and, 101

Jobs's death and, 13, 108

Jobs's second leave and, 105–6

Ku Klux Klan and, 48–49

made COO, 102–3

manufacturing of lighter laptops, 104

Manus × Machina exhibit and, 302–3

Nike and, 103

operations team and, 93–95, 99–100, 101

outsourcing of production by, 97–98

on preparing for opportunities, 55

on purpose of life, 58

reckless driving by, 57–58

relationships with suppliers, 98, 101–2, 103

retention stock grants, 17

sexual orientation and gender identity laws and, 223–24, 228–29

2012 D: All Things Digital conference, 107–8

"Cook Doctrine," 106

Cooper, Anderson, 224

Cooper, Lisa Straka, 49–50

Coster, Danny, xiv, 71, 326, 366, 400

Cotton, Katie, xiii, 155, 156–57, 195

Cramer, Jim, 234, 344

Creative Artists Agency (CAA), 352

creativity

Cook on Jobs's, 13

design dilemma of being functional while radically different, 36

of Ive, 27–28

Jobs and, 19–20

Polaroid and, 15

Sony and, 16

Walt Disney Company and, 15

"Crossing the Canyon" program, 93

Cue, Eddy

Apple in television, 352, 353

Apple Music and, 253, 260, 261, 383

Apple Pay introduction, 212

basic Apple facts about, xiii

Beats Electronics acquisition, 198

in charge of iCloud, iTunes, and Apple Maps, 196–97

Cook's debut as Apple spokesman and, 11

on executive team, 196–97

Cue, Eddy (*cont.*)
  Forstall and, 118
  retention stock grants, 17
Curtis, Richard, 398–99

Darbyshire, Martin, 38, 40
Dauber, Jeff, 181–82, 183, 190–91,
    214, 243
De Anza College (Flint Center for
    the Performing Arts), 72, 188,
    211–16
De Iuliis, Daniele, 66–67, 84, 400
Deneve, Paul, 187–88, 219, 238–39,
    243, 245–46, 279
Designed by Apple in California ad, 90
Dowling, Steve, xiii, 224, 288, 289,
    290
Dre, Dr. (aka Andre Young), 197–98,
    206, 207–8
Drexler, Mickey, 273
Dye, Alan
  background, 135–36
  Chaudhri and, 328–29
  iOS and, 137
  Ive and tenth anniversary iPhone
    changes and, 323, 324
  Ive as part-time and, 248, 249,
    305

Ellison, Larry, 112
engineers, xv
Epic Games, 387, 407–9
Erlicht, Jamie, 353, 383, 388
Esslinger, Hartmut, 69, 74
executive team, xiii–xiv, 17–18

facial recognition technology,
    263–64, 304, 361–63
Fadell, Tony
  basic Apple facts about, xiii

departure from Apple, 106
Forstall and, 118
iPod development and, 80
Nest Labs and, 197
Project Purple and, 86
Farook, Rizwan and wife, 280–81,
    283
FBI, 281–86, 293
Federighi, Craig, 120, 130, 131,
    287–89
Flint Center for the Performing
    Arts (De Anza College), 72,
    188, 211–16
Forlenza, Nick
  production of Power Macs, 78
  smartwatch and, 185, 186
  supply chain control and, 78–79
  in Tokyo with Ive, 76–77
  unit led by, made extension of
    design team, 79
Forstall, Scott
  background, 114–15
  basic Apple facts about, xiii
  Blanchard and, 120
  concerns about Ive's smartwatch
    idea, 112–13
  Cook's debut as Apple
    spokesman and, 11
  Fadell and, 118
  forced resignation of, 126–27
  iPhone apps and, 117
  iPhone mapping system and,
    121–22, 123–25, 126–27
  Ive and, 114, 118–19
  Jobs and, 115–16, 117
  NeXT and, 115–16
  patents of, 119
  Project Purple and, 86–87,
    116–17
  reinvention of TV, 113–14

retention stock grants, 17
  as "Sorcerer's Apprentice," 117–18
Foster + Partners, 177, 202, 267, 271,
     365
Foxconn Technology Group, 97–98,
     101, 103, 237–38, 378
Freud, Matthew, 330, 331
Fry, Stephen, 248–49, 331–32

Gawker, 226, 227
Google, 120–21, 293
Gore, Al, 197
Gottesman, David "Sandy," 316–17
Gou, Terry, 97, 238
Grinyer, Clive, 32–33, 38–39, 40

HomePod speakers, 304, 326–27
homosexuality, 51–52, 223–31
Hönig, Julian
  Apple Watch and, 133, 140, 176
  basic Apple facts about, xiv
  characteristics, 84
Howarth, Richard, 133, 248, 249,
     325–26
Huawei, 378

IBM
  automation at, 56
  Cook at, 54, 56–57, 58–59
  culture and structure of, 55
  PCs, 55, 65
  ThinkPad, 69
  transition to service business, 59
Ibuka, Masaru, 15–16
Icahn, Carl, 164–66
iCloud
  Cue and, 196–97
  privacy of information in, 283, 293
  subscriptions, 325, 339
Ideal Standard sink, 38–39

Iger, Bob, 217
iMacs
  flat-panel version, 97–98
  initial development of, 70–72
  introduction of, 72–73
  mouse, 74
  success of, 9, 73, 74
industrial design team, xiv, 79, 269–70
  collaborative process and, 74–75
  components, 70
  concerns about Cook as CEO, 110
  defections from, 326–28
  Howarth and, 326
  importance of, 21, 83, 84
  industrial, xiv, 79, 269–70
  Ive and, 65, 67, 83, 84, 303–5
  after Ive's departure, 412–13
  Jobs and, 21–22
  lighter laptops, 103–4
  members of, 69, 400
  move to Apple Park, 373–74
  offices, 75
  philosophy, 301
  under Spindler as CEO, 66
industrial engineering, 53–54
Ingram Micro, 60
Intelligent Electronics, 58–61
iOS software, 117, 119, 135, 136–38
Iovine, Jimmy
  Apple in television and, 350–51,
     353
  Apple Music artists' and labels'
     compensation, 256, 261
  Apple Music introduction,
     257–58
  Apple Watch and, 250
  basic facts about, 197–98
  Beats's acquisition by Apple,
     198–99, 205, 206–9
  retirement of, 383

iPads
  debut of, 9
  development of, 89
  suspension of, 326
iPhones
  as characterized by Jobs, 7
  in China, 194–96, 251
  customer loyalty, 122, 316
  debut of, 9, 117
  development of, 11, 367
  facial recognition technology,
    263–64
  first new, after death of Jobs, 144
  glass screens, 104–5
  Google and, 293
  intra-Apple rivalries and, 118–20
  iOS7, 162–64
  Ive and operating system of,
    131–32
  Ive and tenth anniversary
    changes, 323–24
  law enforcement's access to,
    281–95
  manufacturing, 104
  mapping system, 120–22,
    123–27
  multitouch and, 85–86
  popularity of, 87, 182
  Project Purple, 86–87, 116
  sales, 87, 143, 144, 182, 193,
    222, 319, 337, 357, 360, 373,
    378–79, 380
  Samsung competition, 145–48,
    150, 164, 212, 307–9, 315,
    316
  7 series, 311–15, 316
  subscription basis, 316
  "ticktock cycle," 263
  volume buttons, 174
  X series, 360, 361–63, 377–79

iPods
  as characterized by Jobs, 7
  compatibility with Windows, 16
  debut of, 9
  development of, 79–81
  flash memory, 103
  iTunes and, 81
  Ive and, 101
  marketing, 81
  Mini, 205
  Nano, 101, 103, 134–35, 206
  popularity of, 81
  production deadlines, 88
Isaacson, Walter, 274–75
iTunes
  Apple Music and, 264, 340
  Beats and, 205, 206, 256
  in China, 319–20
  compatibility with Windows, 16
  Cue and, 196–97
  iPod sales and, 81
  Jobs and, 204
  music industry move to
    subscriptions, 204
  Songs of Innocence and, 218
  success of, 257, 387
Ive, Heather, 38–40, 64
Ive, Jonathan Paul "Jony"
  Apple Pay and, 35
  Apple Watch and, 112–13, 133–
    35, 138–39, 141–42, 173–74,
    175–77, 179–80, 185–90, 210,
    215, 216–17, 218, 219–21,
    240–44, 246, 277–79
  on Apple's financial troubles, 68
  Apple's new headquarters,
    270–72
  autonomous cars and, 268–69,
    297, 298
  awards and honors, 73

background, 22–25
basic Apple facts about, xiii,
    40–41, 399–400, 401, 411
in Castro neighborhood of San
    Francisco, 64
as chancellor of Royal College of
    Art, 374–75
characteristics, 1, 3, 21, 25–26,
    31, 38, 39, 65, 76, 78, 82, 134,
    210, 274
Chaudhri and, 328–29
Claridge Christmas tree, 305–6
Cook and, 101, 247–48, 274, 366
custom-made Leica camera,
    170–72
De Iuliis and, 66–67
departure of, 401–3, 410
depression, 411
design of second version of tablet
    by, 65–66
design team and, 65, 83–84, 169
as design team head, 67, 82,
    83–84
enchantment with San Francisco,
    33, 37
endurance of Apple after death
    of Jobs, 404–5
fiftieth birthday celebrations,
    329–34
Forstall and, 118–19, 129–32
growth of projects and
    responsibilities, 405
hearing aids for children, 33–34
in high school, 25–26, 27–29
iMac success and, 73
initial development of iMac, 70–72
as internal Apple politics player,
    81–83
after introduction of Apple
    Watch, 232–33

iOS7 and, 162–63
iPad development and, 89
iPhone 11 and, 367
iPhone operating system, 131–32,
    135, 136–38
iPhone tenth anniversary changes
    and, 323–24
iPods and, 79–81, 101
Juggernaut project, 39–40
knighted, 90–91
loss of Jobs's feedback, 90
LoveFrom and, 403, 412
Macintosh computer and, 34
Macintosh redesign by, 66–67
Manus × Machina exhibit and,
    299–303
marketing and, 31, 186–90,
    240–44
materials' usage, 174–75
Media Arts "Intention"
    campaign, 160–61
Media Arts "Leave It Better
    Than You Found It" pitch,
    159–60, 161
meeting with employees after
    death of Jobs, 19–20
mouseless computer idea and,
    85–86
new Apple headquarters at Apple
    Park and, 177–78
new headquarters at Apple Park
    and, 369–70
New Yorker profile of, 211
at Newcastle Polytechnic, 29–32,
    33–36
as part-time, 247–49, 266–67,
    272–74, 303–5, 363, 365–67,
    405–6
pay package, 363–64
personal plane, 266

Ive, Jonathan Paul "Jony" (*cont.*)
  phone designed for Royal Society
    of Arts competition, 35–36
  portable projector design, 27–28
  power of, 83, 84
  production of Power Macs, 77–78
  Project Purple and, 86–87
  retention stock grants, 17
  Riccio and, 298
  at Roberts Weaver, 32–33, 37
  Rubinstein and, 81
  Satzger and, 82–83
  stamps drawn by, 30
  Steve Jobs Theater opening,
    358–59
  supply chain scrutiny by, 76–78
  thought of leaving Apple, 87–88
  tribute to Jobs, 393–94, 395,
    396
  yellow Saab convertible and, 64
  See also Jobs and Ive relationship
Ive, Michael John (father of
    Jonathan)
  Apple computers and, 34
  basic facts about, 22–23, 24–25
  relationship with Jony, 24, 27, 28,
    91, 394–95, 397
  stroke, 394, 397
Ive, Pamela Mary Walford (mother
    of Jonathan), 22, 24–25

Jackson, Lisa, 341, 342
Jobs, Laurene Powell
  Apple Watch and, 235–36
  biopic about Steve, 274–75
  on bond between Jobs and Ive,
    332–33
  endurance of Apple after death
    of Jobs, 404–5

Jobs's good-byes to colleagues,
    11, 12
Manus × Machina exhibit and,
    302
  at Royal College of Art, 374–75
Jobs, Steve
  App Store and, 340
  background, 7–8
  balance of art with engineering
    under, 371
  bureaucracy and, 14
  characteristics, 6, 7, 13, 22, 72, 75,
    76, 116, 341, 359
  computers for masses, 8
  Cook made COO by, 102–3
  Cook made successor, 6, 106–7
  Cook's hiring and, 62–63
  Cook's house purchase and, 99
  Cook's obsession with work and,
    100
  death of, 12–13, 42–43, 90, 108
  design team and, 21–22
  empowerment of lieutenants by,
    102
  on ethos of Apple, 18, 358
  Forstall and, 115–16, 117
  illness of
    bedridden in Palo Alto, 5–6
    doubt about future of Apple
      and, 88
    kept secret, 6
    liver transplant for cancer, 88
    second leave, 105
    unable to go to work, 89
  iMac and, 70, 71, 72–73
  importance of design to, 72
  iPad and, 89
  iPhone apps, 117
  iPhone glass screens, 104–5

iPod Mini and, 205–6
as irreplaceable, 112
iTunes and, 204
lighter laptops and, 103–4
lobbying in Washington and, 157
marketing under, 148–49
as master of reinvention and
    innovation, 406
memorial service for, 18–20
mouseless computers and, 85–86
NeXT and, 9
"one more thing" and, 213
ousting of, 8, 65, 102
personal plane, 266
Pixar and, 102
popularity of personal computers
    and, 8, 55
product quadrants and, 176
Project Purple and, 86, 116
Rare Light and, 111
rehiring of former employees
    and, 181–82
reinvention of TV idea of, 113
retention stock grants to
    executive team and, 17
return to Apple with NeXT, 68
Sculley and, 8, 65, 102
structure of Apple and, 16–17,
    130–31
tribute to, 393–94, 395–97
uniform worn by, 9
See also Jobs and Ive relationship
Jobs and Ive relationship
as balancing each other, 75–76
cementing of, 74
death of Jobs and, 12, 274–76
direct line of communication
    between, 81–82
iPod development, 80
    Ive's ability to put Jobs at ease,
        72–73
    Laurene Jobs on deepness of
        bond, 332–33
    Jobs's first public iPhone call to
        Ive and, 88–89
    Jobs's return to Apple and, 68–70
    overlapping design sensibilities, 75
    power of Ive and, 83
    product greatness and, 75
    success metrics and, 87
Johnson, Ron, 106
"Jony's tears," 369
Joswiak, Greg, xiii, 132
Juggernaut project, 39–40
just-in-time ordering processes, 56

Keane, Roy, 96–97
Keats, Jason, 171, 172
Kerr, Duncan, 85
Khan, Sabih, 100
Kim, Eugene, 181
Koh, D. J., 307
Ku Klux Klan, 48–49
Kudlow, Larry, 354, 355, 356
Kushner, Jared, 341, 342

Lady Gaga, 397–98
Lagerfeld, Karl, 219, 220, 221
Lamiraux, Henri, 118
Land, Edwin, 13, 15
Levin, Carl, 152–55, 157, 158–59
Li Keqiang, 321, 349
LoveFrom, 403, 412
Lynch, Loretta, 283–84

MacBook Airs, 7, 104
Macintosh computers
    Brunner and, 39

Macintosh computers (*cont.*)
   introduction of, 71
   Ive's introduction to, 34
   Power Macs, 77–78
   redesign by Ive, 66–67
Maestri, Luca
   basic Apple facts about, xiii
   Cook and, 274
   iPhone access for law
      enforcement and, 287–90
   promoted to CFO, 199
   review of contracts with third-
      party suppliers, 200
Mansfield, Bob
   autonomous electric cars and,
      318–19
   basic Apple facts about, xiii
   as possible successor to Jobs, 106
   resignation of, 400
   retention stock grants, 17
   smartwatch and, 181
manufacturing design team, 79
Manus × Machina exhibit
      (Metropolitan Museum of
      Art), 299–303
Marcom, 149, 150
Marcom team, 159
marketing
   Apple Watch, 186–90, 240–44
   Clow and, 159
   Cook and, 186, 188–89
   "Designed by Apple in
      California" campaign, 164
   "Genius" campaign, 149–50
   "Get a Mac" campaign, 147, 149
   Ive and, 186–90 under Jobs,
      148–49
   Media Arts "Leave It Better
      Than You Found It"
      campaign, 159–60, 161

Media Arts "The Walk" pitch,
   155–57
Samsung Galaxy campaign,
   146–48, 150
under Schiller, 149–52
SuperBowl "1984" spot, 8, 159
team members, xv
"Think Different" campaign, 74,
   155, 159
McDonough, Denis, 283–84
Media Arts Lab, 149, 150, 152,
   155–57, 159–61, 162
Memphis (Italian design
   movement), 31
Menkes, Suzy, 217–18, 241–42
Metropolitan Museum of Art,
   299–303
Meyerhoffer, Thomas, 82
Milanese loop, 174
Milner, Duncan, 155–57
Milunovich, Steve, 236, 245
Morita, Akio, 13, 15–16
Mossberg, Walt, 108, 264
Muir, David, 217, 290–92
multitouch, development of, 85–86

Nest Labs, 197
Netflix, 351, 352
Newcastle Polytechnic, 29–32,
   33–36
Newson, Marc
   Apple Watch and, 141–42,
      219–21, 241
   Claridge Christmas tree, 305–6
   custom-made Leica camera,
      170–72
   hired by Apple, 172–73
   LoveFrom and, 403
   at Royal College of Art, 374–75
Newton MessagePad, 65–66

NeXT, 9, 68, 115–16
Nike, 103

Obama administration, 283–84, 336
O'Brien, Deidre
  background, 99–100
  basic Apple facts about, xiv
  characteristics, 382–83
  under Cook, 100
operations team
  Cook and, 93–95, 99–100, 101
  members, xv
Oppenheimer, Peter
  basic Apple facts about, xiv
  board of directors and, 274
  retention stock grants, 17
  retirement of, 199
"Orator," 35–36, 37
Ording, Bas, 85
O'Sullivan, Joe, 93–94, 95, 96–97, 99

Paris Fashion Week, 219–21, 235
Parsey, Tim, 66, 83
Passif Semiconductor, 312–13
Pegg, Heather, 26, 27, 30
  See also Ive, Heather
Pendleton, Todd, 144–46
personal computers (PCs)
  Apple II computer sales, 8
  IBM PS/ValuePoint, 58
  popularity of, 8, 55
  price war between Compaq and
    IBM, 65
Petsch, Greg, 60–62, 63
Planet of the Apps, 351
Podolny, Joel, 16
Polaroid, 13, 15
Power Macs, 77–78
PrimeSense, 263
Project North Star, 352, 353

Project Purple, 86–87, 116
Project Titan, 203–4, 251–52,
  267–70, 298
  developmental problems, 297–98
  Mansfield and, 318–19

Quanta Computer, 237
QuesTek Innovations, 175
"Quickboard," 180

racism and Cook, 48–49
Rams, Dieter, 31
Rare Light, 111, 139, 182
Reynolds Aluminum, 53
Reznor, Trent, 253, 260
Riccio, Dan
  autonomous cars, 269
  basic Apple facts about, xiv
  electric cars, 297
  Ive and, 298
Roberts Weaver, 28, 29
Roberts Weaver Group, 32–33, 37
Robertsdale, Alabama, 45–49
Royal Society of Arts, 35–36
Rubinstein, Jon
  basic Apple facts about, xiv,
    79–80
  Cook's house purchase and, 99
  departure from Apple, 106
  iPod development and, 79–80
  Ive and, 81
  Jobs and, 72
  Palm and, 81

Saint John, Bozoma, 254
Samsung smartphones, 145–48, 150,
  164, 212, 307–9, 315, 316
  Barwick, Joni and John, 308
  Samsung Galaxy campaign,
    146–48, 150

San Francisco, 33, 37, 64
Satzger, Doug
    on design team's collaborative
        process, 74–75
    iMac and, 71, 74
    iPod development and, 81
    Ive and, 82–83
    Jobs's return to Apple and, 69
Schiller, Phil
    AirPods, 312, 313
    App Store and, 279
    Apple Watch and, 187, 279
    basic Apple facts about, xiv
    Cook's debut as Apple
        spokesman and, 11
    iPhone access for law
        enforcement, 287–88
    iPhone mapping system, 121–22
    iPhone operating system, 132
    iPhone X, 362
    iPod development, 80
    Mac Pro preview, 163
    marketing and, 149–52, 187
    Media Arts "The Walk" and, 155,
        156–57
    as possible successor to Jobs, 106
    retention stock grants, 17
    Vincent and, 149, 151–52
Schusser, Oliver, 383
Sculley, John, 8, 65, 66, 102
72andSunny, 146
Sewell, Bruce
    basic Apple facts about, xiv
    Cook's debut as Apple
        spokesman and, 10
    death of Jobs and, 12–13
    iPhone access for law
        enforcement, 281–83, 286–87,
        288–90, 292, 293–95
    retention stock grants, 17

sexual orientation, 51–52, 223–31
    Out, 227
Silicon Valley headquarters, 372–73
Siri, 11, 327, 387
skeumorphism, 131–32
smart speakers
    Cook and, 262
    Echo, 262
    HomePod speakers, 304,
        326–27
smartphones
    Huawei, 378
    privacy and security of, 287
    Samsung, 145–48, 150, 164,
        212, 307–9, 315, 316
    See also iPhones
smartwatches, 138
    See also Apple Watch
software team, xiv–xv, 137
Songs of Innocence, 216, 218
Sony Walkman, 13, 15–16, 79, 80
Sottsass, Ettore, 31
Spielberg, Steven, 388–89
Spindler, Michael, 8–9, 66, 67
Spotify, 199, 204, 255, 260, 350, 387
Stern, Peter, 325, 339
Steve Jobs Theater, 358–62
Stiller, Ben, 332
Stockdale, Charlotte, 141
Stringer, Chris
    basic Apple facts about, xiv
    HomePod and, 327
    iMac and, 71
    Jobs's return to Apple and, 69
    resignation of, 366, 400
success metrics, 87
Swift, Taylor, 258–60, 261

tablets, 9, 65–66, 89, 326
Tangerine, xvi, 38–40

taptic engines, 180, 238
TBWA\Media Arts Lab, 81
Tevanian, Avie, 106
"This Emperor Needs New
    Clothes" (Friedman), 235
Tidal, 204
TomTom, 121–22, 126
Top 100 event, 122–23
Trump, Donald
    attacks on China, businesses
        outsourcing jobs, and
        immigration ban, 336,
        342–43, 345, 348
    Cook and, 321, 346–47, 354–
        56, 357
    election of, 321–22
    inauguration, 335–36
    tax laws, 346–47
    trade with China, 348, 349,
        354–56
Trump, Ivanka, 341, 342
TVs, reinvention of, 113–14
Tyrangiel, Josh, 224–26, 230

U2, 215–16, 218

Van Amburg, Zack, 353, 383, 388
Vincent, James, 81, 149, 150–52
Vogue, 217–18, 219–20

Walkman, 13, 15–16
Walt Disney Company, 13, 14, 15
    Disney, Walt, 13, 14, 15
Walton High School, 27–29
wearable technology, 112–13
wearables business, 312–15, 411
    See also Apple Watch
Weaver, Barrie, 32, 33
Weschler, Ted, 316–17
Whang, Eugene "Eug," 84–85, 412

Williams, Brian, 148
Williams, Jeff
    Apple Watch and, 180, 181, 182,
        184, 191, 237–38, 245–46
    Apple Watch Series 2 and, 311
    background, 99
    basic Apple facts about, xiv
    as Cook's consigliere, 274
    as Cook's number two, 99
    demand for Nanos, 103
    departure of Ive and, 403
    iPhone glass screens, 104–5
    iPhone mapping system, 126
    retention stock grants, 17
    similarities to Cook, 180–81
Williamson, Richard, 121, 122, 123,
    124–25
Winfrey, Oprah, 353–54, 389–90
Wintour, Anna, 189–90, 219–20,
    277–78, 300
Witherspoon, Reese, 353, 389
World News Tonight, 217, 290
Wozniak, Steve
    background, 8
    popularity of personal computers
        and, 8, 55
    on Steve Jobs Theater, 360

Xi Guohua, 194–96
Xi Jinping, 319–20, 409

Yesterday, 398–99

Zorkendorfer, Rico
    AirPods and, 312–13
    Apple Watch and, 133, 140,
        176
    basic Apple facts about, xiv
    resignation of, 400
Zuckerman, Andrew, 90, 364–65